Biological Liquid-Liquid Phase Separation, Biomolecular Condensates, and Membraneless Organelles

Biological Liquid-Liquid Phase Separation, Biomolecular Condensates, and Membraneless Organelles

Editor

Vladimir N. Uversky

MDPI

Basel • Beijing • Wuhan • Barcelona • Belgrade • Novi Sad • Cluj • Manchester

Editor
Vladimir N. Uversky
Department of Molecular Medicine
University of South Florida
Tampa
United States

Editorial Office
MDPI
St. Alban-Anlage 66
4052 Basel, Switzerland

This is a reprint of articles from the Special Issue published online in the open access journal *International Journal of Molecular Sciences* (ISSN 1422-0067) (available at: www.mdpi.com/journal/ijms/special_issues/BLL).

For citation purposes, cite each article independently as indicated on the article page online and as indicated below:

Lastname, A.A.; Lastname, B.B. Article Title. *Journal Name* **Year**, *Volume Number*, Page Range.

ISBN 978-3-0365-8931-2 (Hbk)
ISBN 978-3-0365-8930-5 (PDF)
doi.org/10.3390/books978-3-0365-8930-5

Contents

About the Editor

Vladimir N. Uversky

Vladimir N. Uversky, PhD, DSc, FRSB, FRSC, obtained his Ph.D. in biophysics from the Moscow Institute of Physics and Technology (1991) and D.Sc. in biophysics from the Institute of Experimental and Theoretical Biophysics, Russian Academy of Sciences (1998). He spent his early career investigating protein folding at the Institute of Protein Research and the Institute for Biological Instrumentation (Russian Academy of Sciences). In 1998, he moved to the University of California Santa Cruz, and in 2004, he was invited to join the Center for Computational Biology and Bioinformatics at the Indiana University–Purdue University Indianapolis. Since 2010, he has been with the Department of Molecular Biology at the University of South Florida, where he started as an Associate Professor and was promoted to the rank of Full Professor. At the University of South Florida, Dr. Uversky's work focuses on various aspects of the protein intrinsic disorder phenomenon and on the analysis of protein folding and misfolding. Dr. Uversky has authored over 1150 scientific publications. According to the Thomson Reuters' Web of Science, his work has been cited over 66,000 times, and he has an H-index of 126. Dr. Uversky has been included in the prestigious Clarivate list of Highly Cited Researchers™ every year from 2014 to 2020. In 2021, he was elected as a Fellow of the Royal Society of Biology and a Fellow of the Royal Society of Chemistry. Over the years, he has delivered more than 240 invited talks at various national and international conferences and invited seminars at various universities and research institutions around the globe. Dr. Uversky has collaborated with more than 12,500 colleagues from more than 2,750 research organizations in 89 countries/territories. He is an Editor of numerous scientific journals and has edited several books and book series on protein structure, function, folding, and misfolding.

International Journal of
Molecular Sciences

MDPI

Editorial

Biological Liquid–Liquid Phase Separation, Biomolecular Condensates, and Membraneless Organelles: Now You See Me, Now You Don't

Vladimir N. Uversky

Department of Molecular Medicine and Byrd Alzheimer's Research Institute, Morsani College of Medicine, University of South Florida, Tampa, FL 33612, USA; vuversky@usf.edu

Liquid–liquid phase separation (LLPS, also known as biomolecular condensation) and the related biogenesis of various membraneless organelles (MLOs) and biomolecular condensates (BMCs) are now considered fundamental molecular mechanisms governing the spatiotemporal organization of the intracellular space [1–5]. Physically, LLPS represents a special case of phase transition taking place in a homogeneous solution, i.e., a system including a solvent (in this case, water) and homogenously distributed solutes (e.g., proteins and nucleic acids). Under specific conditions, this homogeneous solution undergoes a process of spontaneous separation into two (or more) distinct immiscible liquids reflected in the emergence of dense and dilute phases that, respectively, contain more and less specific solutes. Although as a result of this phase transition, the chemical nature of solvent is not changed, and dense and dilute phases are still filled with water, the resulting separated phases are characterized by noticeable changes in the solvent properties of water [6]. Therefore, in addition to the consideration of the peculiarities of the interaction between different solutes undergoing LLPS, one should keep in mind that water plays a crucial role in biological LLPS and in the biogenesis of various MLOs [6].

When LLPS of biopolymers takes place in a cell, it leads to the emergence of MLOs or BMCs, which, being liquid droplets, represent specific compartments within a cell that are not enclosed by a lipid membrane [7–10]. Therefore, the biogenesis of MLOs is linked to the ability of biopolymers (e.g., intrinsically disordered proteins (IDPs) [11–13] and nucleic acids [14,15]) to separate into phases under specific conditions. LLPS is known to be controlled by various external factors and environmental cues, such as changes in the temperature, pH, ionic strength of the solution, posttranslational modifications [16], small molecules [17], and a number of other causes [18]. Furthermore, most LLPS processes are reversible, and many MLOs exist transiently and show "now you see me, now you don't" behavior, rapidly emerging when conditions are changed and rapidly disintegrating as soon as the original conditions are restored.

MLOs/BMCs are many (to date, about a hundred different MLOs are known [19]) and can be found in eukaryotes, archaea, bacteria, and, likely, viruses. They exist as liquid droplets (or cellular bodies, puncta, etc.) in the cytoplasm, nucleoplasm, mitochondrial matrix, and stroma of chloroplasts. Figure 1 displays the dramatic increase in the appreciation of the importance of MLOs in recent literature. Despite being almost unknown to researchers until quite recently, MLOs are rapidly becoming mainstream in modern cellular research. In line with this ever-increasing interest in MLOs, this Special Issue includes seven research articles and three reviews considering different aspects related to the LLPS, MLOs, and BMCs. These articles are briefly outlined below.

Citation: Uversky, V.N. Biological Liquid–Liquid Phase Separation, Biomolecular Condensates, and Membraneless Organelles: Now You See Me, Now You Don't. *Int. J. Mol. Sci.* **2023**, *24*, 13150. https://doi.org/10.3390/ijms241713150

Received: 10 August 2023
Accepted: 18 August 2023
Published: 24 August 2023

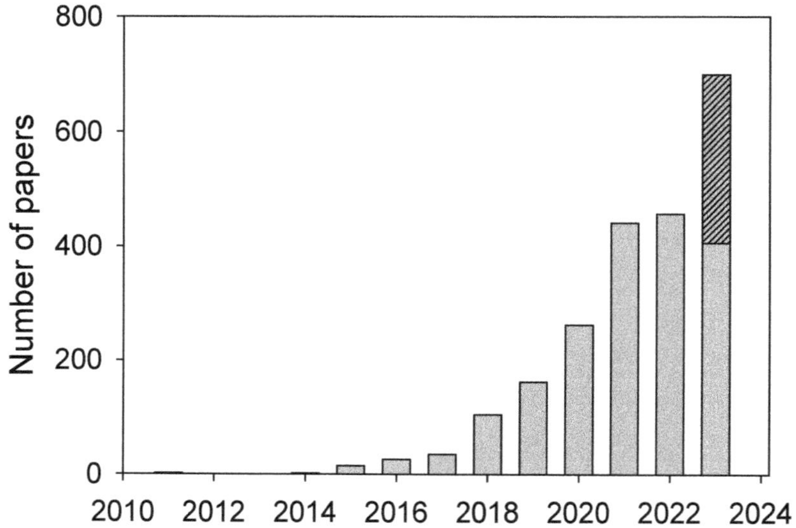

Figure 1. Increase in the number of publications dealing with the different aspects of membraneless organelles. Plot represents the output of a PubMed search for "membraneless organelles" conducted on 9 August 2023. The actual annual publication rate is shown by gray bars, whereas the hatched gray bar shows extrapolated data based on the number of papers published in 2023 by the date of analysis.

In their research article, Mohtadin Hashemi, Siddhartha Banerjee, and Yuri L. Lyubchenko investigated the effects of membrane and free cholesterol on the early stages of the aggregation process of amyloid β (Aβ) [20]. The authors showed that the aggregation of this protein, which is related to the formation of neurotoxic species in Alzheimer's disease, is dramatically enhanced via direct Aβ-membrane interactions. Importantly, this interaction promotes oligomer assembly on the lipid bilayer at physiologically low concentrations of the Aβ monomer. The process is strongly dependent on the membrane composition, and the presence of cholesterol in the membranes significantly enhances the aggregation kinetics. Furthermore, in agreement with previous studies where the presence of free cholesterol in amyloid plaques was reported, the authors found that free cholesterol can further accelerate the aggregation process and promote fast formation of aggregates of significantly larger sizes (as evidenced by an in-solution time-lapse Atomic Force Microscopy (AFM) analysis). Furthermore, free cholesterol accelerated the dissociation of Aβ oligomers from the surface and their accumulation in bulk solution [20].

Juliet F. Nilsson, Hakima Baroudi, Frank Gondelaud, Giulia Pesce, Christophe Bignon, Denis Ptchelkine, Joseph Chamieh, Hervé Cottet, Andrey V. Kajava, and Sonia Longhi also studied the formation of amyloid fibrils, but analyzed the aggregation behavior of an interesting viral protein, the phosphoprotein (P protein) from the Nipah and Hendra viruses (NiV and HeV), which is an essential polymerase cofactor [21]. A peculiar feature of this system is that the gene encoding the P protein in both NiV and HeV also encodes the V and W proteins. These P, V, and W proteins have unique C-terminal domains (CTD) but share an intrinsically disordered N-terminal domain (NTD). A short PTN3 region found within the shared NTD is capable of amyloid-like structure formation. The HeV PNT3 containing the amyloidogenic motif (EYYY) was used to analyze the relevance of each of three contiguous tyrosine residues to the fibrillation process. This analysis revealed that the ability to form fibrils is dramatically reduced by the removal of a single tyrosine independently of its position. Furthermore, the authors showed that PNT3 fibrillation can be regulated by the C-terminal half of this protein, showing an inhibitory effect on fibril formation [21].

Aleksandra E. Badaczewska-Dawid, Vladimir N. Uversky, and Davit A. Potoyan reported the development of a convenient web platform, Bioinformatics Analysis of LLPS Sequences (BIAPSS) [22]. The need for such a tool is based on the premises that despite a broad acceptance of the importance of biological LLPS, MLOs, and BMCs, there is a remarkable gap in the current knowledge which prevents a complete understanding of the sequence "codes" of phase separation required for the design of new phase-separating sequences of fundamental, medical, and technological importance. Therefore, the goals of this tool were to enhance apprehension of the interplay between the primary sequences of proteins and their capability to undergo spontaneous LLPS and thereby to uncover the sequence-encoded signals of the LLPS potential and related biogenesis of numerous functional MLOs and cellular bodies [22]. Researchers can use this web server in on-the-fly analysis as BIAPSS provides a useful tool for the visualization and interpretation of the physicochemical and structural features for the superset of curated LLPS proteins [22].

Yoon-Jeong Choi, Yujin Lee, Yuxi Lin, Yunseok Heo, Young-Ho Lee, and Kiwon Song dedicated their research article to the investigation of a P-body-associated protein from *Saccharomyces cerevisiae,* Nst1 [23]. The authors emphasized that not all highly promiscuous proteins found within MLOs are made equal, with only a few of them acting as key scaffolds. Being capable of inducing condensation of the components of P-bodies (PBs), the multidomain scaffolding protein Nst1 serves as an important subject for targeted analysis of the roles of different domains in biomolecular condensation processes. To this end, a series of Nst1 domain deletion mutations were prepared and investigated [23]. This analysis revealed several important features. For example, deletion of the aggregation-prone domain (APD) significantly inhibited self-condensation, whereas the Nst1 mutant with the deleted multivalent polyampholyte domain (PD) within the intrinsically disordered region ($Nst1_{\Delta PD}$) was able to form self-condensates but failed to interact and condensate Dcp2, a decapping protein and PB component. Furthermore, the $Nst1_{\Delta PD}$ deletion mutant was also unable to condensate other PB components, such as Xrn1, Dhh1, and Edc3, indicating that the PD of the IDR in Nst1 functions as a hub domain interacting with other PB components [23].

The research team of Tamami Miyagi, Rio Yamazaki, Koji Ueda, Satoshi Narumi, Yuhei Hayamizu, Hiroshi Uji-i, Masahiko Kuroda, and Kohsuke Kanekura investigated how the patterns of charged amino acids and the net charge within proteins undergoing LLPS determine their targeting of the specific MLOs, such as nuclear speckles and the nucleolus [24]. This was an important exercise aimed at understanding the basic mechanisms underlying the distributions of the LLPS-prone proteins with charged low-complexity domains (LCDs) to specific MLOs. The authors used proteins with Arg-enriched mixed-charge domains (R-MCDs) primarily composed of R and Asp (D) and known to accumulate in nuclear speckles via LLPS. They demonstrated that the distribution of R-MCD can be shifted from nuclear speckles to the nucleolus by redistributing their R and D residues from an alternately sequenced pattern to uneven blocky sequences [24]. Furthermore, it was established that the blocky R-MCD peptide showed affinity to RNA, acidic poly-Glu, and the acidic nucleolar protein nucleophosmin and was capable of efficient phase separation. On the other hand, the R-MCD peptide with alternating amino acids did not undergo LLPS. Furthermore, localization of the R-MCDs to the MLOs and their accumulation in the nucleolus were promoted by the incorporation of the basic residues into the R-MCDs. Based on these observations, it was concluded that the proximal positioning of D and R linked to the mutual neutralization of their charges is required for the distribution of proteins to nuclear speckles [24].

A series of two papers from the group headed by Ji-Long Liu is dedicated to the analysis of cytoophidia, filamentous structures formed by the CTP synthase (CTPS) [25,26]. Cytoophidia, or "cellular snakes", are evolutionary conserved MLOs found in the cells of many species in all three domains of life and represent an interesting case of metabolic regulation via enzyme filamentation and resulting compartmentalization [27,28]. In the first paper of this miniseries, Yi-Fan Fang, Yi-Lan Li, Xiao-Ming Li, and Ji-Long Liu used fluorescence recovery after photobleaching (FRAP) to study the dynamic characteristics of cytoophidium in human cell lines and also utilized stimulated emission depletion (STED) microscopy to analyze the super-resolution structure of these cellular snakes [25]. This analysis revealed that cytoophidia are dynamic and reticular, with the reticular structure of CTPS cytoophidia potentially providing space for other components [25], such as inosine monophosphate dehydrogenase (IMPDH), another enzyme capable of cytoophidia formation [29]. Curiously, in their structural analysis, the authors also observed CTPS granules with tentacles [25].

In the second paper of the miniseries, Qiao-Qi Wang, Dong-Dong You, and Ji-Long Liu used the female reproductive system of Drosophila as a model for studying the physiological function of cytoophidia [26]. Utilization of a CTPSH355A mutant with a diminished cytoophidium-forming ability revealed the ingression and increased heterogeneity of follicle cells in the CTPSH355A egg chambers, indicating that cytoophidia may play a role in upholding the integrity of the follicle epithelium [26].

This Special Issue concludes with three interesting reviews regarding important biological implementations of LLPS. Lin Zhang, Shubo Wang, Wenmeng Wang, Jinming Shi, Daniel B. Stovall, Dangdang Li, and Guangchao Sui considered the associations between aberrant LLPS, misbehaving MLOs, and the development of various pathological conditions [30]. The authors systemized the properties of different MLOs and BMCs and summarized the multiple LLPS-regulated biological processes. They also emphasized that although normally functioning LLPS controls the biogenesis and composition of dozens of MLOs and BMCs in the cell, the onset and progression of various diseases, including neurodegenerative disorders and cancers, may be associated with (or even driven by) altered physiological conditions or genetic mutations, because of which phase-separated condensates may undergo aberrant formation, maturation, or gelation [30].

Woei Shyuan Ng, Hendrik Sielaff, and Ziqing Winston Zhao focused their review on phase-separation-mediated chromatin organization and dynamics [31]. The authors show how cells can utilize specific physico-chemical properties of chromatin-based phase condensates with various regulatory functions in a spatially and temporally controlled manner. In addition to presenting some key recent findings on the mechanistic roles of phase separation in regulating the organization and dynamics of chromatin-based molecular processes and illuminating the complex phase-separation-mediated interplay between chromatin and diverse chromatin-interacting molecular species, the authors also focused on the quantitative characterizations of these condensates using advanced imaging-based approaches. They emphasized that such phase-separation-mediated chromatin organization defines an emerging multifaceted, multimodal, and multiscale landscape responsible for the hierarchical regulation of the genome. The authors also discussed some deficiencies in existing studies and emphasized the need for multiparametric approaches for the in-depth characterization of chromatin-based phase separation in close-to-native cellular contexts [31].

Finally, Olga I. Povarova, Iuliia A. Antifeeva, Alexander V. Fonin, Konstantin K. Turoverov, and Irina M. Kuznetsova discussed the important roles of LLPS in the spatiotemporal regulation of the cytoskeleton assembly/disassembly, including the formation of actin filaments [32]. The authors argued that the LLPS leading to the formation of coacervates of actin-binding proteins can increase local concentration of G-actin and thereby initiate its polymerization. Furthermore, coacervates can acts as biological reactors, wherein, in addition to this local increase in the G-actin concentration, integration of proteins controlling actin polymerization, such as N-WASP, Arp2/3, and Cortactin, into such MLOs enhances the activity of these actin-binding proteins, thereby providing additional means for the efficient formation of actin filaments [32].

Funding: This research received no external funding.

Institutional Review Board Statement: Not applicable.

Informed Consent Statement: Not applicable.

Data Availability Statement: No new data were created or analyzed in this study.

Conflicts of Interest: The author declares no conflict of interest. The funders had no role in the design of the study; in the collection, analyses, or interpretation of data; in the writing of the manuscript; or in the decision to publish the results.

References

1. Hyman, A.A.; Weber, C.A.; Jülicher, F. Liquid-liquid phase separation in biology. *Annu. Rev. Cell Dev. Biol.* **2014**, *30*, 39–58. [CrossRef] [PubMed]
2. Shin, Y.; Brangwynne, C.P. Liquid phase condensation in cell physiology and disease. *Science* **2017**, *357*, eaaf4382. [CrossRef] [PubMed]
3. Antifeeva, I.A.; Fonin, A.V.; Fefilova, A.S.; Stepanenko, O.V.; Povarova, O.I.; Silonov, S.A.; Kuznetsova, I.M.; Uversky, V.N.; Turoverov, K.K. Liquid-liquid phase separation as an organizing principle of intracellular space: Overview of the evolution of the cell compartmentalization concept. *Cell Mol. Life Sci.* **2022**, *79*, 251. [CrossRef] [PubMed]
4. Bratek-Skicki, A.; Van Nerom, M.; Maes, D.; Tompa, P. Biological colloids: Unique properties of membraneless organelles in the cell. *Adv. Colloid Interface Sci.* **2022**, *310*, 102777. [CrossRef]
5. van Tartwijk, F.W.; Kaminski, C.F. Protein Condensation, Cellular Organization, and Spatiotemporal Regulation of Cytoplasmic Properties. *Adv. Biol. (Weinh)* **2022**, *6*, e2101328. [CrossRef]
6. Zaslavsky, B.Y.; Uversky, V.N. In Aqua Veritas: The Indispensable yet Mostly Ignored Role of Water in Phase Separation and Membrane-less Organelles. *Biochemistry* **2018**, *57*, 2437–2451. [CrossRef]
7. Brangwynne, C.P. Phase transitions and size scaling of membrane-less organelles. *J. Cell Biol.* **2013**, *203*, 875–881. [CrossRef]
8. Shorter, J. Membraneless organelles: Phasing in and out. *Nat. Chem.* **2016**, *8*, 528–530. [CrossRef]
9. Banani, S.F.; Lee, H.O.; Hyman, A.A.; Rosen, M.K. Biomolecular condensates: Organizers of cellular biochemistry. *Nat. Rev. Mol. Cell Biol.* **2017**, *18*, 285–298. [CrossRef]
10. Rangachari, V. Biomolecular condensates—Extant relics or evolving microcompartments? *Commun Biol.* **2023**, *6*, 656. [CrossRef]
11. Turoverov, K.K.; Kuznetsova, I.M.; Fonin, A.V.; Darling, A.L.; Zaslavsky, B.Y.; Uversky, V.N. Stochasticity of Biological Soft Matter: Emerging Concepts in Intrinsically Disordered Proteins and Biological Phase Separation. *Trends Biochem. Sci.* **2019**, *44*, 716–728. [CrossRef] [PubMed]
12. Murthy, A.; Fawzi, N.L. The (un)structural biology of biomolecular liquid-liquid phase separation using NMR spectroscopy. *J. Biol. Chem.* **2020**, *295*, 2375–2384. [CrossRef] [PubMed]
13. Abyzov, A.; Blackledge, M.; Zweckstetter, M. Conformational Dynamics of Intrinsically Disordered Proteins Regulate Biomolecular Condensate Chemistry. *Chem. Rev.* **2022**, *122*, 6719–6748. [CrossRef]
14. Roden, C.; Gladfelter, A.S. RNA contributions to the form and function of biomolecular condensates. *Nat. Rev. Mol. Cell Biol.* **2021**, *22*, 183–195. [CrossRef] [PubMed]
15. André, A.A.M.; Spruijt, E. Rigidity Rules in DNA Droplets: Nucleic Acid Flexibility Affects Model Membraneless Organelles. *Biophys. J.* **2018**, *115*, 1837–1839. [CrossRef]
16. Li, J.; Zhang, M.; Ma, W.; Yang, B.; Lu, H.; Zhou, F.; Zhang, L. Post-translational modifications in liquid-liquid phase separation: A comprehensive review. *Mol. Biomed.* **2022**, *3*, 13. [CrossRef]
17. Li, S.; Wang, Y.; Lai, L. Small molecules in regulating protein phase separation. *Acta Biochim. Biophys. Sin. (Shanghai)* **2023**, *55*, 1075–1083. [CrossRef]
18. Brangwynne, C.P.; Tompa, P.; Pappu, R.V. Polymer physics of intracellular phase transitions. *Nat. Phys.* **2015**, *11*, 899–904. [CrossRef]
19. Darling, A.L.; Uversky, V.N. Known types of membrane-less organelles and biomolecular condensates. In *Droplets of Life: Membrane-Less Organelles, Biomolecular Condensates, and Biological Liquid-Liquid Phase Separation*, 1st ed.; Uversky, V.N., Ed.; Elsevier: Amsterdam, The Netherlands, 2023; pp. 271–335. [CrossRef]
20. Hashemi, M.; Banerjee, S.; Lyubchenko, Y.L. Free Cholesterol Accelerates Abeta Self-Assembly on Membranes at Physiological Concentration. *Int. J. Mol. Sci.* **2022**, *23*, 2803. [CrossRef]
21. Nilsson, J.F.; Baroudi, H.; Gondelaud, F.; Pesce, G.; Bignon, C.; Ptchelkine, D.; Chamieh, J.; Cottet, H.; Kajava, A.V.; Longhi, S. Molecular Determinants of Fibrillation in a Viral Amyloidogenic Domain from Combined Biochemical and Biophysical Studies. *Int. J. Mol. Sci.* **2022**, *24*, 399. [CrossRef]
22. Badaczewska-Dawid, A.E.; Uversky, V.N.; Potoyan, D.A. BIAPSS: A Comprehensive Physicochemical Analyzer of Proteins Undergoing Liquid-Liquid Phase Separation. *Int. J. Mol. Sci.* **2022**, *23*, 6204. [CrossRef] [PubMed]
23. Choi, Y.J.; Lee, Y.; Lin, Y.; Heo, Y.; Lee, Y.H.; Song, K. The Multivalent Polyampholyte Domain of Nst1, a P-Body-Associated Saccharomyces cerevisiae Protein, Provides a Platform for Interacting with P-Body Components. *Int. J. Mol. Sci.* **2022**, *23*, 7380. [CrossRef] [PubMed]
24. Miyagi, T.; Yamazaki, R.; Ueda, K.; Narumi, S.; Hayamizu, Y.; Uji, I.H.; Kuroda, M.; Kanekura, K. The Patterning and Proportion of Charged Residues in the Arginine-Rich Mixed-Charge Domain Determine the Membrane-Less Organelle Targeted by the Protein. *Int. J. Mol. Sci.* **2022**, *23*, 7658. [CrossRef] [PubMed]

25. Fang, Y.F.; Li, Y.L.; Li, X.M.; Liu, J.L. Super-Resolution Imaging Reveals Dynamic Reticular Cytoophidia. *Int. J. Mol. Sci.* **2022**, *23*, 11698. [CrossRef]
26. Wang, Q.Q.; You, D.D.; Liu, J.L. Cytoophidia Maintain the Integrity of Drosophila Follicle Epithelium. *Int. J. Mol. Sci.* **2022**, *23*, 15282. [CrossRef]
27. Liu, J.L. The Cytoophidium and Its Kind: Filamentation and Compartmentation of Metabolic Enzymes. *Annu. Rev. Cell Dev. Biol.* **2016**, *32*, 349–372. [CrossRef]
28. Aughey, G.N.; Liu, J.L. Metabolic regulation via enzyme filamentation. *Crit. Rev. Biochem. Mol. Biol.* **2015**, *51*, 282–293. [CrossRef]
29. Ji, Y.; Gu, J.; Makhov, A.M.; Griffith, J.D.; Mitchell, B.S. Regulation of the interaction of inosine monophosphate dehydrogenase with mycophenolic Acid by GTP. *J. Biol. Chem.* **2006**, *281*, 206–212. [CrossRef]
30. Zhang, L.; Wang, S.; Wang, W.; Shi, J.; Stovall, D.B.; Li, D.; Sui, G. Phase-Separated Subcellular Compartmentation and Related Human Diseases. *Int. J. Mol. Sci.* **2022**, *23*, 5491. [CrossRef]
31. Ng, W.S.; Sielaff, H.; Zhao, Z.W. Phase Separation-Mediated Chromatin Organization and Dynamics: From Imaging-Based Quantitative Characterizations to Functional Implications. *Int. J. Mol. Sci.* **2022**, *23*, 8039. [CrossRef]
32. Povarova, O.I.; Antifeeva, I.A.; Fonin, A.V.; Turoverov, K.K.; Kuznetsova, I.M. The Role of Liquid-Liquid Phase Separation in Actin Polymerization. *Int. J. Mol. Sci.* **2023**, *24*, 3281. [CrossRef] [PubMed]

International Journal of
Molecular Sciences

MDPI

Review

The Role of Liquid–Liquid Phase Separation in Actin Polymerization

Olga I. Povarova †, Iuliia A. Antifeeva †, Alexander V. Fonin, Konstantin K. Turoverov and Irina M. Kuznetsova *

Laboratory of Structural Dynamics, Stability and Folding of Proteins, Institute of Cytology, Russian Academy of Sciences, 4 Tikhoretsky Ave., 194064 St. Petersburg, Russia
* Correspondence: imk@incras.ru
† These authors contributed equally to this work.

Abstract: To date, it has been shown that the phenomenon of liquid–liquid phase separation (LLPS) underlies many seemingly completely different cellular processes. This provided a new idea of the spatiotemporal organization of the cell. The new paradigm makes it possible to provide answers to many long-standing, but still unresolved questions facing the researcher. In particular, spatiotemporal regulation of the assembly/disassembly of the cytoskeleton, including the formation of actin filaments, becomes clearer. To date, it has been shown that coacervates of actin-binding proteins that arise during the phase separation of the liquid–liquid type can integrate G-actin and thereby increase its concentration to initiate polymerization. It has also been shown that the activity intensification of actin-binding proteins that control actin polymerization, such as N-WASP and Arp2/3, can be caused by their integration into liquid droplet coacervates formed by signaling proteins on the inner side of the cell membrane.

Keywords: actin; actin polymerization; actin-binding proteins; liquid–liquid phase separation (LLPS); coacervate; membrane; signaling proteins

Citation: Povarova, O.I.; Antifeeva, I.A.; Fonin, A.V.; Turoverov, K.K.; Kuznetsova, I.M. The Role of Liquid–Liquid Phase Separation in Actin Polymerization. *Int. J. Mol. Sci.* **2023**, *24*, 3281. https://doi.org/10.3390/ijms24043281

Academic Editor: Pavel Draber

Received: 5 January 2023
Revised: 30 January 2023
Accepted: 2 February 2023
Published: 7 February 2023

1. Introduction

Actin exists in nature in a monomeric globular form (G-actin) and a polymeric fibrillar form (F-actin). In the cytoplasm of cells, F-actin, which is the main part of the cytoskeleton, has a functional role. The actin cytoskeleton is necessary for a variety of processes in cells, including the establishment of cell polarity, the activation of cell migration, the launch of cytokinesis, and the positioning of intracellular organelles. Despite extensive research during the last 100 years, there is no complete understanding of the dynamic nature of the cytoskeleton and its regulation in cells and tissues. Recent work has shown a significant role of biomolecular condensates in the regulation of the formation, functioning, and assembly/disassembly of the cytoskeleton [1–5].

The presence of coacervates in protoplasm was also observed a long time ago [6]; however, the significance and universality of this phenomenon of the most diverse intracellular processes became clear only after the work of Brangwynne et al. [7]. The study of LLPS, and its role in cell life, is one of the most rapidly developing areas of molecular and cellular biology. It becomes obvious that the phase separation of biomacromolecules plays a significant role in almost all intracellular processes.

It is believed that intracellular liquid–liquid phase separation leads to the formation of coacervates (also called droplets, membraneless organelles, bodies, and granules), which do not have a membrane separating them from the rest of the cellular space, and which include proteins and nucleic acids. A key role in the formation of membraneless organelles is played by proteins containing intrinsically disordered regions (IDRs), which are mainly scaffold proteins of these structures [8,9].

The IDRs of proteins are important for the fluid-like behavior of coacervates. Studies have shown that proteins rich in IDRs are more likely to separate into phases that retain the dynamic properties of the IDPs forming them [10]. Post-translational modifications (PTMs), such as phosphorylation and SUMOylation, also have a significant impact on LLPS, allowing cells to dynamically regulate LLPS in response to various cellular signals [11–13].

A high degree of multivalence, also inherent in IDPs, is necessary for recruiting other components into membraneless organelles and thus increasing their local concentration [14]. Liquid–liquid phase separation (LLPS) has been recognized as a general cellular mechanism of control for the spatiotemporal dynamics of many important signaling pathways [14,15].

The most important, functionally significant features of membraneless organelles are that only weak interactions are necessary for their formation, they are easily assembled and disassembled after performing their function, the absence of a rigid membrane allows them to freely exchange molecules with the environment, and weak impacts can significantly change their properties. At the same time, mature, membraneless organelles can be complex structures that include hundreds of protein molecules, including globular proteins and RNA [16]. The components of condensates are subdivided into scaffold proteins responsible for phase separation and client proteins, which are included in membraneless organelles to fulfill their function. In coacervates, the concentration of client proteins, including actin, can significantly exceed the concentration in the environment surrounding the organelle.

It is well known that a high concentration of actin is required in a solution to initiate the polymerization of actin monomers. This is primarily due to the fact that the formation of a polymerization nucleus requires the interaction of at least three actin monomers [17]. Coacervates working as biological reactors can contribute to an increase in the likelihood of this event.

2. Model Coacervates as Reactors of Actin Polymerization Initiation

The possibility of actin polymerization through increasing its concentration in model coacervates was shown for the first time in prior work [18], which aimed to demonstrate the possibility of creating a bioreactor that provides a significant increase in the reaction rate. It is known that charged homopolymers (polyelectrolytes) form liquid phases through complex coacervation [19] and can include charged proteins [20–22] and small molecules. To demonstrate the operation of a bioreactor based on poly-L-lysine (pLK) polycation and poly-(L, D)-glutamic acid (pRE) polyanion, two proteins were used: monomeric G-actin and bovine serum albumin (BSA) [18]. G-actin and BSA monomers are globular proteins; they are similar in size (42 and 66 kDa, respectively) and carry a comparable negative charge (isoelectric points of 5.23 and 5.60, respectively).

The proteins were visualized using fluorescently labeled protein-TMR-actin (green) and Alexa-647-BSA (orange). It turned out that both proteins accumulated in the created coacervates model; moreover, actin formed fibrils that were localized at the phase boundary, and BSA filled the coacervates evenly (Figure 1). Using fluorescently labeled phalloidin (Alexa Fluor 647-phalloidin)—a small, uncharged toxin known to bind specifically to F-actin—fibrils, formed by actin, near the coacervate surface were shown to be F-actin (Figure 1D). The extent to which the coacervate microenvironment influenced the rate of actin polymerization was investigated by recording pyrene fluorophore fluorescence.

A study of the polymerization of 1.5 mM of actin in a solution showed the presence of a characteristic lag phase, indicating a slow stage of filament nucleation [23] followed by a phase of rapid growth and then saturation after reaching a steady state [24]. At an actin concentration of 1.5 mM, the initial lag phase is usually 10 min, and the steady state is reached after 120 min. The presence of pLK/pRE coacervates eliminates the lag phase, and a steady state is established within 10 min. Thus, actin incorporation in pLK/pRE coacervates significantly stimulated its polymerization.

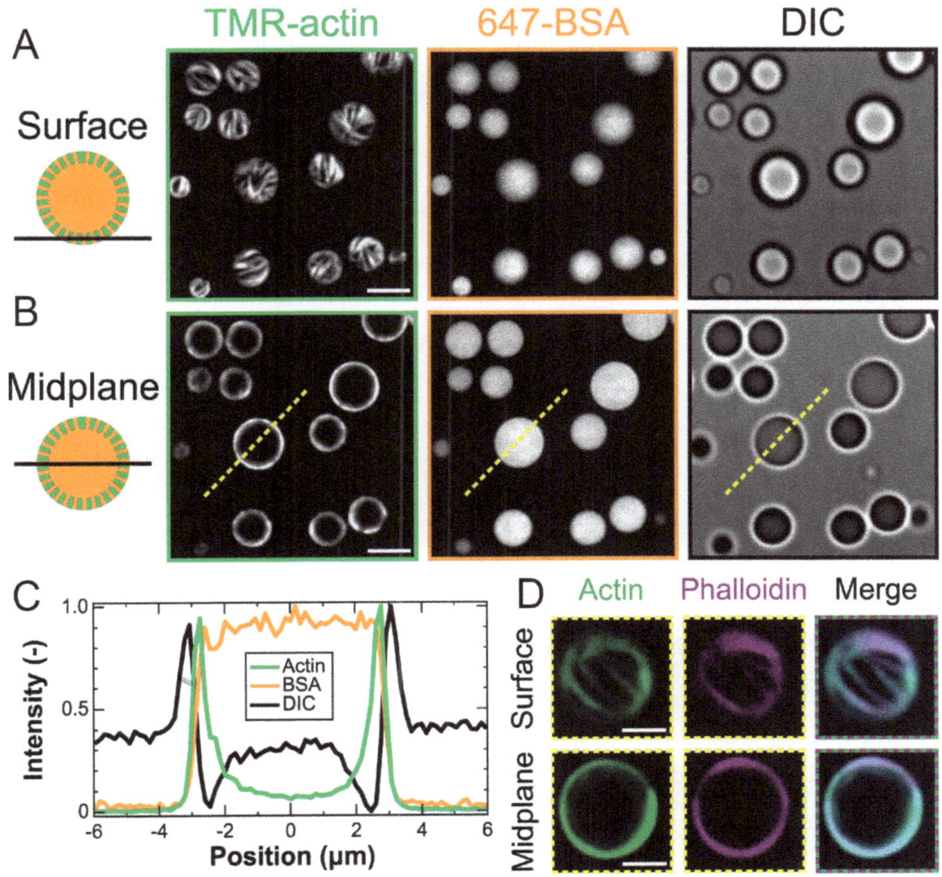

Figure 1. Actin polymerization in the coacervates formed by poly-L-lysine (pLK) and the polyanion poly-(L,D)-glutamic acid (pRE) at the phase boundary. (**A**,**B**) Microphotographs, obtained using confocal fluorescence microscopy (left and middle) and differential interference contrast (DIC) microscopy (right), show polypeptide coacervates containing TMR-actin (green) and Alexa 647-BSA (orange). The focal plane is at the interface of the coacervates and the substrate (surface (**A**)) or near the midplane of the droplet (**B**), indicated by the dashed yellow line. Scale bar, 5 mm. (**C**) Normalized intensity line scans are provided along the dashed yellow lines, as indicated in (**B**). (**D**) False-colored fluorescence images are provided in (**A**,**B**) from the surface (upper row) and midplane (bottom row). The figure represents panels (A), (B), and (D) of Figure 1 and panels (C) and (D) of Figure 2 from the article McCall et al., Biophysical Journal 114, 1636–1645, © 2018 [18].

The authors showed that the mechanism underlying the increased assembly rate is an increase in the local concentration of actin in coacervates. The threshold monomer concentration/critical concentration required for the polymerization of Mg-ATP-actin is 0.1 mM [25]. If actin is concentrated 30-fold in coacervate droplets, then the expected assembly of actin within coacervates should be observed at total actin concentrations of 0.003 mM. However, F-actin formation in coacervates was experimentally observed at total actin concentrations of 0.05 mM. Interestingly, a peripheral shift of F-actin in pLK/pRE coacervates was observed at all actin concentrations studied—even at the lowest concentration (0.01 mM), at which the filaments are almost indistinguishable.

A little later, actin polymerization was examined in a model of the two-phase poly(ethylene glycol) (PEG)/dextran system [26]. It was shown that globular actin was distributed evenly between the phases, while F-actin was concentrated inside the droplets and F-actin bundles were distributed along the periphery of the droplets, deforming them. In addition, it was shown in this work that, in the two-phase PEG/dextran system, the threshold concentration of KCl (which causes actin polymerization) decreased by about one order of magnitude, and the threshold concentration of $MgCl_2$ (which causes the formation of F-actin bundles) decreased by a third [26].

The authors of [18] consider three physical mechanisms of F-actin localization on the periphery of coacervates that do not exclude each other: the bulging of filaments, the depletion of macromolecules, and interfacial adsorption. However, it is noted that the results available to date do not provide a definitive answer for what determines the peripheral localization of F-actin; this could be an interesting problem for future research.

3. Initiation of Actin Polymerization from abLIM1 Coacervates

For the first time, the phase separation of the actin-binding protein, abLIM1, which leads to the formation of coacervates (i.e., reactors for the polymerization of cortical actin) was shown in work by Yang et al. [27]. The nonerythroid actin-binding protein, abLIM1, is known to be critical for stable interactions between the plasma membrane and the actin cortex under mechanical stress. This actin-binding protein has a C-terminal region that is 40% identical to dematin, an erythroid protein [28,29], which consists of a large, N-terminal, intrinsically disordered region (IDR) and a short actin-binding region. This region is identical to the villin headpiece (VHP) domain [30–33]. It is known that abLIM1 is expressed as three isoforms; namely, the long (abLIM-L, which is very rare), medium (abLIM-M, the widely expressed main isoform), and short (abLIM-S, a dematin-like isoform) isoforms [29]. These isoforms differ in a number of N-terminal LIM domains that mainly serve in protein–protein interactions [34]. The proteins abLIM1 and dematin are important for the stable attachment of the cortex to the plasma membrane in non-erythroid cells [28] and erythrocytes [31,35,36], respectively (Figure 2A).

Previously, it was shown that bacterially expressed His-tagged GFP-ΔLIM (3 μM) induces the formation of dense networks of bound F-actin in vitro [28], such as dematin [30], and asters ~10 μm in diameter. Usually, asters contain an amorphous core involving His-GFP-ΔLIM, from which numerous actin filaments radiate, decorated with GFP-ΔLIM. It was shown that the asters' sizes depended upon ΔLIM concentration, and aster fibers contained densely packed actin filaments (Figure 2C). This allowed the authors of the work [27] to suggest that ΔLIM condensates could serve as the centers of the actin filament asters' organization in vitro. To prove this idea, it was necessary to rule out the presence of F-actin prior to aster formation (Figure 2C).

To this end, G-actin in a storage buffer (containing 0.2 mM of CaCl2) was mixed with GFP-ΔLIM (6 μM) followed by an actin polymerization buffer. Then, TRITC phalloidin (tetramethylrhodamine B isothiocyanate) was used to visualize F-actin. It appeared that GFP-ΔLIM-decorated actin bundles, especially GFP-ΔLIM punctate condensates, were shown to gradually emerge after about 15 min and expand rapidly with time (Figure 2C).

The network (web) of actin fibrils turned out to be quite stable and remained preserved for hours without signs of disassembly. The web had a thickness of 6 to 7 μm and was located above an array of His-GFP-ΔLIM condensates at the bottom of the substrate, while numerous F-actin asters of various sizes floated in the solution along with sporadic bundles. Thus, it was shown that ΔLIM was capable of the self-organization of asters and webs of actin bundles in vitro [27] (Figure 2F).

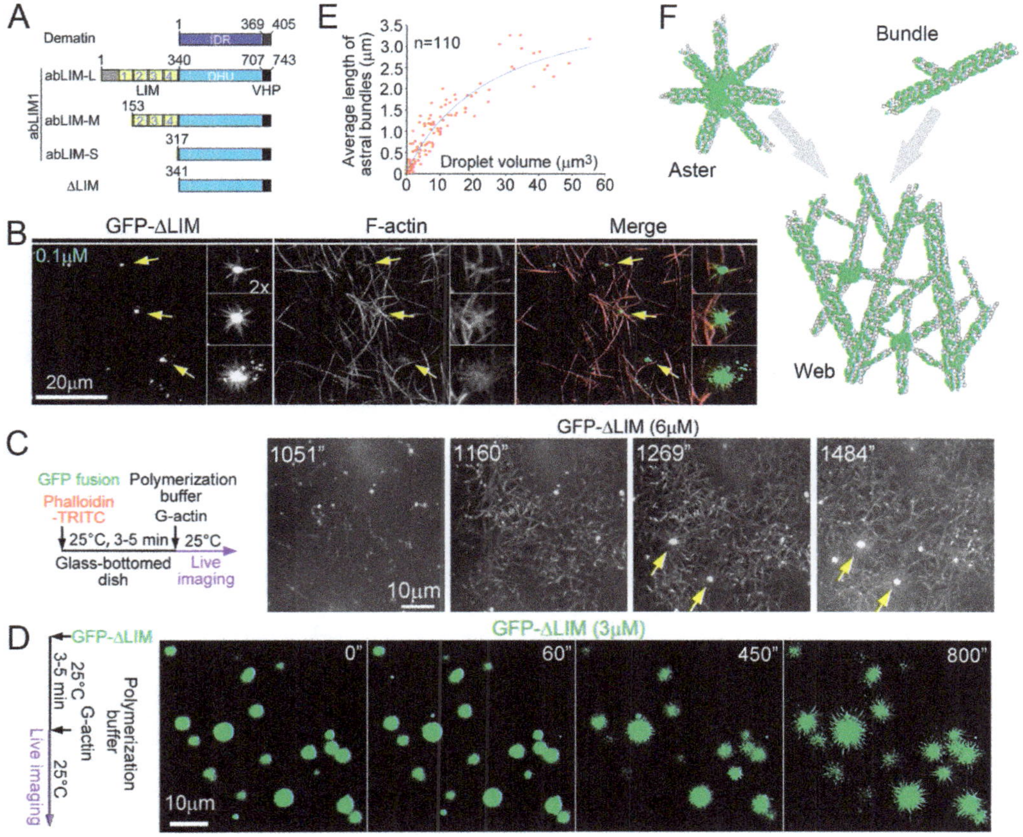

Figure 2. ΔLIM induces asters and webs of F-actin bundles in vitro. (**A**) Isoforms of human abLIM1. (**B**) Effects of GFP-ΔLIM on F-actin organizations in vitro. Arrows indicate representative astral structures. This presents the results of experiments in which the GFP-ΔLIM concentration was 0.1 μM. (**C**) The scheme of the experiment and the results of live imaging. The final concentration of G-actin was 6 μM. Phalloidin-TRITC (final concentration: 4 μM) was used to label F-actin. Representative time-lapse images showing the formation of His-GFP-ΔLIM-induced actin webs. Z-stack images, at 1 μm intervals, were captured for the GFP autofluorescence to cover a depth of 6 μm close to the bottom of the substratum via spinning disk microscopy at ~3.6 s intervals. The time started immediately after the addition of G-actin. Arrows denote two asters integrated into the web. (**D**) Massive actin polymerization from liquid droplets of GFP-ΔLIM. The experiments were performed without phalloidin-TRITC to show that the aster formation was not due to the presence of phalloidin. (**E**) Positive size–length correlations between the GFP-ΔLIM droplets and astral F-actin bundles. The size (volumes) of each droplet and the mean length of its astral bundles were measured from time-lapse images. (**F**) Illustrations showing ΔLIM (green)-induced F-actin-based asters, bundles, webs, and their relationships. The figure represents panels A of Figure 1; line 5 of panel B; panels E, F, and H of Figure 2; and panels A and C of Figure 4 from Yang et al., PNAS 119 (29), e2122420119 [27]. Copyright © 2022 the Yang et al., published by PNAS. This article is distributed under Creative Commons Attribution-NonCommercial-NoDerivatives License 4.0 (CC BY-NC-ND).

A comprehensive study of the properties of ΔLIM condensates in astral centers has been made (Figure 2B), and it has been proven that they have the properties of coacervates. To this end, in particular, the change in the turbidity of the solution with a change in temperature was monitored [27]. Increasing the temperature of the GFP-ΔLIM samples from 0 °C to 25 °C over 5 min resulted in an increase in turbidity. Moreover, the turbidity was reversible [27]. In addition, droplets were shown to increase in both size and density depending on the concentration of GFP-ΔLIM and could be clearly observed at a concentration of 1 μM. They could easily merge into larger ones, which also confirms their liquid properties. Fluorescence recovery after photobleaching (FRAP) assays revealed a rapid recovery of droplet fluorescence, indicating a dynamic exchange of ΔLIM molecules with the environment. Finally, in the presence of polyethylene glycol (PEG), a crowding agent known to promote the separation of protein phases [37,38], droplet formation was observed even at 8 nM [27]. The ability to achieve the phase separation of various modifications of His-GFP-DHU-VHP, including His-GFP-DHU and His-GFPHis-GFP-VHP, was considered. The results of these studies concluded that the IDR region homologous to dematin mediated LLPS.

Finally, since aromatic amino acids (Fs and Ys) are known to be critical for the LLPS of some IDRs [14,27,39], DHU mutants that had a serine (S) residue instead of aromatic residues were investigated, and it was shown that that the resulting mutants, DHU25S and ΔLIM25S, did not form coacervates. Thus, it was proven that ΔLIM can, indeed, be separated into liquid droplets due to its IDR. In order to prove that ΔLIM liquid droplets can function as actin polymerization sites for the direct production of de novo radial actin filaments, GFP-ΔLIM liquid droplets were examined through live imaging immediately after the addition of G-actin. Numerous hair-like protrusions were shown to continuously grow from droplets (Figure 2D,E), indicating the massive nucleation and elongation of F-actin. To prove that the rays were formed by F-actin, the slides were stained with 4 μM of phalloidin-TRITC. Although the presence of phalloidin can potentially stabilize actin filaments, it does not affect the rate of actin polymerization [40] and, thus, does not interfere with the process of aster formation. Indeed, phalloidin-TRITC rapidly congressed to the outer surface of the droplets within 30 s after the G-actin addition, after which, GFP-positive F-actin arrays gradually appeared around all of the droplets (Figure 2A,D). It is noteworthy that some of the drops gradually became hollow during the growth of the aster, which suggests that the elongation of the astral bundle absorbs the contents of the drops. Accordingly, larger droplets produced longer bundles of astral actin (Figures 2D and 3A).

It was also shown that aster formation was not observed in the presence of Cytochalasin D (CytoD) or LatrunculinA (LatA), drugs that inhibit actin polymerization and prevent actin assembly in the plus-end [24] or sequestering G-actin [41].

To find out whether the formation of aster was simply the result of a directed arrangement of abLIM1 molecules on a spherical surface and not their phase separation, a control experiment was carried out with magnetic beads. To perform this experiment, HEK293T cells were overexpressed with GFP-ΔLIM and concentrated with anti-GFP, antibody-conjugated magnetic beads ~1.5 μm in diameter. Thus, the ΔLIM molecules on the beads were oriented so that their actin-binding VHP domain was directed radially from the center of the bead. Direct imaging showed that the fluorescence intensity of the GFP beads was comparable to that of similarly sized liquid GFP-ΔLIM droplets (Figure 3B), suggesting a similar density of GFP-ΔLIM molecules on the surface of the beads and the droplets. However, in the presence of G-actin, the beads could not produce visible actin filaments, while liquid droplets induced aster formation. Therefore, the formation of ΔLIM coacervates is necessary for aster formation [27].

Thus, the authors of the work [27] proved that ΔLIM, consisting mainly of a disordered fragment of DHU, can form coacervates, which are actin polymerization reactors, due to LLPS. At the same time, it was shown that both DHU and VHP are necessary for actin aster formation since no aster formation was observed in experiments with GFP-DHU liquid droplets.

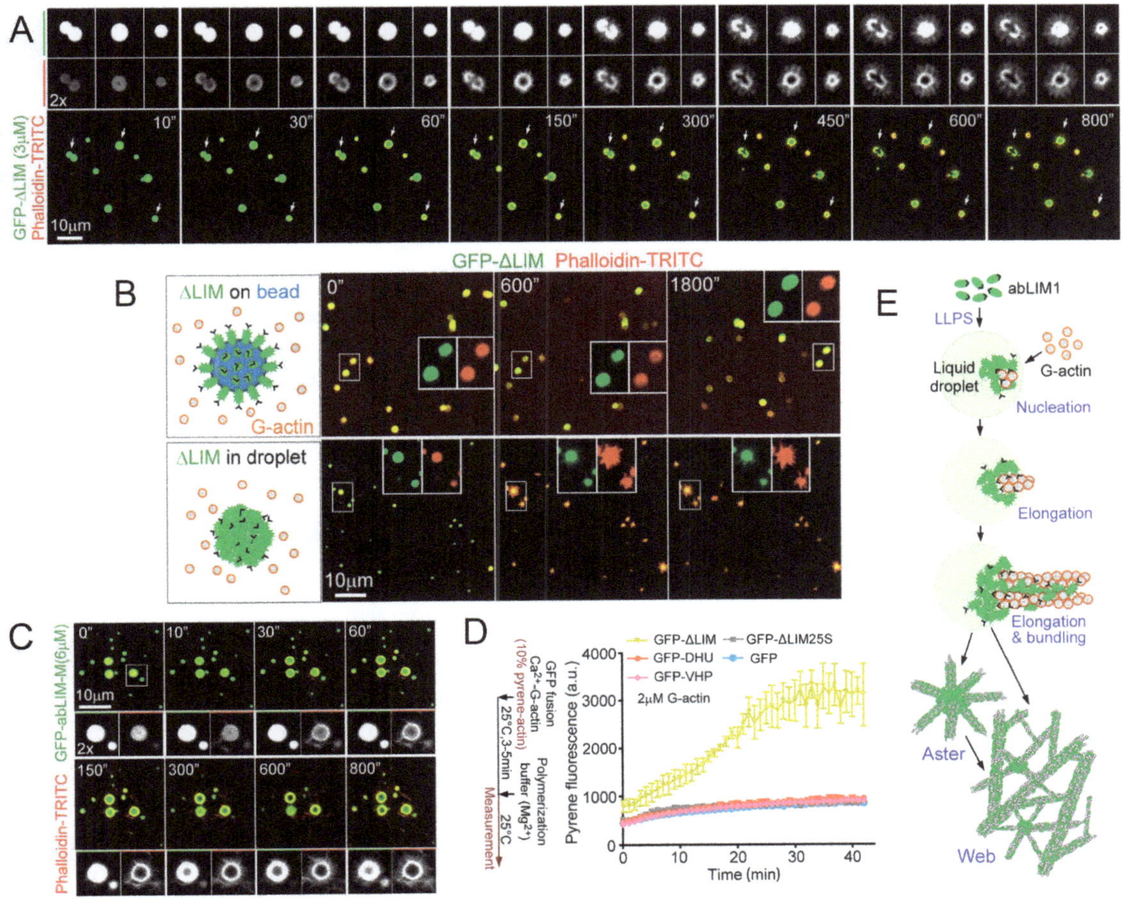

Figure 3. Nucleation, elongation of polymerization, formation of asters, and F-actin filaments in the presence of ΔLIM or abLIM-M liquid droplets. (**A**) Intense actin polymerization from GFP-ΔLIM liquid droplets in real time. The initial moment of time was taken as the time when G-actin was introduced into the actin polymerization buffer containing preliminarily formed GFP-ΔLIM liquid drops. (**B**) F-actin does not polymerize from GFP-ΔLIM immobilized with an anti-GFP antibody on magnetic beads (upper panel); however, in the presence of preformed His-GFP-ΔLIM droplets, actin filaments grow rapidly (lower panel). (**C**) abLIM-M liquid droplets can also generate asters. Although the LIM domain inhibits droplet formation, GFP-abLIM-M forms droplets in the presence of 1% PEG. Drops of GFP-abLIM-M, as well as GFP-ΔLIM, induce aster formation after the addition of G-actin. (**D**) G-actin polymerization requires the presence of GFP-ΔLIM or GFP-abLIM-M droplets. Individually, neither GFP-DHU (IDR domain) nor GFP-VHP (actin-binding domain) affects the rate of actin polymerization. (**E**) A model illustrating the role of abLIM1 in actin polymerization. At least, its abLIM-S or abLIM-M isoform undergoes LLPS to form condensates that promote actin concentration, initiate polymerization, and cross-link actin filaments to form asters and sporadic bundles that further develop into the F-actin web. The figure represents panels B, E, and G of Figure 4 and panels C and F of Figure 5 of Yang et al., PNAS 119 (29), e2122420119, 2022, [27]. Copyright © 2022 Yang et al., published by PNAS. This article is distributed under Creative Commons Attribution-NonCommercial-NoDerivatives License 4.0 (CC BY-NC-ND).

Since actin-binding proteins are the entire abLIM family, it was necessary to find out how the ordered LIM fragment affects LLPS and whether these proteins could form coacervates. The tendency to precipitate His-GFP-abLIM-L [28] made it difficult to perform experiments with this form. However, His-GFP-abLIM-M, a widely expressed major isoform (Figure 1A) [29], was soluble, so the study of whether its liquid droplets could also generate aster was carried out with this isoform. It was found that GFP-abLIM-M did not phase separate, even at 105 μM, consistent with the inhibitory role of the LIM region. However, it was subjected to LLPS at 6 μM in the presence of 1% PEG (Figure 3C).

Live imaging showed that, after the addition of G-actin, liquid droplets mediated aster formation in a similar manner (Figure 3C). Therefore, the separated abLIM-M and abLIM-S phases were able to facilitate local actin polymerization and binding to form complex aster.

In [27], the rate of actin polymerization in the presence of abLIM1 was also estimated. For this, the authors used the fluorescence of pyrene. An essential increase in the actin polymerization rate was shown in the presence of abLIM1. It is known that the stage limiting rate of actin polymerization is the formation of nuclei, i.e., the formation of initial actin trimers [42,43]. To quantify how abLIM1 promotes actin nucleation, the G-actin concentration was chosen to be 2 μM, which excludes spontaneous actin nucleation [44,45]. This made it possible to evaluate the activity of nucleation mediated by nucleators [46,47]. It turned out that while pyrene fluorescence in GFP samples only slightly increased over time; it steadily increased in 3 μM GFP-ΔLIM samples and stabilized after about 25 min (Figure 3D). The steady-state pyrene intensity was more than five times higher than that of the GFP samples (Figure 3D). However, separately, neither GFP-DHU nor GFP-VHP had a clear effect on actin polymerization kinetics compared to GFP (Figure 3D), confirming the need for both regions. As expected, the GFPΔLIM25S that did not undergo LLPS was unable to stimulate actin polymerization (Figure 3D).

Thus, based on all in vitro studies, it can be concluded that LLPS-induced ΔLIM condensates are capable of strongly stimulating actin nucleation. The resulting actin filaments are then stitched together to form stable asters and sporadic beams that further develop into stable networks (Figure 3E).

Condensates of abLIM1 likely move along actin filaments to regionally nucleate and cross-link actin filaments together to self-organize elastic, interconnected networks of F-actin cortical bundles so that cells properly resist mechanical stress. The tendency of F-actin polymerizing arrays to consume liquid droplets and the positive correlation between the length of actin astral bundles and the size of liquid droplets suggest fluid fluidity of ΔLIM and abLIM-M along F-actin in vitro. In [27], it was concluded that the absence of the F-actin association of the phase-defective ΔLIM25S and abLIM25S-M mutants, the appearance of liquid droplets of abLIM-M and ΔLIM during actin depolymerization in cells, their resorption into repolymerizing actin cytoskeletons, and their constant propagation along intracellular F-actin bundles confirmed that the intracellular molecules, abLIM-M and abLIM-S, associated with F-actin were also form coacervates. Moreover, previous studies demonstrate that dense, interconnected cortical actin networks in RPE1 and U2OS cells become sparse and rich in thick linear filaments when abLIM1 is depleted, leading to membrane swelling during cell proliferation or migration [28]. In addition, because abLIM1 is highly enriched in the Z-disc of sarcomeres [29,48], it may also assist in the construction of actin networks in striated muscle cells through the nucleation and cross-linking of actin filaments in the Z-disc. This is also consistent with the ΔLIM-induced efficient polymerization of Ca^{2+}-G-actin in the absence of Mg^{2+}.

4. LLPS of Signaling Proteins Leads to the Formation of Coacervates Concentrating Actin-Binding Proteins, N-WASP, Arp2/3, and Cortactin

A recent review on the role of phase transitions in the formation of the cytoskeleton [3] analyzed the role of phase transitions in actin polymerization and briefly mentioned several studies showing that the phase separation of signaling proteins modulates the functioning of proteins associated with actin polymerization.

It was shown in [1] that the phase transition of the neuronal Wiskott–Aldrich syndrome protein (N-WASP)—which interacts with its biological partners, Nck and phosphorylated nephrin—leads to the formation of coacervates, which include complexes (actin-related proteins), Arp2/3, which are actin nucleation factors. It has been shown that actin polymerization increases dramatically when coacervates of signaling proteins are formed and decreases when the LLPS process or the interaction of Arp2/3 with N-WASP is disrupted. In [49], a mechanism was proposed for the intensification of the Arp2/3 complex in coacervates. It has been shown that the phase separation of Nephrin/Nck/N-WASP signaling proteins on lipid bilayers leads to an increase in the residence time of N-WASP and the Arp2/3 complex on the membrane and, therefore, intensifies actin assembly.

Another example of condensates that enhance actin polymerization is activated T-cell receptors [2]. The authors biochemically recreated a 12-component signaling pathway on model membranes, starting with T-cell receptor (TCR) activation and ending with actin assembly. When TCR phosphorylation was triggered, signaling proteins spontaneously separated into fluid-like clusters that facilitated signaling both in vitro and in human Jurkat T cells. The reconstituted clusters were enriched in kinases, but excluded phosphatases and enhanced the assembly of actin filaments by recruiting and organizing actin polymerization regulators, N-WASP-Arp2/3. Thus, it was demonstrated that the LLPS of proteins can create separate physical and biochemical compartments that facilitate signaling.

Another example of this phenomenon was considered in [50], where signaling in synapses was considered. Obviously, the correct formation and rapidity of responses to synaptic stimulation are fundamental for the functions of the mammalian brain; however, the molecular basis that governs the formation and modulation of separated synaptic ensembles has, so far, remained unclear. Using a biochemical reconstruction approach, the authors showed that post-synaptic scaffold proteins at physiological concentrations can form highly condensed, self-assembling postsynaptic density protein (PSD)-like assemblies via liquid–liquid phase separation (LLPS). Such PSD scaffold condensates can cluster glutamate receptors, incorporate synaptic enzymes, and promote the formation of actin bundles; however, they do not allow gephyrin, which is the main scaffold protein for almost all inhibitory synapses, to pass through.

When creating model coacervates, the authors used the main PSD scaffold proteins, including PSD-95, GKAP, Shank, and Homer, which serve to connect ion channels/receptors on the postsynaptic plasma membrane to the actin cytoskeleton in the PSD cytoplasm. It is known that Shank directly interacts with regulatory proteins of the actin cytoskeleton, such as cortactin [51] and subunits of the Arp2/3 complex [52].

Cortactin has an N-terminal acidic domain that binds to and stimulates the Arp2/3 complex, followed by tandem cortactin repeats (CR) that bind to F-actin, and a C-terminal SH3 domain that binds to Shank3. It was shown that when cortactin, the Arp2/3 complex, and G-actin were introduced into the 63 PSD system assembled on lipid bilayers, polymerized structures of F-actin began to appear in the condensed PSD assemblies 15 min after the onset of the phase transition. Actin bundles colocalized with PSD condensates gradually became thicker and longer. Phalloidin staining confirmed that the bundles were actin filaments. Without the addition of the five PSD components but with the presence of cortactin, the Arp2/3 complex, and G-actin, actin bundles were not observed. Finally, it has been shown that PSD can also promote the formation of actin bundles on lipid bilayers in the absence of Arp2/3. Thus, it has been shown that PSD condensates can promote F-actin assembly by increasing the concentration of G-actin and cortactin.

5. Conclusions

To date, it is becoming more and more obvious that the LLPS of proteins is of a universal nature and, apparently, mediates all known cellular processes [9]. The process of the formation of F-actin, which is one of the main components of the cytoskeleton, is no exception.

Figure 4 shows the coacervates that are known, to date, to arise from LLPS, which plays a crucial role in actin polymerization. This model shows:

- coacervates of LAT signaling proteins capable of integrating actin-binding proteins, which, as noted in the literature, leads to the intensification of their work (see Section 3);
- coacervates of the actin-binding protein, abLIM1, which integrates G-actin monomers and thereby increases their local concentration, thus leading to the initiation of polymerization and the growth of F-actin fibrils in the form of asters (see Section 2);
- VASP coacervates, which trigger the spontaneous, self-sustaining growth of actin bundles and are analyzed in detail in the yet-unpublished work of Graham et al. [53]. Actin polymerizes inside the VASP droplets, and elongating filaments are distributed along the periphery of the droplet, forming an actin-rich ring inside the droplet. As actin polymerizes and the ring thickens, its rigidity increases and eventually overcomes the surface tension of the drop, deforming the drop and transforming it into a linear bundle. The resulting bundles contain long, parallel actin filaments growing from their tips (Figure 4). The length of the final F-actin bundle can be many times greater than the size of the initial droplet.

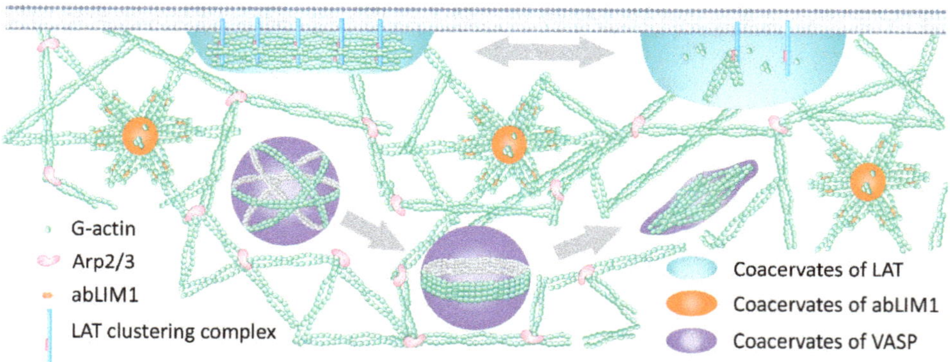

Figure 4. The currently discovered, coacervate-dependent models of actin polymerization and bundling. The figure was created based on the results of the works [2,27,53].

Actin filaments are known to be associated with the membrane. Within the framework of new ideas of actin polymerization, it becomes clear that this is energetically favorable since the threshold concentration of coacervate formation on the membrane is an order of magnitude lower than in free space [54]. New ideas about actin polymerization reveal new directions in the study of the cytoskeleton and formulate new challenges for researchers. It remains to be seen whether some other actin-binding proteins can form coacervates during phase separation and under which scenario they will work, similar to how the abLIM1 or VASP coacervates interact with formed filaments [3]. It is interesting to evaluate the existence of proteins that can form coacervates similar to signaling proteins in response to internal cell stimuli.

Together with earlier work on the mechanism of microtubule formation through the concentration of tubulin in centrosomes [4], recent works on actin polymerization highlight the universal role of LLPS in the formation of the cytoskeleton. Researchers are now challenged to explore the dynamic cytoskeletal organization within the frame of this new paradigm.

Author Contributions: Writing—original draft preparation O.I.P., I.A.A., A.V.F., K.K.T. and I.M.K.; writing—review and editing, O.I.P., I.A.A., A.V.F., K.K.T. and I.M.K. All authors have read and agreed to the published version of the manuscript.

Funding: This research received no external funding.

Institutional Review Board Statement: Not applicable.

Informed Consent Statement: Not applicable.

Data Availability Statement: Not application.

Conflicts of Interest: The authors declare no conflict of interest.

References

1. Li, P.; Banjade, S.; Cheng, H.C.; Kim, S.; Chen, B.; Guo, L.; Llaguno, M.; Hollingsworth, J.V.; King, D.S.; Banani, S.F.; et al. Phase transitions in the assembly of multivalent signalling proteins. *Nature* **2012**, *483*, 336–340. [CrossRef]
2. Su, X.; Ditlev, J.A.; Hui, E.; Xing, W.; Banjade, S.; Okrut, J.; King, D.S.; Taunton, J.; Rosen, M.K.; Vale, R.D. Phase separation of signaling molecules promotes t cell receptor signal transduction. *Science* **2016**, *352*, 595–599. [CrossRef] [PubMed]
3. Wiegand, T.; Hyman, A.A. Drops and fibers—How biomolecular condensates and cytoskeletal filaments influence each other. *Emerg. Top. Life Sci.* **2020**, *4*, 247–261. [PubMed]
4. Woodruff, J.B.; Ferreira Gomes, B.; Widlund, P.O.; Mahamid, J.; Honigmann, A.; Hyman, A.A. The centrosome is a selective condensate that nucleates microtubules by concentrating tubulin. *Cell* **2017**, *169*, 1066–1077.e10. [CrossRef] [PubMed]
5. Shin, Y.; Brangwynne, C.P. Liquid phase condensation in cell physiology and disease. *Science* **2017**, *357*, eaaf4382. [CrossRef] [PubMed]
6. Wilson, E.B. The structure of protoplasm. *Science* **1899**, *10*, 33–45. [CrossRef] [PubMed]
7. Brangwynne, C.P.; Eckmann, C.R.; Courson, D.S.; Rybarska, A.; Hoege, C.; Gharakhani, J.; Jülicher, F.; Hyman, A.A. Germline p granules are liquid droplets that localize by controlled dissolution/condensation. *Science* **2009**, *324*, 1729–1732. [CrossRef]
8. Uversky, V.N. Intrinsically disordered proteins in overcrowded milieu: Membrane-less organelles, phase separation, and intrinsic disorder. *Curr. Opin. Struct. Biol.* **2017**, *44*, 18–30. [CrossRef]
9. Fonin, A.V.; Antifeeva, I.A.; Kuznetsova, I.M.; Turoverov, K.K.; Zaslavsky, B.Y.; Kulkarni, P.; Uversky, V.N. Biological soft matter: Intrinsically disordered proteins in liquid-liquid phase separation and biomolecular condensates. *Essays Biochem.* **2022**, *66*, 831–847.
10. Boeynaems, S.; Alberti, S.; Fawzi, N.L.; Mittag, T.; Polymenidou, M.; Rousseau, F.; Schymkowitz, J.; Shorter, J.; Wolozin, B.; Van Den Bosch, L.; et al. Protein phase separation: A new phase in cell biology. *Trends Cell Biol.* **2018**, *28*, 420–435. [CrossRef]
11. Hofweber, M.; Dormann, D. Friend or foe-post-translational modifications as regulators of phase separation and rnp granule dynamics. *J. Biol. Chem.* **2019**, *294*, 7137–7150. [CrossRef] [PubMed]
12. Monahan, Z.; Ryan, V.H.; Janke, A.M.; Burke, K.A.; Rhoads, S.N.; Zerze, G.H.; O'Meally, R.; Dignon, G.L.; Conicella, A.E.; Zheng, W.; et al. Phosphorylation of the fus low-complexity domain disrupts phase separation, aggregation, and toxicity. *EMBO J.* **2017**, *36*, 2951–2967. [CrossRef]
13. Snead, W.T.; Gladfelter, A.S. The control centers of biomolecular phase separation: How membrane surfaces, ptms, and active processes regulate condensation. *Mol. Cell* **2019**, *76*, 295–305. [CrossRef] [PubMed]
14. Banani, S.F.; Lee, H.O.; Hyman, A.A.; Rosen, M.K. Biomolecular condensates: Organizers of cellular biochemistry. *Nat. Rev. Mol. Cell Biol.* **2017**, *18*, 285–298. [CrossRef] [PubMed]
15. Alberti, S.; Gladfelter, A.; Mittag, T. Considerations and challenges in studying liquid-liquid phase separation and biomolecular condensates. *Cell* **2019**, *176*, 419–434. [CrossRef]
16. Gomes, E.; Shorter, J. The molecular language of membraneless organelles. *J. Biol. Chem.* **2019**, *294*, 7115–7127. [CrossRef] [PubMed]
17. Dominguez, R. Actin filament nucleation and elongation factors–structure-function relationships. *Crit. Rev. Biochem. Mol. Biol.* **2009**, *44*, 351–366. [CrossRef]
18. McCall, P.M.; Srivastava, S.; Perry, S.L.; Kovar, D.R.; Gardel, M.L.; Tirrell, M.V. Partitioning and enhanced self-assembly of actin in polypeptide coacervates. *Biophys. J.* **2018**, *114*, 1636–1645. [CrossRef]
19. van der Gucht, J.; Spruijt, E.; Lemmers, M.; Cohen Stuart, M.A. Polyelectrolyte complexes: Bulk phases and colloidal systems. *J. Colloid Interface Sci.* **2011**, *361*, 407–422. [CrossRef]
20. Black, K.A.; Priftis, D.; Perry, S.L.; Yip, J.; Byun, W.Y.; Tirrell, M. Protein encapsulation via polypeptide complex coacervation. *ACS Macro Lett.* **2014**, *3*, 1088–1091. [CrossRef]
21. Lindhoud, S.; Claessens, M.M. Accumulation of small protein molecules in a macroscopic complex coacervate. *Soft Matter* **2016**, *12*, 408–413. [CrossRef] [PubMed]
22. Martin, N.; Li, M.; Mann, S. Selective uptake and refolding of globular proteins in coacervate microdroplets. *Langmuir* **2016**, *32*, 5881–5889. [CrossRef] [PubMed]
23. Bubb, M.R.; Govindasamy, L.; Yarmola, E.G.; Vorobiev, S.M.; Almo, S.C.; Somasundaram, T.; Chapman, M.S.; Agbandje-McKenna, M.; McKenna, R. Polylysine induces an antiparallel actin dimer that nucleates filament assembly: Crystal structure at 3.5-a resolution. *J. Biol. Chem.* **2002**, *277*, 20999–21006. [CrossRef]
24. Cooper, J.A. Effects of cytochalasin and phalloidin on actin. *J. Cell Biol.* **1987**, *105*, 1473–1478. [CrossRef]
25. Pollard, T.D. Rate constants for the reactions of atp- and adp-actin with the ends of actin filaments. *J. Cell Biol.* **1986**, *103*, 2747–2754. [CrossRef]

26. Nakatani, N.; Sakuta, H.; Hayashi, M.; Tanaka, S.; Takiguchi, K.; Tsumoto, K.; Yoshikawa, K. Specific spatial localization of actin and DNA in a water/water microdroplet: Self-emergence of a cell-like structure. *Chembiochem.* **2018**, *19*, 1370–1374. [CrossRef]
27. Yang, S.; Liu, C.; Guo, Y.; Li, G.; Li, D.; Yan, X.; Zhu, X. Self-construction of actin networks through phase separation-induced ablim1 condensates. *Proc. Natl. Acad. Sci. USA* **2022**, *119*, e2122420119. [CrossRef] [PubMed]
28. Li, G.; Huang, S.; Yang, S.; Wang, J.; Cao, J.; Czajkowsky, D.M.; Shao, Z.; Zhu, X. Ablim1 constructs non-erythroid cortical actin networks to prevent mechanical tension-induced blebbing. *Cell Discov.* **2018**, *4*, 42. [CrossRef]
29. Roof, D.J.; Hayes, A.; Adamian, M.; Chishti, A.H.; Li, T. Molecular characterization of ablim, a novel actin-binding and double zinc finger protein. *J. Cell Biol.* **1997**, *138*, 575–588. [CrossRef]
30. Chen, L.; Jiang, Z.G.; Khan, A.A.; Chishti, A.H.; McKnight, C.J. Dematin exhibits a natively unfolded core domain and an independently folded headpiece domain. *Protein Sci.* **2009**, *18*, 629–636. [CrossRef] [PubMed]
31. Khanna, R.; Chang, S.H.; Andrabi, S.; Azam, M.; Kim, A.; Rivera, A.; Brugnara, C.; Low, P.S.; Liu, S.C.; Chishti, A.H. Headpiece domain of dematin is required for the stability of the erythrocyte membrane. *Proc. Natl. Acad. Sci. USA* **2002**, *99*, 6637–6642. [CrossRef] [PubMed]
32. Rana, A.P.; Ruff, P.; Maalouf, G.J.; Speicher, D.W.; Chishti, A.H. Cloning of human erythroid dematin reveals another member of the villin family. *Proc. Natl. Acad. Sci. USA* **1993**, *90*, 6651–6655. [CrossRef] [PubMed]
33. Vardar, D.; Chishti, A.H.; Frank, B.S.; Luna, E.J.; Noegel, A.A.; Oh, S.W.; Schleicher, M.; McKnight, C.J. Villin-type headpiece domains show a wide range of f-actin-binding affinities. *Cell Motil. Cytoskelet.* **2002**, *52*, 9–21. [CrossRef] [PubMed]
34. Kadrmas, J.L.; Beckerle, M.C. The lim domain: From the cytoskeleton to the nucleus. *Nat. Rev. Mol. Cell Biol.* **2004**, *5*, 920–931. [CrossRef]
35. Koshino, I.; Mohandas, N.; Takakuwa, Y. Identification of a novel role for dematin in regulating red cell membrane function by modulating spectrin-actin interaction. *J. Biol. Chem.* **2012**, *287*, 35244–35250. [CrossRef]
36. Lu, Y.; Hanada, T.; Fujiwara, Y.; Nwankwo, J.O.; Wieschhaus, A.J.; Hartwig, J.; Huang, S.; Han, J.; Chishti, A.H. Gene disruption of dematin causes precipitous loss of erythrocyte membrane stability and severe hemolytic anemia. *Blood* **2016**, *128*, 93–103. [CrossRef]
37. Bergeron-Sandoval, L.P.; Safaee, N.; Michnick, S.W. Mechanisms and consequences of macromolecular phase separation. *Cell* **2016**, *165*, 1067–1079. [CrossRef]
38. Jiang, H.; Wang, S.; Huang, Y.; He, X.; Cui, H.; Zhu, X.; Zheng, Y. Phase transition of spindle-associated protein regulate spindle apparatus assembly. *Cell* **2015**, *163*, 108–122. [CrossRef]
39. Wang, J.; Choi, J.M.; Holehouse, A.S.; Lee, H.O.; Zhang, X.; Jahnel, M.; Maharana, S.; Lemaitre, R.; Pozniakovsky, A.; Drechsel, D.; et al. A molecular grammar governing the driving forces for phase separation of prion-like rna binding proteins. *Cell* **2018**, *174*, 688–699.e16. [CrossRef]
40. Estes, J.E.; Selden, L.A.; Gershman, L.C. Mechanism of action of phalloidin on the polymerization of muscle actin. *Biochemistry* **1981**, *20*, 708–712. [CrossRef]
41. Coué, M.; Brenner, S.L.; Spector, I.; Korn, E.D. Inhibition of actin polymerization by latrunculin a. *FEBS Lett.* **1987**, *213*, 316–318. [CrossRef] [PubMed]
42. Campellone, K.G.; Welch, M.D. A nucleator arms race: Cellular control of actin assembly. *Nat. Rev. Mol. Cell Biol.* **2010**, *11*, 237–251. [CrossRef] [PubMed]
43. Chhabra, E.S.; Higgs, H.N. The many faces of actin: Matching assembly factors with cellular structures. *Nat. Cell Biol.* **2007**, *9*, 1110–1121. [CrossRef] [PubMed]
44. Tobacman, L.S.; Korn, E.D. The kinetics of actin nucleation and polymerization. *J. Biol. Chem.* **1983**, *258*, 3207–3214. [CrossRef]
45. Zuchero, J.B. In vitro actin assembly assays and purification from acanthamoeba. *Methods Mol. Biol.* **2007**, *370*, 213–226.
46. Derivery, E.; Gautreau, A. Assaying wave and wash complex constitutive activities toward the arp2/3 complex. *Methods Enzymol.* **2010**, *484*, 677–695. [PubMed]
47. Ho, H.Y.; Rohatgi, R.; Lebensohn, A.M.; Kirschner, M.W. In vitro reconstitution of cdc42-mediated actin assembly using purified components. *Methods Enzymol.* **2006**, *406*, 174–190.
48. Frank, D.; Frey, N. Cardiac z-disc signaling network. *J. Biol. Chem.* **2011**, *286*, 9897–9904. [CrossRef]
49. Case, L.B.; Zhang, X.; Ditlev, J.A.; Rosen, M.K. Stoichiometry controls activity of phase-separated clusters of actin signaling proteins. *Science* **2019**, *363*, 1093–1097. [CrossRef]
50. Zeng, M.; Chen, X.; Guan, D.; Xu, J.; Wu, H.; Tong, P.; Zhang, M. Reconstituted postsynaptic density as a molecular platform for understanding synapse formation and plasticity. *Cell* **2018**, *174*, 1172–1187.e16. [CrossRef]
51. Naisbitt, S.; Kim, E.; Tu, J.C.; Xiao, B.; Sala, C.; Valtschanoff, J.; Weinberg, R.J.; Worley, P.F.; Sheng, M. Shank, a novel family of postsynaptic density proteins that binds to the nmda receptor/psd-95/gkap complex and cortactin. *Neuron* **1999**, *23*, 569–582. [CrossRef] [PubMed]
52. Han, K.; Holder, J.L., Jr.; Schaaf, C.P.; Lu, H.; Chen, H.; Kang, H.; Tang, J.; Wu, Z.; Hao, S.; Cheung, S.W.; et al. Shank3 overexpression causes manic-like behaviour with unique pharmacogenetic properties. *Nature* **2013**, *503*, 72–77. [CrossRef] [PubMed]

53. Graham, K.; Chandrasekaran, A.; Wang, L.; Ladak, A.; Lafer, E.M.; Rangamani, P.; Stachowiak, J.C. Liquid-like VASP condensates drive actin polymerization and dynamic bundling. *Nat. Phys.* **2023**, 1–12. [CrossRef]
54. Ditlev, J.A. Membrane-associated phase separation: Organization and function emerge from a two-dimensional milieu. *J. Mol. Cell Biol.* **2021**, *13*, 319–324. [CrossRef]

International Journal of
Molecular Sciences

MDPI

Article

Molecular Determinants of Fibrillation in a Viral Amyloidogenic Domain from Combined Biochemical and Biophysical Studies

Juliet F. Nilsson [1], Hakima Baroudi [1], Frank Gondelaud [1], Giulia Pesce [1], Christophe Bignon [1], Denis Ptchelkine [1], Joseph Chamieh [2], Hervé Cottet [2], Andrey V. Kajava [3] and Sonia Longhi [1,*]

1 Laboratoire Architecture et Fonction des Macromolécules Biologiques (AFMB), UMR 7257, Aix Marseille University and Centre National de la Recherche Scientifique (CNRS), 163 Avenue de Luminy, Case 932, CEDEX 09, 13288 Marseille, France
2 IBMM, Centre National de la Recherche Scientifique (CNRS), Ecole Nationale Supérieure de Chimie de Montpellier (ENSCM), Université Montpellier, 34095 Montpellier, France
3 Centre de Recherche en Biologie Cellulaire de Montpellier (CRBM), UMR 5237, Centre National de la Recherche Scientifique (CNRS), Université Montpellier, 34293 Montpellier, France
* Correspondence: sonia.longhi@univ-amu.fr

Abstract: The Nipah and Hendra viruses (NiV and HeV) are biosafety level 4 human pathogens classified within the *Henipavirus* genus of the *Paramyxoviridae* family. In both NiV and HeV, the gene encoding the Phosphoprotein (P protein), an essential polymerase cofactor, also encodes the V and W proteins. These three proteins, which share an intrinsically disordered N-terminal domain (NTD) and have unique C-terminal domains (CTD), are all known to counteract the host innate immune response, with V and W acting by either counteracting or inhibiting Interferon (IFN) signaling. Recently, the ability of a short region within the shared NTD (i.e., PNT3) to form amyloid-like structures was reported. Here, we evaluated the relevance of each of three contiguous tyrosine residues located in a previously identified amyloidogenic motif (EYYY) within HeV PNT3 to the fibrillation process. Our results indicate that removal of a single tyrosine in this motif significantly decreases the ability to form fibrils independently of position, mainly affecting the elongation phase. In addition, we show that the C-terminal half of PNT3 has an inhibitory effect on fibril formation that may act as a molecular shield and could thus be a key domain in the regulation of PNT3 fibrillation. Finally, the kinetics of fibril formation for the two PNT3 variants with the highest and the lowest fibrillation propensity were studied by Taylor Dispersion Analysis (TDA). The results herein presented shed light onto the molecular mechanisms involved in fibril formation.

Keywords: paramyxoviruses; Hendra virus; intrinsically disordered proteins; amyloid-like fibrils; Taylor Dispersion Analysis (TDA); negative staining Transmission Electron Microscopy (ns-TEM); Polyethylene glycol (PEG) precipitation assays; Congo Red; Small-Angle X-ray Scattering (SAXS)

Citation: Nilsson, J.F.; Baroudi, H.; Gondelaud, F.; Pesce, G.; Bignon, C.; Ptchelkine, D.; Chamieh, J.; Cottet, H.; Kajava, A.V.; Longhi, S. Molecular Determinants of Fibrillation in a Viral Amyloidogenic Domain from Combined Biochemical and Biophysical Studies. *Int. J. Mol. Sci.* **2023**, *24*, 399. https://doi.org/10.3390/ijms24010399

Academic Editor: Ludmilla A. Morozova-Roche

Received: 17 November 2022
Revised: 16 December 2022
Accepted: 19 December 2022
Published: 26 December 2022

1. Introduction

The Hendra virus (HeV), together with the closely related Nipah virus (NiV), is a Biosafety Level 4 (BSL-4) pathogen belonging to the *Henipavirus* genus within the *Paramyxoviridae* family [1]. Henipaviruses are zoonotic viruses responsible in humans for severe encephalitis [1]. They are enveloped viruses with a non-segmented, single-stranded RNA genome of negative polarity [2]. Their genome is wrapped by the nucleoprotein (N) within a helical nucleocapsid that is the template used by the viral polymerase for transcription and replication. The polymerase consists of the L protein, which bears all the enzymatic activities, and of the phosphoprotein (P). P serves as an indispensable polymerase co-factor as not only it tethers the L protein onto the nucleocapsid, but also keeps L in a soluble and competent form for transcription and replication [3–5].

The N and P proteins from Henipaviruses encompass long intrinsically disordered regions (IDRs) [6–8], i.e., regions devoid of stable secondary and tertiary structure [9–12]. The *Henipavirus* P protein consists of a long N-terminal intrinsically disordered domain (NTD) and a C-terminal region that possesses both structured and disordered regions (Figure 1) [7,8,13–18].

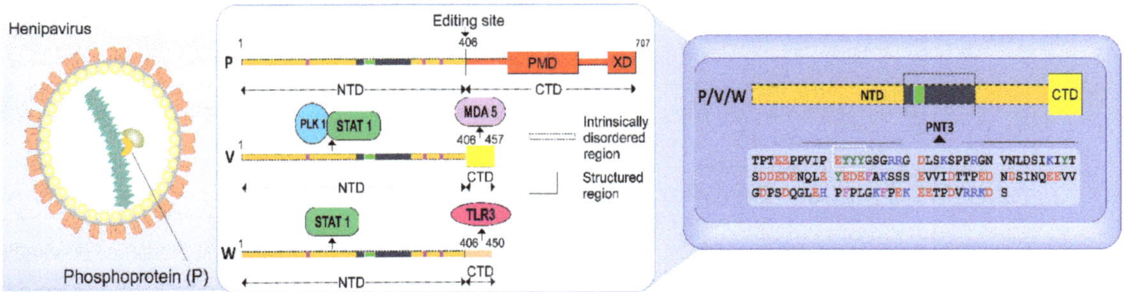

Figure 1. Schematic illustration of the HeV particle, of the organization of the P, V and W proteins and sequence of the HeV PNT3 region shared by the P, V, and W proteins. The left panel displays a scheme of the HeV virion, with the genome encapsidated by the nucleoprotein (in green) and the P (yellow) and the Large (wheat) proteins attached onto the nucleocapsid. The central panel displays a scheme of the P, V and W protein organization, showing that they share a common N-terminal domain (NTD) and have distinct C-terminal domains (CTD). PMD: P multimerization domain; XD: X domain of P. Interaction sites with proteins associated with the host innate immune response are shown. STAT: Signal Transducers and Activators of Transcription; PLK1: Polo-Like Kinase 1; MDA5: Melanoma Differentiation-Associated protein 5; TLR3: Toll-like receptor 3. Pink bars correspond to cysteine residues within NTD. The localization and the sequence of the PNT3 region within NTD is shown in grey and its sequence is displayed in the blue, right panel. The EYYY motif is shown as a green square and is framed within the PNT3 sequence.

As in many paramyxoviruses [19], the P gene from HeV and NiV also encodes the C, V and W non-structural proteins. While the C protein is encoded from an alternative reading frame within the P gene, the V and W proteins (~50 kDa) result from a mechanism of co-transcriptional editing of the P messenger: the addition of either one or two non-templated guanosines at the editing site of the P messenger yields the V and W proteins, respectively (Figure 1). The editing site is located at the end of the NTD-encoding region (Figure 1). The P, V and W proteins therefore share a common NTD but have distinct C-terminal domains (CTDs) (Figure 1). While the CTD of V adopts a zinc-finger conformation [20], the CTD of W is disordered [21].

The V and W proteins are key players in the evasion of the antiviral type I interferon (IFN-I)-mediated response [22–24]. This property relies on their ability to bind to a number of key cellular proteins involved in the antiviral response (Figure 1).

We previously reported the ability of the HeV V protein to undergo a liquid-to-gel transition, with a region within the NTD (referred to as PNT3, aa 200–310) (Figure 1) being identified as responsible for this behavior [25]. In those previous studies, we characterized PNT3 using a combination of biophysical and structural approaches. Congo Red (CR) binding assays, together with negative-staining transmission electron microscopy (ns-TEM) studies, showed that PNT3 forms amyloid-like structures [25]. Noteworthy, Congo red staining experiments provided hints that these amyloid-like fibrils form not only in vitro but also *in cellula* after transfection or infection suggesting a probable functional role. In light of the critical role of the *Henipavirus* V and W proteins in evading the host innate immune response, we previously proposed that in infected cells PNT3-mediated fibrillar aggregates could sequester key cellular proteins involved in the antiviral response. In particular, sequestration of STAT and 14-3-3 proteins would lead to prevention of IFN signaling and abrogation of the NF-κB-induced proinflammatory response [26]. Consistent with the presence of PNT3 within their NTD, the *Henipavirus* W proteins were shown to be able to form amyloid-like fibrils as well [21].

Within PNT3, a motif encompassing three contiguous tyrosines (EYYY) was predicted as an amyloidogenic region [25]. The ArchCandy predictor [27,28] also predicted a fibril architecture in which the three contiguous tyrosines of the motif are part of the first β-strand of a β-strand-loop-β-strand motif (see Figure 4C in [25]). ArchCandy identifies amyloidogenic regions based on their ability to form β-arcades. Indeed, the core structural element of a majority of naturally occurring and disease-related amyloid fibrils is a β-arcade representing a parallel and in register stacks of β-strand-loop-β-strand motifs called β-arches [27]. Substitution of the three contiguous tyrosine residues with three alanine residues yielded a variant (referred to as $PNT3^{3A}$) that was shown to possess a dramatically reduced fibrillation ability, thus providing direct experimental evidence for the predicted involvement of the EYYY motif in building up the core of the fibrils [25].

Here, with the aim of achieving a better understanding of the molecular determinants of HeV PNT3 fibrillation, we designed and characterized a set of additional PNT3 variants that were conceived to either further investigate the EYYY amyloidogenic motif or probe the contribution of the C-terminal half of the protein to the fibrillation process. Results, as obtained by combining various biophysical and structural approaches, show that removal of one out of the three tyrosines of the motif, irrespective of position, is sufficient to lead to a significantly reduced fibrillation ability. In addition, our results revealed that the C-terminal half of PNT3 acts as a natural dampener of the fibrillation process.

2. Results and Discussion

2.1. Influence of pH on the Formation of HeV PNT3 Amyloid-like Fibrils

We previously documented the ability of the PNT3 region of the HeV V protein to form amyloid-like structures [25]. A number of studies reported an impact of pH on both fibril structure and kinetics [29–31]. As a first step towards an in-depth characterization of PNT3 fibrils, and with the aim of selecting appropriate conditions to investigate the kinetics of fibril formation, we sought at assessing the possible impact of pH on the fibrillation process. To this end, we used a previously described method based on the titration of polyethylene glycol (PEG), a crowding agent, to quantitatively assess the relative solubility of proteins [32]. Hence, after optimization of this method (see Materials and Methods), we performed PEG precipitation assays to evaluate the relative solubility of HeV PNT3 at three different pH values, namely 6.5, 7.2 and 8.0 (Figure 2). From this assay, the $PEG_{1/2}$ value, which corresponds to the PEG concentration at which 50% of the protein is still soluble, can be obtained and allows comparing protein aggregation propensities under different conditions [32]. Results display that HeV PNT3 at pH 6.5 shows less relative solubility compared to pH 7.2, indicating a higher aggregation propensity at the lower pH (Figure 2A). This behavior might be at least partly rationalized based on the isoelectric point of the protein (pI = 4.6), as proteins are well known to display minimal solubility at pH values close to their pI. The impact of pH on fibril formation, and the correlation with the $PEG_{1/2}$ value, was confirmed by ns-TEM (Figure 2B). The obtained micrographs show that at pH 6.5 fibrillar aggregates are present even at time 0, while at pH 7.2 equivalent fibrillar aggregates are only observed after an incubation of 96 h at 37 °C (i.e., no fibrillar aggregates can be detected at time 0, Figure 2B). This trend is further confirmed at pH 8, a condition where PNT3 displays the lowest propensity to form fibrillar aggregates (Figure 2). In line with expectations, at pH 4, a value closer to the pI of the protein, the sample was found to exhibit a strongly reduced solubility (Supplementary Figure S1), and ns-TEM studies showed mainly amorphous aggregates rather than fibrillar aggregates (Figure 2B). Notably, in addition to the presence of fibrils, the micrographs obtained at all the pH values also show the presence of amorphous aggregates. These results, beyond advocating for a role of electrostatics in the aggregation process, prompted us to define pH 7.2 as the standard pH value for further studies: the rationale for this choice was that we wanted to be able to monitor the appearance and growth of fibrils over time, while at pH 6.5 fibrils were detected as early as at time zero.

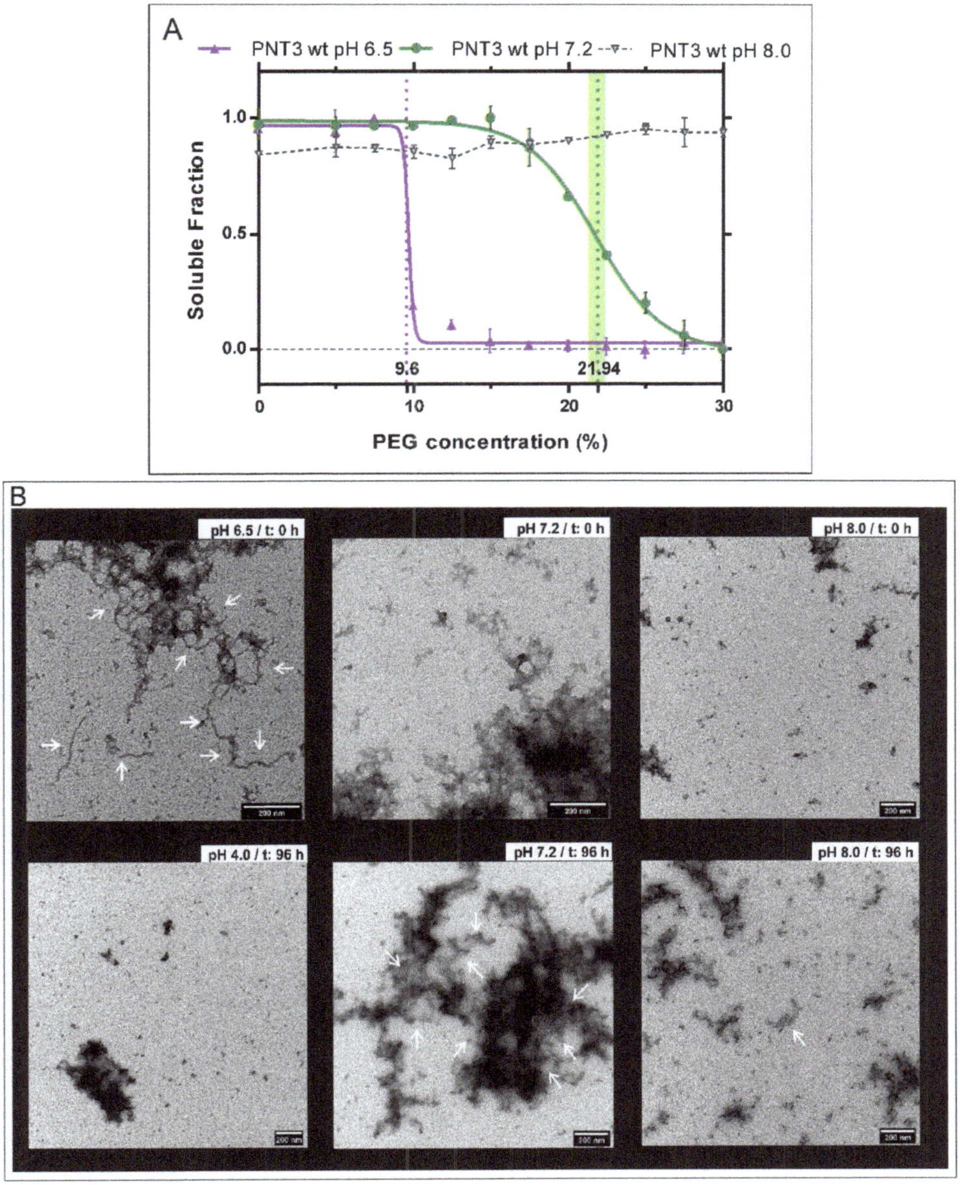

Figure 2. Fibrillation propensity of HeV PNT3 *wild-type* (*wt*) at different pHs. (**A**) PEG assay and relative solubility of HeV PNT3 *wt* at pH 7.2 (green), 6.5 (violet) and 8.0 (light gray). The vertical lines correspond to $PEG_{1/2}$ values (with their 95% confidence intervals) as obtained after a normalization and fitting step to a sigmoid function. Note that the curve obtained at pH 6.5 exhibits poor fitting thus preventing calculation of the confidence interval. Data points at pH 8.0 could obviously not be fitted. (**B**) Ns-TEM of HeV PNT3 *wt* fibrils at four different pH values: 6.5 (at time 0, t: 0 h), 7.2, 8.0 (at time 0 and after 96 h of incubation at 37 °C, t: 96 h) and 4.0 (after 96 h of incubation at 37 °C, t: 96 h). White arrows indicate fibrils.

2.2. Rational Design and Generation of PNT3 Variants

2.2.1. Design of PNT3 Variants Targeting the EYYY Motif

Within the HeV PNT3 region, we previously identified an amyloidogenic motif encompassing three contiguous tyrosine residues (EYYY) [25]. The relevance of this motif for fibril formation was experimentally confirmed through the generation of the PNT3^{3A} variant, in which the three tyrosine residues were replaced with three alanine residues, which showed a reduced ability to form amyloid-like fibrils [25]. With the goal of further investigating this amyloidogenic motif and of unveiling whether all the three tyrosine residues were critical for fibril formation or whether only a subset of them was so, we rationally designed three single-site PNT3 variants where each one of the three contiguous tyrosine residues was replaced with one alanine (PNT3^{A1}, PNT3^{A2} and PNT3^{A3}, Figure 3A). In this context, a construct encoding the corresponding PNT3 region from the NiV V protein (NiV PNT3) was also generated taking advantage of the fact that the corresponding NiV PNT3 motif encompasses only 2 contiguous tyrosine residues (EHYY) (Figure 3A and Supplementary Figure S2).

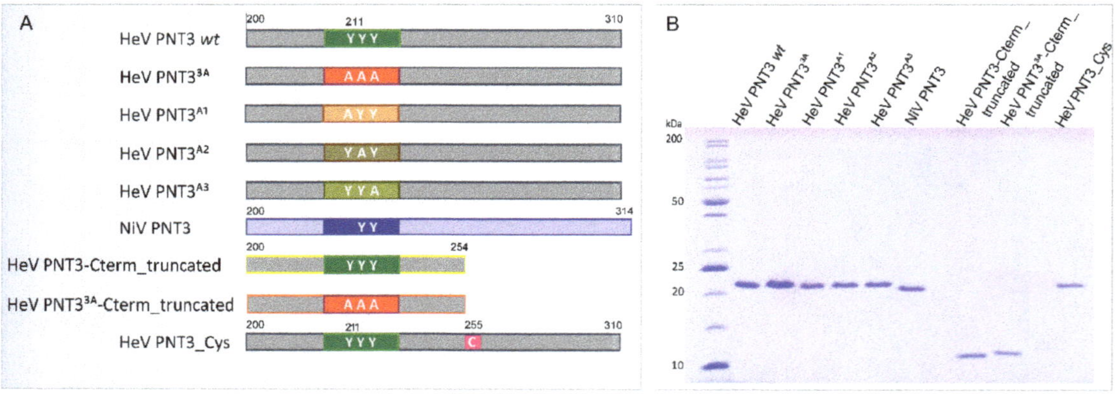

Figure 3. (**A**) Schematic diagram of HeV PNT3 variants designed and constructed in this work. The EYYY motif of the HeV PNT3 is shows in green. (**B**) SDS-PAGE analysis of the purified proteins.

2.2.2. Design of HeV PNT3 Truncated Variants Devoid of the C-Terminal Region

As mentioned above, the EYYY motif plays a significant role in fibril formation. However, the fact that the substitution of the triple tyrosine motif only reduces but does not fully abrogate the ability of PNT3 to form amyloid-like fibrils [25] indicates that other motifs and/or sequence attributes, that remained to be identified, contribute to the fibrillation process. Hence, to deepen the characterization and to assess the possible contribution of the PNT3 C-terminal region, we designed a C-terminally truncated HeV PNT3 variant (PNT3 C-term truncated) that lacks the second half of the protein (Figure 3A). In addition, we also designed a C-terminally truncated HeV PNT3 variant where the three contiguous tyrosines of the EYYY motif were replaced with three alanines (PNT3^{3A} C-term truncated) (Figure 3A).

2.2.3. Design of a HeV PNT3 Variant Bearing a Unique Cysteine

The N-terminal domain, shared by the HeV P, V, and W proteins, has 3 cysteines distributed along the sequence (Figure 1). Recently, Pesce & Gondelaud et al. suggested that disulfide bridges could be involved in preventing aggregation of the W protein [21]. Thus, in this context, we reasoned that a HeV PNT3 variant bearing a cysteine residue could be useful to investigate the possible impact of disulfide bridge-mediated protein dimerization on the fibrillation abilities of PNT3. We targeted for cysteine substitution the unique alanine residue of PNT3 (Ala255) to yield PNT3 variant A255C (PNT3_Cys)

(Figure 3A). The rationale for choosing an alanine, rather than a serine residue which would have enabled a more isosteric substitution, was to introduce as much a conservative as possible substitution, while preserving the content in OH groups, which might play a role in the establishment of stabilizing inter-chain interactions in the core of the fibrils.

2.2.4. Expression and Purification of the PNT3 Variants

All the proteins were expressed in *E. coli* as hexahistidine tagged forms with no solubility tag. The proteins were purified from the total fraction of the bacterial lysate under denaturing conditions by Immobilized Metal Affinity Chromatography (IMAC) and size exclusion chromatography (SEC). The purity of the final purified products was assessed by SDS-PAGE (Figure 3B). The identity of all the variants was confirmed by mass spectrometry (MS) analysis of peptides resulting from the tryptic digestion of each protein (Supplementary Figure S3).

2.2.5. Conformational Characterization of the PNT3 Variants

First, we evaluated the hydrodynamic properties of all the variants through analytical SEC. Table 1 shows the Stokes radius (R_S) value inferred for each protein, along with the Compaction Index (CI) associated with each variant. By comparing the mean measured Stokes radius (R_S^{OBS}) with the theoretical Stokes radii expected for the various conformational states (i.e., R_S^{NF}: natively folded protein; R_S^{PMG}: premolten globule, PMG; R_S^{U}: fully unfolded form; R_S^{IDP}: IDP), all the proteins were found to have R_S values consistent with a PMG state [33]. Their compaction indexes are relatively close to each other, with the notable exception of the HeV PNT3_C-term_truncated variant that is much more compact. Strikingly, the introduction of the triple alanine motif in the context of the truncated variant leads to a more extended conformation, a phenomenon already observed, although with a borderline significance, in the context of the full-length PNT3 protein (*cf.* HeV PNT3 *wt* and HeV PNT3^{3A} in Table 1). These results suggest that the second half of the protein is a determinant of chain expansion, with this effect being counteracted by the presence of the triple alanine motif.

Table 1. Stokes radii (R_S^{OBS}, Å) of the PNT3 variants as inferred from the elution volume of the major SEC peak. Shown are also the expected values for the various conformational states, along with the ratios between the R_S^{OBS} and each R_S state, and compaction index (CI) values.

Proteins	Mass	R_S^{OBS}	R_S^{NF}	R_S^{PMG}	R_S^{U}	R_S^{IDP}	R_S^{OBS}/R_S^{NF}	R_S^{OBS}/R_S^{PMG}	R_S^{OBS}/R_S^{U}	R_S^{OBS}/R_S^{IDP}	CI
HeV PNT3 *wt*	15198	**27.0 ± 0.3**	19.4	**27.9**	33.9	30.0	1.39	0.97	0.80	0.90	0.47 ± 0.02
HeV PNT3^{3A}	14922	**27.5 ± 0.4**	19.3	**27.7**	33.5	30.0	1.42	0.99	0.82	0.92	0.42 ± 0.03
HeV PNT3^{A1}	15106	**28.4 ± 0.4**	19.4	**27.9**	33.8	30.0	1.36	0.94	0.78	0.88	0.37 ± 0.03
HeV PNT3^{A2}	15106	**28.1 ± 0.3**	19.4	**27.9**	33.8	30.0	1.31	0.91	0.76	0.85	0.39 ± 0.02
HeV PNT3^{A3}	15106	**27.5 ± 0.2**	19.4	**27.9**	33.8	30.0	1.42	0.99	0.81	0.92	0.44 ± 0.01
NiV PNT3	14928	**27.3 ± 0.7**	19.3	**27.7**	33.5	30.5	1.41	0.99	0.82	0.90	0.44 ± 0.05
HeV PNT3 C-term_truncated	9020	**19.0 ± 0.1**	**16.1**	**22.6**	25.8	22.7	1.17	0.84	0.73	0.83	0.71 ± 0.01
HeV PNT3^{3A} C-term_truncated	8743	**20.1 ± 0.6**	16.0	**22.4**	25.4	22.7	1.26	0.90	0.79	0.89	0.56 ± 0.06
HeV PNT3_Cys	15230	**28.0 ± 0.4**	19.5	**28.0**	33.9	30.0	1.44	1.00	0.83	0.93	0.41 ± 0.03

R_S^{OBS}: experimentally observed Stokes radius (mean value and s.d. from three independent experiments); R_S^{NF}: R_S expected for a natively folded (NF) form; R_S^{PMG}: R_S expected for a pre-molten globule (PMG); R_S^{U}: R_S expected for a fully unfolded form; R_S^{IDP}: R_S expected for an IDP based on the simple power law model; all radii are given in Å. Mass: molecular mass (Daltons) calculate d from the amino acid sequence of the recombinant protein. Compaction index (CI) mean values and s.d., as obtained from three independent experiments.

In order to evaluate the secondary structure content of each variant, we performed a Circular Dichroism (CD) analysis in the far ultraviolet (UV) region. All the variants present a spectrum typical of a disordered protein lacking any stable organized secondary structure, as judged from the large negative peak centered at 200 nm, and from the low ellipticity in the 220–230 nm region and at 190 nm (see [34] and references therein cited) (Supplementary Figure S4). These results indicate that the introduced substitutions and/or the truncation impact only marginally, if at all, the secondary structure content of the protein. They also indicate that NiV PNT3 has a secondary structure content very close to that of its HeV counterpart. The finding that the CD spectra of PNT3 variants bearing the triple alanine motif are virtually superimposable onto those of variants bearing either the naturally occurring triple tyrosine motif or just one Tyr to Ala substitution, rules out the possibility that the expansion effect driven by the triple alanine motif, as observed in SEC studies, may arise from the presence of a transiently populated α-helix encompassing the motif. Therefore, the mechanism underlying the counteracting effect exerted by the triple alanine motif on chain compaction remains to be elucidated.

To achieve a more quantitative description of the conformational properties of the variants, we carried out Small-Angle X-ray Scattering (SAXS) studies coupled to SEC (SEC-SAXS). We selected a set of representative variants (i.e., HeV $PNT3^{3A}$, NiV PNT3, HeV PNT3_C-term_truncated, and HeV $PNT3^{3A}$_C-term_truncated) along with HeV PNT3 *wt*. Although we already previously reported SEC-SAXS studies of HeV PNT3 *wt* [25], this sample was herein again investigated under exactly the same conditions used for the other variants, so as to enable meaningful comparisons. For all the five PNT3 proteins, linearity of the Guinier region in the resulting scattering curves (Supplementary Figure S5A) allowed meaningful estimations of the radius of gyration (R_g) (Table 2). The R_g values obtained for HeV $PNT3^{3A}$, NiV PNT3, and HeV PNT3 *wt* are very close to each other's clustering in a group with R_g around 37–39 Å. As expected, the two truncated variants have smaller, and close to each other's, R_g values (Table 2).

Table 2. R_g and D_{max} as obtained from SEC-SAXS studies and expected values for the various conformational states.

Proteins	I(0) cm^{-1}	R_g (Å) (Guinier)	D_{max} (Å)	R_g^{IDP} (Å)	R_g^{U} (Å)
HeV PNT3 *wt*	$0.030 \pm 1.8 \times 10^{-4}$	36.67 ± 0.41	140	32.6	35.9
HeV $PNT3^{3A}$	$0.022 \pm 8.1 \times 10^{-5}$	39.48 ± 0.33	144	32.6	35.9
NiV PNT3	$0.030 \pm 8.2 \times 10^{-5}$	37.37 ± 0.21	147	33.1	36.5
HeV PNT3 C-term_truncated	$0.018 \pm 4.1 \times 10^{-5}$	27.53 ± 0.14	115	24.5	25.9
HeV $PNT3^{3A}$ C-term_truncated	$0.015 \pm 5.6 \times 10^{-5}$	27.34 ± 0.21	119	24.5	25.9

I(0): Intensity at zero angle as determined from Guinier approximation; R_g Guinier: R_g values as obtained from Guinier approximation; D_{max}: maximal intramolecular distance from P(r). R_g^{IDP}: R_g expected for an IDP based on the simple power-law model. R_g^{U}: theoretical R_g value expected for a chemically denatured (U) protein.

Notably, all obtained experimental R_g values are close to the theoretical R_g^{U}, corresponding to chemically denatured (U) proteins, and hence reflecting a highly extended conformation. Because of this, the R_g-based CI could not be computed, the numerator ($R_g^{U} - R_g^{OBS}$) in Equation (10) being ≤ 0 (see Section 3). These results are in contrast with the previous SEC results, where the variants were found to adopt a PMG conformation. A possible explanation for this might be related to the differences in the buffer used in the two techniques. As expected, the five variants were all found to be disordered as judged from the presence of a plateau in the normalized Kratky (Supplementary Figure S5B) and Kratky-Debye plots (Supplementary Figure S5C). However, the two truncated variants, and particularly the HeV PNT3 C-term truncated one, showed a slight deviation from the pure random-coil regime as observed in the normalized Kratky plot. This deviation is consistent with a slightly more compact conformation, in line with the R_S-based CI values discussed above.

2.3. *Relevance of the PNT3 EYYY Motif in Fibrillation Abilities*

2.3.1. Aggregation Propensity of the EYYY Motif PNT3 Variants

With the aim of elucidating the contribution of each tyrosine in the EYYY motif to the fibrillation process, we first assessed the aggregation propensity of the set of variants bearing alanine substitutions within the EYYY amyloidogenic motif. To this end, we took advantage of the same PEG solubility assay described above [32]. For each of the five EYYY variants (PNT3^{3A}, PNT3$^{A1, A2, A3}$ and NiV PNT3), we therefore carried out PEG solubility assays which enabled ranking them based on their estimated $PEG_{1/2}$ value.

The PNT3^{3A} variant shows a significant increase in the relative solubility compared to PNT3 HeV *wt* (Figure 4A,B), a result in agreement with previous findings that pointed out a much lower propensity to form amyloid-like fibrils for PNT3^{3A} [25]. The three variants where only one Tyr was replaced, however, show no significant difference in the $PEG_{1/2}$ values compared to HeV PNT3 *wt* (Figure 4B), suggesting that the removal of one tyrosine does not affect the aggregation propensity of PNT3. By contrast, and interestingly, NiV PNT3 (EHYY) displays an intermediate relative solubility between HeV PNT3 *wt* and HeV PNT3^{3A} (Figure 4A). In light of the results obtained with the HeV PNT3 variants bearing two tyrosines (i.e., PNT3^{A1}, PNT3^{A2} and PNT3^{A3}), the intermediate aggregation propensity of NiV PNT3 more likely arise from differences in the amino acid context rather than from the absence of just one tyrosine.

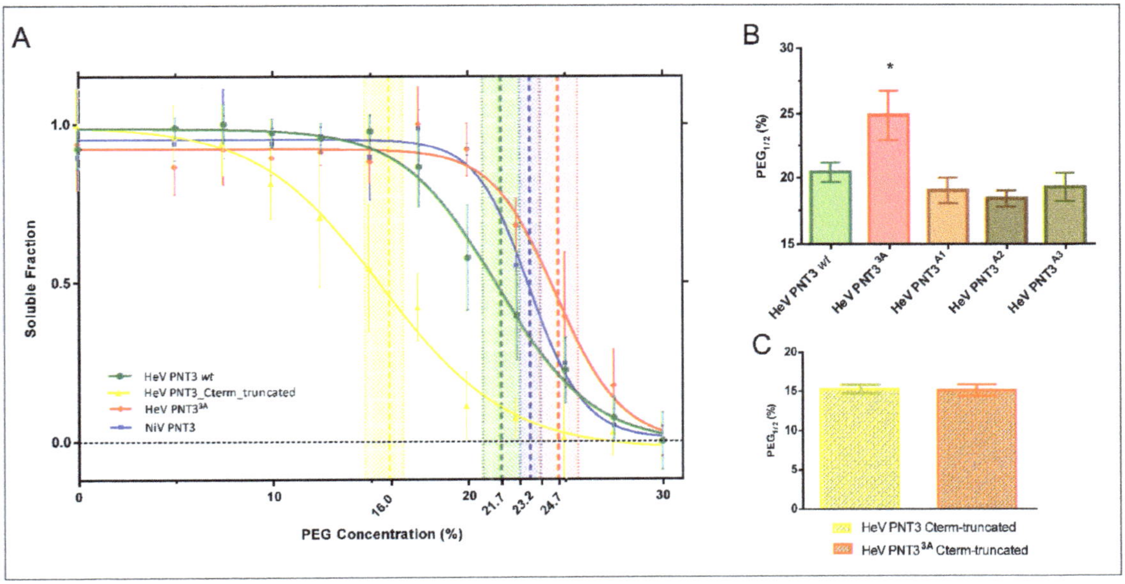

Figure 4. PEG solubility assays for the set of PNT3 variants. (**A**) Soluble fraction of each variant at different PEG concentrations. HeV PNT3 *wt* (green), HeV PNT3^{3A} (red), HeV PNT3_C-term_truncated (yellow), NiV PNT3 (blue). Vertical lines represent $PEG_{1/2}$ values including their 95% confidence intervals obtained after a normalization and fitting step to a sigmoid function. (**B**) $PEG_{1/2}$ values obtained for the *wt*, the triple alanine variant and the single alanine HeV PNT3 variants. (**C**) $PEG_{1/2}$ values obtained for the C-terminally truncated variants. The asterisk indicates statistically significant differences (One-way Anova test, p-value < 0.05).

2.3.2. Congo Red Binding Abilities of PNT3 EYYY Motif Variants

Congo Red is a widely used dye to document the presence of amyloids: binding of this dye to cross β-sheet structures in fact leads to hyperchromicity and a red shift in the absorbance maximum of the CR spectrum. Hence, to further characterize PNT3 EYYY motif variants, we took advantage of CR binding assays. We compared the binding abilities of the PNT3^{A1}, PNT3^{A2}, PNT3^{A3} and NiV variants to those of both PNT3 *wt* and PNT3^{3A}. We spectrophotometrically measured the red shift (from 497 nm to 515 nm) in the absorbance maximum of the CR spectrum of each sample following a four or seven days incubation at 37 °C. Results shown in Figure 5 indicate that all the variants promote a shift in the CR spectrum whose amplitude increases with incubation time, suggesting that all the variants are able to progressively form amyloid-like fibrils or at least structures able to bind CR.

Our previous findings that pointed out that the HeV PNT3^{3A} variant has a reduced ability to bind CR compared to HeV PNT3 *wt* [25] were confirmed here (Figure 5). Two of the three single alanine variants (PNT3^{A1}, PNT3^{A3}) show an intermediate behavior between HeV PNT3 *wt* and HeV PNT3^{3A}, but without significant differences with either the *wt* or the PNT3^{3A} variant, while HeV PNT3^{A2} displays an ability to bind CR significantly higher than that of HeV PNT3^{3A} and similar to that of HeV PNT3 *wt* (Figure 5). These results therefore suggest that the central tyrosine in the motif could be less relevant to the fibrillation process. Notably, the NiV PNT3 variant shows a significantly decreased ability to bind CR compared to PNT3 *wt*, similar to that of the triple alanine variant. Thus, as already observed for the aggregation propensity, the reduced CR binding ability of NiV PNT3 likely results from its amino acid context rather than from the fact that it lacks a tyrosine in the motif.

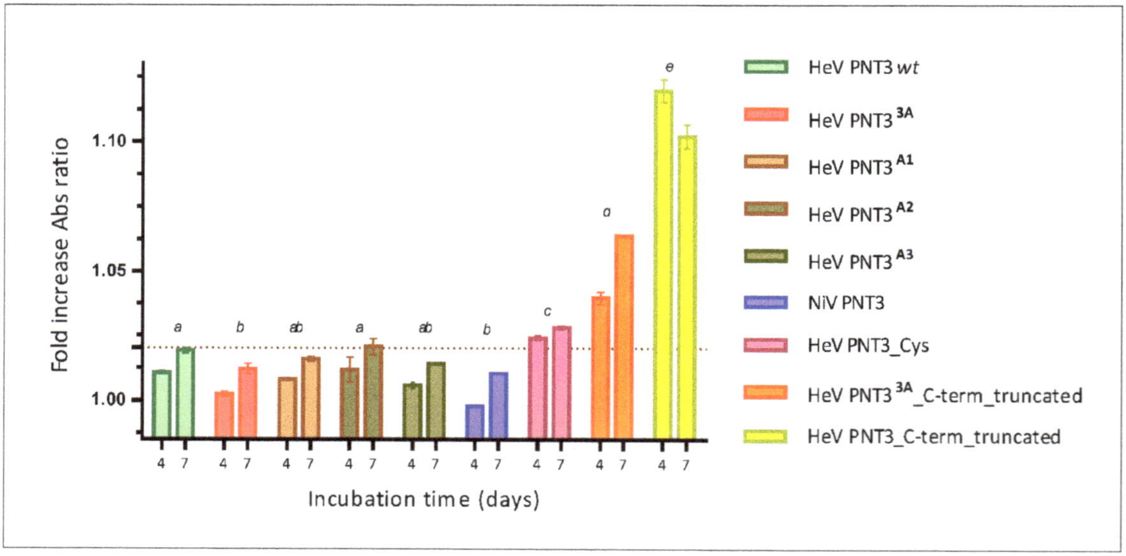

Figure 5. Congo Red binding assay of the set of PNT3 variants. The ability to bind CR is represented by the fold increase in the ratio between the absorbance at 515 and at 497 nm, with respect to a sample containing CR alone, of PNT3 samples at 20 μM after 4 and 7 days of incubation at 37 °C. The error bar corresponds to the standard deviation, with n = 3. Different letters indicate statistically significant differences ($p < 0.05$) (Two-way ANOVA test; *ab* means lack of statistically significant differences with respect to *a* or *b*).

2.3.3. Propensity and Time-Dependance of Fibrillation of the PNT3 EYYY Motif Variants Using Negative-Staining Transmission Electron Microscopy (ns-TEM)

To directly document fibril formation by the PNT3 EYYY variants as a function of time and to obtain orthogonal experimental evidence corroborating the CR binding assay results, we next carried out ns-TEM studies. These analyses were performed for each of the variants at 0, 24 and 96 h of incubation at 37 °C (Figure 6). As shown Figure 6, although HeV PNT3 *wt* in the selected conditions does not form fibrils at time 0, after 24 h of incubation it forms short fibrils, which evolve to long fibrils after 96 h. In line with the CR binding results, ns-TEM studies confirmed that all PNT3 EYYY variants, including NiV PNT3, are able to form amyloid-like fibrils but with a significantly decreased ability compared to the *wt*. Specifically, after 24 h short fibrils can be observed in most variants, except for HeV PNT3^{3A}. In an attempt at identifying possible significant differences among the EYYY variants in spite of their overall similar behavior, we performed a comprehensive analysis of the number and length of fibrils detected for each of them (Figure 7). This analysis revealed no significant differences in the length of the fibrils obtained after 96 h of incubation among the 3 variants with a single alanine (PNT3^{A1}, PNT3^{A2} and PNT3^{A3}). However, a slight, though significant, increase is observed in the PNT3^{A3} variant in the number of fibrils found *per* picture compared to the other two single alanine variants (Figure 7). This slightly higher number of short fibrils observed for the PNT3^{A3} variant may indicate that the last tyrosine in the motif contributes less to the nucleation process compared to the two other tyrosines. Remarkably, the same analysis showed that the NiV PNT3 variant forms significantly longer fibrils compared to the PNT3^{A1}, PNT3^{A2} and PNT3^{A3} variants, therefore confirming that other sequence attributes, beyond the amyloidogenic motif, contribute to the fibrillation process. In other words, fibril formation would rely not only on stabilizing contacts mediated by the E(H/Y)YY motif but also on additional stabilizing interactions established by other protein regions.

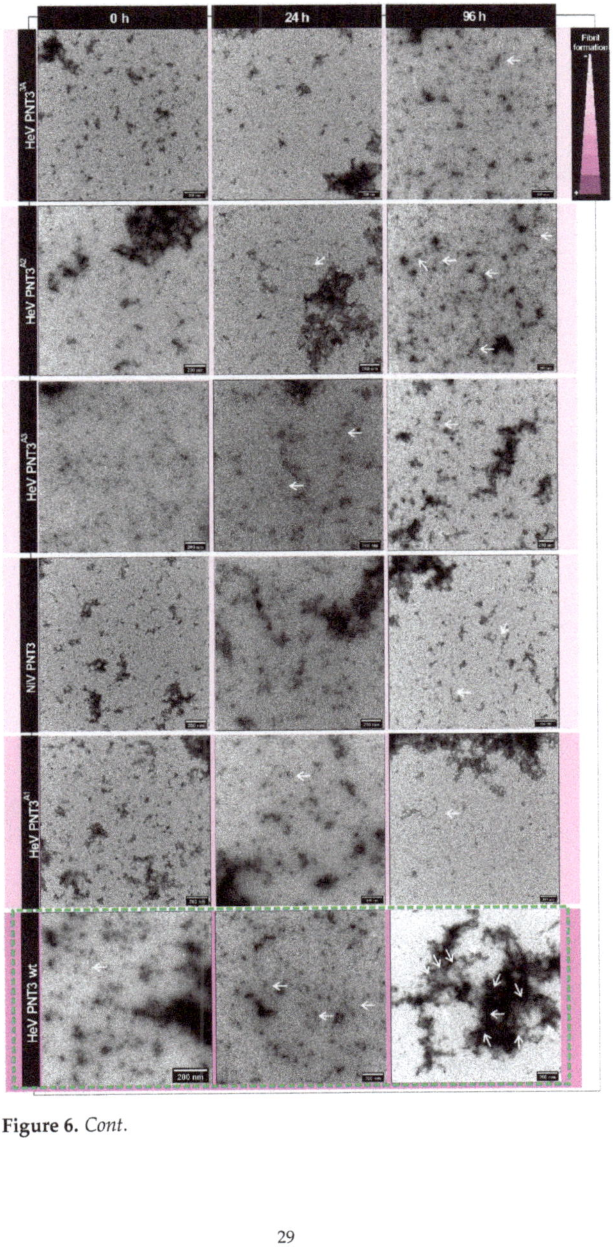

Figure 6. *Cont.*

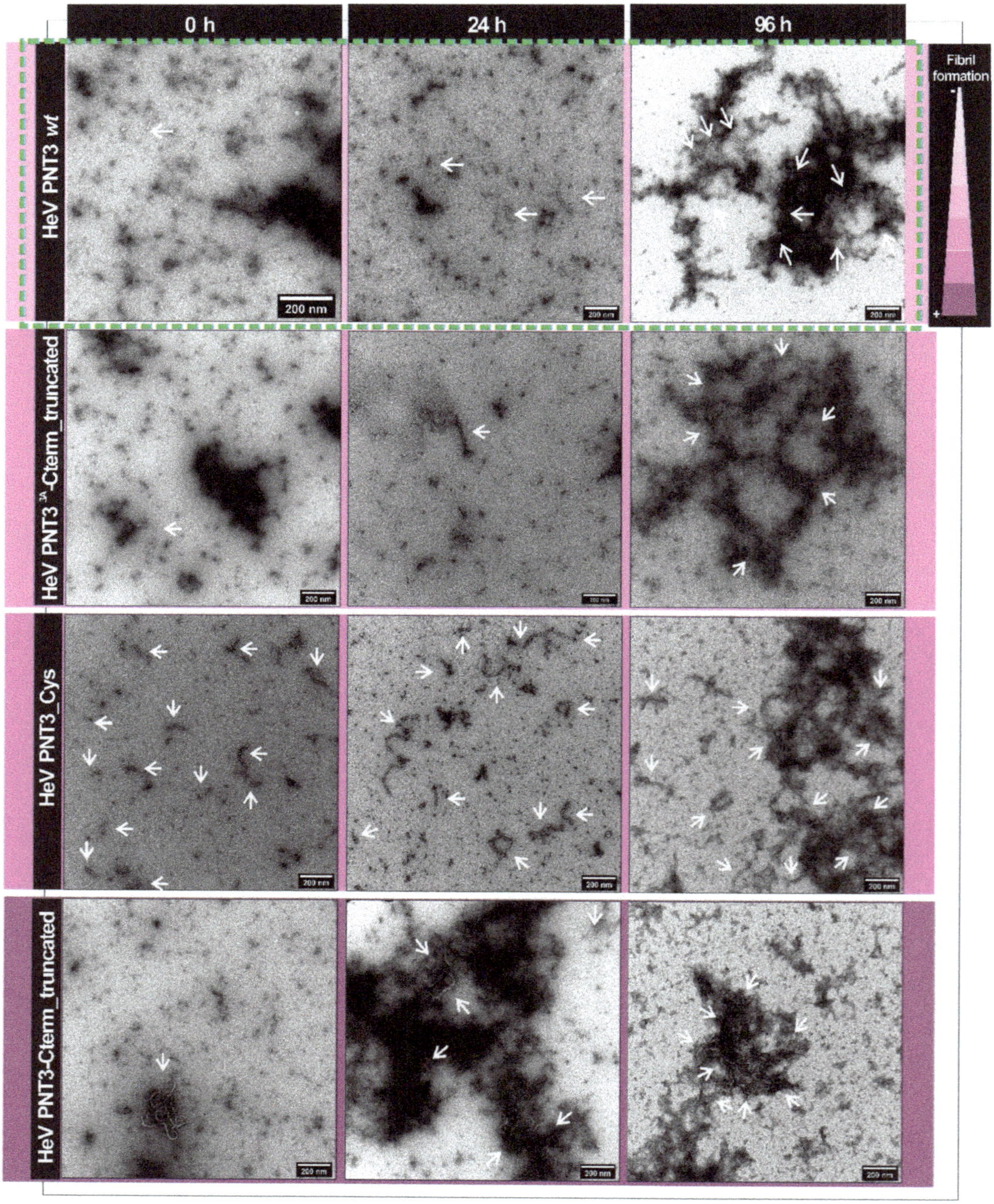

Figure 6. Fibril formation as a function of time by ns-TEM. Ns-TEM analysis of PNT3 variants (200 μM) at time zero and after 24 h or 96 h of incubation at 37 °C. Note that in all cases, samples were diluted to 40 μM prior to being deposited on the grid. White arrows indicate fibrils. The purple gradient represents the propensity to form fibrils. The green dashed box indicates HeV PNT3 *wt* results.

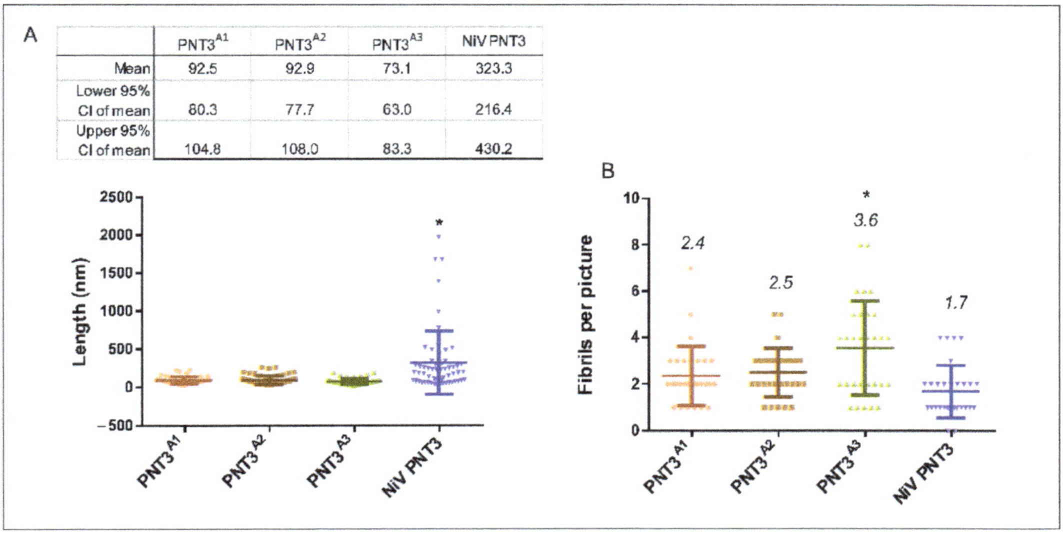

	PNT3A1	PNT3A2	PNT3A3	NiV PNT3
Mean	92.5	92.9	73.1	323.3
Lower 95% CI of mean	80.3	77.7	63.0	216.4
Upper 95% CI of mean	104.8	108.0	83.3	430.2

Figure 7. Fibril length of EYYY variants. (**A**) For each variant, statistics on fibril length were obtained from analysis of the contour length of 60 fibrils after 96 h of incubation at 37 °C. The upper table shows the main statistical results. (**B**) Number of fibrils detected *per* picture for each variant. Upper number indicates the mean value of each variant. The analysis was done using the ImageJ software. The asterisk indicates statistically significant differences (One-way Anova test, p-value < 0.05). Circles: PNT3^{A1}; squares: PNT3^{A2}; triangles: PNT3^{A3}, inverted triangles: NiV PNT3.

Altogether, CR binding assays and ns-TEM enabled the documenting of slight differences among the PNT3^{A1}, PNT3^{A2} and PNT3^{A3} variants. Although other regions beyond the amyloidogenic motif seemingly contribute to the fibrillation process, results strongly suggest that the absence of a single tyrosine in the EYYY motif leads to a significant decrease in the ability to form fibrils irrespective of the position. In light of the finding that the PNT3^{A1}, PNT3^{A2} and PNT3^{A3} variants are able to form short fibrils, whose length does not increase under the experimental conditions herein used, we can assume that the three tyrosines mainly play a role in the elongation phase. The present results support the involvement of tyrosines in the formation of amyloid-like fibrils, with π-π stacking and H-bonding interactions between tyrosines likely allowing to form cross-β-like architectures, as previously suggested [35].

2.4. Impact of the HeV PNT3 C-Terminal Region in Fibrillation Abilities

2.4.1. Aggregation Propensity of C-Terminally Truncated PNT3 Variants

Taking into account the results presented above that lend support to a scenario where other motifs and/or sequence attributes beyond the amyloidogenic motif could be involved in the fibrillation process, we decided to investigate the impact of the PNT3 C-terminal region. We first studied the aggregation propensity of both PNT3 C-terminally truncated variants. Figure 4A shows that the PNT3 C-term truncated variant has the lowest $PEG_{1/2}$ value, indicating a strikingly decreased relative solubility compared to all the full-length PNT3 variants. Notably, there are no significant differences in the $PEG_{1/2}$ values between the PNT3 C-term truncated and its triple alanine mutant (PNT3^{3A} C-term truncated) (Figure 4C). These results indicate that removal of the C-terminal region has a strong impact on the aggregation propensity, with this effect being insensitive to the sequence context of the EYYY motif.

2.4.2. CR Binding Ability of C-Terminally Truncated PNT3 Variants

Motivated by the results obtained by the PEG solubility assays pointing to a much higher aggregation propensity of both truncated variants, we next carried out CR binding assays. As shown in Figure 5, the HeV PNT3 C-terminal truncated variant displays a significantly increased ability to bind CR compared to full-length HeV PNT3 *wt*. In striking contrast with PEG solubility assays that detected no differences in terms of aggregation propensities between the two truncated variants, CR binding assays revealed significant differences between the two variants. In particular, the PNT3^{3A} truncated variant has a much-decreased ability to bind CR with respect to the truncated variant bearing a native EYYY motif, hence displaying an intermediate behavior between the full- length and truncated HeV PNT3 *wt* (Figure 5). Therefore, it can be concluded that the C-terminal region, far from being inert, negatively affects the ability of the protein to form CR-binding structures. Albeit the PNT3^{3A} truncated variant binds more CR than the *wt* variant, the contribution of the EYYY motif is evident when the two truncated variants are compared. These results, however, suggest that the EYYY motif would have only a marginal role in driving the formation of CR-binding structures in the context of C-terminally truncated form.

2.4.3. Propensity and Kinetics of Fibrillation of C-Terminally Truncated PNT3 Variants Using ns-TEM Studies

In order to ascertain whether the increased binding to CR and decreased solubility of the truncated variants is actually reflected in a higher fibrillation ability, we analyzed them by ns-TEM studies. In line with expectations, Figure 6 clearly shows a significantly higher abundance of fibrils, as well as an increased fibril length, in the PNT3 C-terminal truncated variant at short incubation times, thus confirming its increased fibrillation potential. Notably, the PNT3^{3A} truncated variant displays a decreased fibrillation ability compared to its *wt* counterpart, similar to the full-length HeV PNT3 *wt* (Figure 6). These findings suggest that the removal of the C-terminal region results in an acceleration of the fibrillation rate, reflecting an interaction between this region and the rest of the sequence that negatively affects the kinetics of fibril formation. Remarkably, similar results were previously documented in the case of the aggregation of α-synuclein (α-syn), an extensively characterized protein associated with neurodegeneration and whose transition from a soluble to a fibrillar form is thought to contribute to pathogenesis [36]. Compelling experimental evidence indicates that C-terminal truncation of α-syn promotes in vitro oligomer and fibril formation (see [37] and references therein cited). The middle region of α-syn, referred to as "non-amyloid component" (NAC) domain, forms the core of α-syn filaments. The C-terminal region of α-syn can adopt conformations in which the C-terminus contacts the hydrophobic NAC domain thus shielding it from pathological templating interactions [37]. The negative charge of the C-terminal region has been proposed to contribute to this self-chaperoning activity via the establishment of electrostatic interactions [37].

In an attempt at rationalizing the observed self-inhibitory effect of the C-terminal region of PNT3 on fibril formation, we analyzed the charge distribution within the PNT3 sequence using the CIDER server (http://pappulab.wustl.edu/CIDER/, accessed on 24 October 2022) [38] (Supplementary Figure S6 and Table S1). Although full-length PNT3, and its constituent N-terminal and C-terminal regions fall in very close positions in the phase diagram plot, the C-terminal region has a higher fraction of negatively charged residues compared to the N-terminal region (Supplementary Figure S6 and Table S1). In addition, the full-length form of PNT3 and the truncated variant strongly differ in their net charge at pH 7.0. Taking into account the strong impact of pH on PNT3 fibrillation, where a decrease to pH 6.5 strongly promotes fibrillation and leads to a behavior similar to that of the truncated PNT3 variant at pH 7.2, it is conceivable that the results obtained with the truncated variant could be, at least partly, accounted for by electrostatics, as in the case of α-syn. A plausible alternative scenario for the self-inhibitory effect of the C-terminal region of PNT3 on fibril formation could be the following: the disordered region downstream the fibril core may hamper fibril formation by slowing the disorder-to-order transition expected to take place in the core of the fibril, through either a purely entropic effect or through a combination of enthalpy and entropy as already documented in the case of fuzzy appendages adjacent to molecular recognition elements (for examples see [39,40]).

Altogether, these findings advocate for a key role of the C-terminal region in regulating the fibrillation properties of PNT3. In particular, the C-terminal region may act either as a molecular shield, as in the case of α-syn [37], or by slowing down the rate of folding of the core of the fibrils, with this property, irrespective of the underlying mechanisms, being also possibly relevant to biological function in vivo. Definite answers on the precise molecular mechanisms and on the possible biological relevance await future studies.

2.5. Impact of a Cysteine in the HeV PNT3 Sequence on Fibrillation Abilities

With the aim of elucidating the possible impact of a disulfide bridge-mediated PNT3 dimerization on its fibrillation abilities, we first studied the ability of the HeV PNT3 variant bearing a cysteine residue (PNT3_Cys) to bind CR. As shown in Figure 5, this variant shows a significant increased ability to bind CR compared to HeV PNT3 *wt*. Subsequently, the ability of this variant to form fibrils was assessed by ns-TEM in the same conditions used for PNT3 *wt*. Figure 6 shows that the PNT3_Cys variant is able to form fibrils even at time 0, indicating a higher fibrillation propensity compared to HeV PNT3 *wt*. Notably, the PNT3_Cys was the unique variant displaying an enrichment in shortened fibrils (Figure 6). These findings suggest that the presence of one cysteine in the sequence mainly impacts the nucleation phase. We reasoned that the peculiar fibrillation behavior of this variant could result from disulfide bridge-mediated protein dimerization. However, the addition of DTT was found to have a negligible impact, as judged from the presence at time 0 of very short fibrils and of long fibrils at 96 h (Supplementary Figure S7), resulting in an intermediate behavior between PNT3 *wt* and PNT3_cys under non-reducing conditions. Thus, the peculiar behavior of this variant cannot be ascribed to protein dimerization and rather stems from other intrinsic properties that remain to be elucidated.

2.6. Characterization of the Aggregation Process by Taylor Dispersion Analysis

We next sought at shedding light onto the fibrillation kinetics using Taylor Dispersion Analysis (TDA) [41]. TDA is a new technique in the field of protein aggregation that has the notable advantages of being able to (i) capture intermediate species, (ii) quantify early and late-stage aggregates, and (iii) provide both kinetic and equilibrium constants. In addition, TDA is not dominated by aggregates (as opposite to scattering techniques), and is thus ideally suited to study the molecular mechanisms of protein fibrillation. Recently, this technique successfully allowed obtaining a complete quantitative picture of the aggregation process of both Aβ(1-40) and Aβ(1-42), including the size of the oligomers and protofibrils, the kinetics of monomer consumption, and the quantification of different early- and late-formed aggregated species [41–43].

In light of the fibrillation properties of all the PNT3 variants herein investigated, as revealed by the ensemble of studies described above, we decided to focus on the two variants with the most extreme phenotype, namely the variant with the highest fibrillation propensity, i.e., HeV PNT3_C-term_truncated, and the least fibrillogenic variant, i.e., $PNT3^{3A}$. Figure 8A,C show a three-dimensional overview of the obtained taylorgrams during the aggregation process of the two selected variants. In the case of the truncated variant, two major species were detected over time: the monomer (hydrodynamic radius, *Rh*, of ~3 nm) (Figure 8B) and large aggregates with an $Rh \geq 400$ nm (see spikes at the beginning of the run in Figure 8A). Figure 8B shows that the *Rh* value of the soluble fraction increases with time, reaching about double its initial value at the end of the incubation, when this species is present in low proportions. Figure 8B also shows that the monomeric population slowly decreases with time. The decrease of the peak area (*Y*) of the monomeric population could be fitted using a first order exponential decay:

$$Y = A_1 \times \exp\left(-\,t/t_1\right) + Y_0 \tag{1}$$

where $Y_0 = 6.13 \pm 0.20$; $A_1 = 7.63 \pm 0.34$ and $t_1 = 12.06 \pm 1.32$ h. This resulted in a good quality fit as judged from the R^2 value of 0.9394. From the fit, the kinetics of aggregation could be deduced, with a characteristic aggregation time of about $t_1 = 12$ h. From Figure 8B, the evolution in the spikes area gives an estimation of the quantity of large aggregated species entering the capillary. The proportion of these aggregated species increases with time until about 48 h, before decreasing because their size becomes too large to enter the capillary or because of precipitation in the sample vial. The presence of spikes at very short times of incubation suggests a fast aggregation process of the truncated form, while the absence of significant intermediate species suggests that the monomers add to the already present aggregates and elongate them.

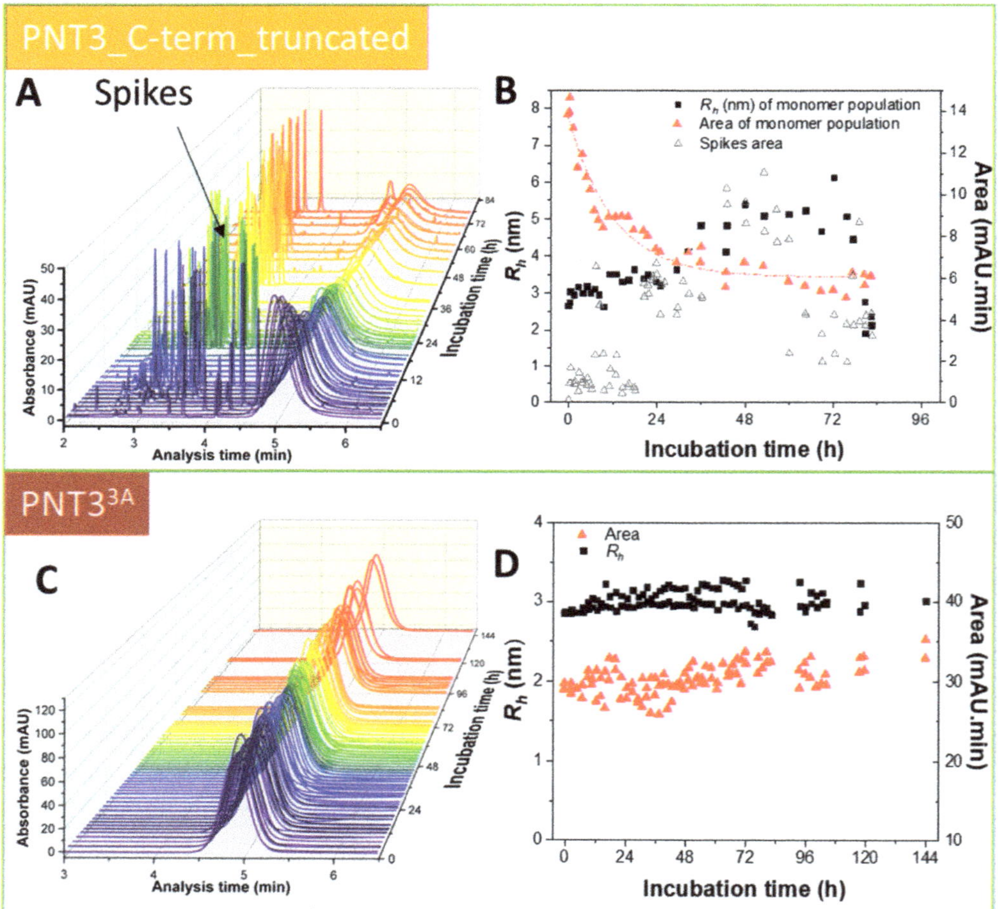

Figure 8. Kinetics of fibril formation by Taylor Dispersion Analysis (TDA). Three-dimensional overview of the obtained taylorgrams during the aggregation process of HeV PNT3-Cterm_truncated (**A**) and HeV PNT3^{3A} (**C**) at different incubation times. The rainbow color code indicates the progress of the incubation time from dark violet to red. Analyses were performed using a protein concentration of 200 μM in 50 mM phosphate buffer, pH 7.2 at 37 °C. Peak area and hydrodynamic radius (*Rh*) evolution of the monomeric species and of the spikes area during the aggregation process of HeV PNT3-Cterm_truncated (**B**) and HeV PNT3^{3A} (**D**).

By contrast, and in agreement with its dramatically reduced fibrillogenic abilities as unveiled by the other approaches herein used, the PNT3^{3A} variant does not show any changes during the incubation time, with both *Rh* and peak area remaining constant in this period (Figure 8C,D).

In conclusion, the data support a much higher fibrillation ability of the truncated variant, as corroborated by the presence of spikes in the elution profiles and by the fast consumption of the monomeric species as compared to PNT3^{3A}, where no evolution in size nor in area was observed. The aggregation process of the truncated variant follows a first order kinetics, without significant formation of intermediate species between the monomeric and the fibrillar species. In PNT3^{3A} the fibrillation process could not be detected, indicating that the fibrils observed by ns-TEM represent a very poorly populated species within the system.

3. Materials and Methods

3.1. Generation of the Constructs

The pDEST17OI/PNT3 and pDEST17OI/PNT3^{3A} expression plasmids, driving the expression of a hexahistidine tagged form of the protein of interest, have already been described [25]. For the construction of expression plasmids encoding the PNT3 variants bearing single alanine substitutions (PNT3^{A1}, PNT3^{A2}, PNT3^{A3}), the pDEST17OI/PNT3 construct was used as template in two separate PCR amplifications using either primers attB1 and specific R_ala*N*-PNT3 (PCR1), or primers F_ala*N*-PNT3 and attB2 (PCR2), where *N* varies from 1 to 3 (see Supplementary Table S2). Primers were purchased from Eurofins Genomics Germany GmbH (Ebersberg, Germany). After DpnI treatment (New England Biolabs, Ipswich, MA, USA), 1 µL of PCR1 and 1 µL of PCR2 were used as overlapping megaprimers along with primers attB1 and attB2 in a third PCR. After purification, the third PCR product was inserted into the pDEST17OI bacterial expression vector using the Gateway® technology (Invitrogen, Carlsbad, CA, USA). This vector allows expression of the recombinant protein under the control of the T7 promoter. The resulting protein is preceded by a stretch of 22 vector-encoded residues (MSYYHHHHHHLESTSLYKKAGF) encompassing a hexahistidine tag. The DNA fragment encoding NiV PNT3 (i.e., residues 200–314 of the NiV P/V/W protein) was PCR-amplified using the pDEST17OI/NiV W construct as template [21] and primers NiV PNT3-AttB1 and NiV PNT3-AttB2. After DpnI treatment (New England Biolabs, Ipswich, MA, USA), the resulting amplicon was cloned into pDEST17OI as described above (Invitrogen, Carlsbad, CA, USA).

The expression construct encoding the truncated HeV PNT3 variant (PNT3_C-term_truncated) was generated by PCR using pDEST17OI/PNT3 [25] as template and attB1 and Trunc_PNT3_B2 as primers. After DpnI treatment (New England Biolabs, Ipswich, MA, USA), the resulting amplicon was cloned in pDEST17OI. The same procedure was used to obtain the HeV PNT3 truncated variant bearing the triple alanine substitution (PNT3^{3A}_C-term_truncated) except that the pDEST17OI/PNT3^{3A} construct [25] was used as template.

The construct encoding the HeV PNT3 variant bearing a cysteine (PNT3_Cys) was obtained using the pDEST17OI/PNT3 construct [25] as template in two separate PCR amplifications using either primers attB1 and specific PNT3_C255_R (PCR1), or primers PNT3_C255_F and attB2 (PCR2). After DpnI treatment, 1 µL of PCR1 and 1 µL of PCR2 were used as overlapping megaprimers along with primers attB1 and attB2 in a third PCR. After purification, the third PCR product was inserted into pDEST17 (Invitrogen, Carlsbad, CA, USA). The list and sequence of primers used to generate the above-described constructs is provided in Supplementary Table S2. Primers were purchased from Eurofins Genomics. All the constructs were verified by DNA sequencing (Eurofins Genomics Germany GmbH, Ebersberg, Germany) and found to conform to expectations.

3.2. Proteins Expression and Purification

The *E. coli* strain T7pRos was used for the expression of all the recombinant proteins upon transformation of bacterial cells with each of the bacterial expression plasmids described above. Cultures were grown over-night to saturation in LB medium containing 100 μg mL^{-1} ampicillin and 34 μg mL^{-1} chloramphenicol. An aliquot of the overnight culture was diluted 1/20 into 1 L of TB medium and grown at 37 °C with shaking at 200 rpm. When the optical density at 600 nm (OD_{600}) reached 0.5–0.8, isopropyl β-D-thiogalactopyanoside (IPTG) was added to a final concentration of 0.5 mM, and the cells were grown at 35 °C overnight. The induced cells were harvested, washed and collected by centrifugation (5000× *g*, 15 min). The resulting pellets were resuspended in buffer A (50 mM Tris/HCl pH 7.5, 1 M NaCl, 20 mM imidazole) containing 6 M guanidium hydrochloride (GDN). The suspension was sonicated to disrupt the cells (using a 750 W sonicator and 3 cycles of 30 s each at 45% power output) and then centrifuged at 14,000× *g* for 30 min at 20 °C. The supernatant was first purified by IMAC by mixing it with 5 mL Nickel resin (GE Healthcare, Uppsala, Sweden) pre-equilibrated in buffer A. The affinity resin was washed with 20 column volumes (CV) of buffer A. Proteins were eluted with ~3 CV of buffer A supplemented with 250 mM imidazole. The fractions eluted were pooled and concentrated in the presence of 6 M GDN up to 1 mM using Centricon concentrators, and the proteins were then frozen at −20 °C. All PNT3 variants were subsequently subjected to SEC (HiLoad 16/600 Superdex 75 pg column, Cytiva, Marlborough, MA, USA), where the SEC column was equilibrated with buffer B (sodium phosphate 50 mM pH 7.2, 100 mM NaCl, 5 mM EDTA). The fractions from SEC, were pooled, supplemented with 6M GDN and concentrated (up to ~750 μM) and stored at −20 °C. In the case of the PNT3 cysteine variant, the GDN-containing sample was also supplemented with 10 mM DTT. Prior to each subsequent analysis, the samples were loaded onto a Sephadex G-25 medium column (Cytiva, Marlborough, MA, USA) to exchange the buffer. The proteins were eluted from G-25 columns using sodium phosphate 50 mM buffer at a pH of 7.2 unless differently specified. IMAC and SEC were performed at room temperature (RT).

Protein concentrations were estimated using the theoretical absorption coefficients at 280 nm as obtained using the program ProtParam from the EXPASY server (http://web.expasy.org/protparam/, accessed on 20 January 2021).

The purity of the final purified products was assessed by SDS-PAGE (Figure 3B). The identity of all the purified PNT3 variants generated in this work was confirmed by mass spectrometry analysis of tryptic fragments obtained after digestion of the purified protein bands excised from SDS-polyacrylamide gels (Supplementary Figure S3). The excised bands were analyzed by the mass spectrometry facility of Marseille Proteomics in the same way as previously done for the PNT3 *wt* variant [25]. Briefly, gel pieces were washed and destained using 100 mM NH4HCO3/acetonitrile (50/50). Destained gel pieces were shrunk with acetonitrile and were re-swollen in the presence of 100 mM ammonium bicarbonate in 50% acetonitrile and dried at room temperature. Protein bands were then rehydrated and cysteines were reduced using 10 mM DTT in 100 mM ammonium bicarbonate pH 8.0 for 45 min at 56 °C before alkylation in the presence of 55 mM iodoacetamide in 100 mM ammonium bicarbonate pH 8.0 for 30 min at room temperature in the dark, then washed twice with 25 mM ammonium bicarbonate pH 8.0 and digested with high-sequencing-grade trypsin (Promega, Madison, WI, USA). Mass spectrometry analysis were carried out by LC-MSMS using a Q Exactive Plus Hybrid Quadrupole-Orbitrap online with a nanoLC Ultimate 3000 chromatography system (Thermo Fisher Scientific™, San Jose, CA, USA). For each biological sample, 5 microliters corresponding to 25% of digested sample were injected in duplicate on the system. After pre-concentration and washing of the sample on an Acclaim PepMap 100 column (C18, 2 cm × 100 μm i.d. 100 A pore size, 5 μm particle size), peptides were separated on a LC EASY-Spray column (C18, 50 cm × 75 μm i.d., 100 A, 2 μm, 100A particle size) at a flow rate of 300 nL/min with a two steps linear gradient (2–22% acetonitrile/H20; 0.1% formic acid for 100 min and 22–32% acetonitrile/H20; 0.1% formic acid for 20 min). For peptides ionization in the EASYSpray source, spray voltage was set at 1.9 kV and the capillary temperature at 250 °C. All samples were measured in a data-dependent acquisition mode. Each run was preceded by a blank MS run in order to monitor system background. The peptide masses were measured in a survey full scan (scan range 375–1500 m/z, with 70 K FWHM resolution at m/z = 400, target AGC value of 3.00×10^6 and maximum injection time of 100 ms). Following the high-resolution full scan in the Orbitrap, the 10 most intense data-dependent precursor ions were successively fragmented in HCD cell and measured in Orbitrap (normalized collision energy of 25%, activation time of 10 ms, target AGC value of 1.00×10^3, intensity threshold 1.00×10^4 maximum injection time 100 ms, isolation window 2 m/z, 17.5 K FWHM resolution, scan range 200 to 2000 m/z). Dynamic exclusion was implemented with a repeat count of 1 and exclusion duration of 20 s.

Raw files generated from mass spectrometry analysis were processed with Proteome Discoverer 1.4 .1.14 (Thermo Fisher Scientific, San Jose, CA, USA) to search against a home-made database containing 20,150 human sequences, 4306 *E.coli* sequences implemented with the expected sequences (swissprot – human – reviewed – 170315 _ 20150 _UP_coli_171120_4306_ Patrick220922 _ ID _ Bandes.fast).

Database search with SequestHT were done using the following settings: a maximum of two trypsin miss cleavage allowed, methionine oxidation and N terminal protein acetylation as variable modifications, and cysteine carbamidomethylation as fixed modification. A peptide mass tolerance of 6 ppm and a fragment mass tolerance of 0.8 Da were allowed for search analysis. Only peptides with high Sequest scores were selected for protein identification. False discovery rate was set to 1% for protein identification.

3.3. PEG Precipitation Assay (Relative Solubility)

The relative solubility of each variant was evaluated at different PEG concentrations using an adaptation of the protocol recently described by Oeller et al. [32]. Briefly, PEG solutions from 0 to 30% were prepared from 50% PEG_{6000} (Steinheim, Germany) stock solution. Then, an aliquot of corresponding protein from a stock (at 260 µM) was mixed with each PEG solutions to obtain a final concentration of 66 µM protein in a 100 µL final reaction volume. The assay was performed in 96-well plates sealed with aluminum plate sealers to prevent possible evaporation (Thermo Fisher Scientific, USA). Plates were incubated at 4 °C for 24 h and centrifugated at maximum velocity (4600× *g*) for 2 h. Immediately after, 2 µL of the supernatant were pipetted to quantify the soluble protein concentration using a ND-1000 Nanodrop Spectrophotometer and theoretical absorption coefficients at 280 nm as obtained using the program ProtParam from the EXPASY server (http://web.expasy.org/protparam/, accessed on 20 January 2021). Each condition was made in triplicate. The soluble fractions obtained were normalized and fitted to a sigmoid function to obtain the $PEG_{1/2}$ value that reports on relative solubility (PRISM software). The error on the $PEG_{\frac{1}{2}}$ and the quality of the fit were estimated by a 95% confidence interval analysis (PRISM software version 9.2, CA, USA).

3.4. Far-UV Circular Dichroism

CD spectra were measured using a Jasco 810 dichrograph (Jasco France, Lisses, France), flushed with N_2 and equipped with a Peltier thermoregulation system. Proteins were loaded into a 1 mm quartz cuvette at 0.06 mg/mL (in 10 mM phosphate buffer at pH 7.2) and spectra were recorded at 37 °C. The scanning speed was 20 nm min^{-1}, with data pitch of 0.2 nm. Each spectrum is the average of ten acquisitions. The spectrum of buffer was subtracted from the protein spectrum. Spectra were smoothed using the "means-movement" smoothing procedure implemented in the Spectra Manager package. As already previously documented, a decrease in the signal spectrum was observed with increasing incubation time and ascribed to fibril formation [25]. Because the different variants have different fibrillation propensities, differences in spectra intensity might reflect differences in the fraction of fibrillar species (which are not detected by CD) rather than *bona fide* spectral differences. We therefore normalized spectra using the maximum negative value of intensity as a normalization factor.

Mean molar ellipticity values per residue (MRE) were calculated as

$$[\theta] = 3300 \times m \times \Delta A/(l \times c \times n) \quad (2)$$

where l is the path length in cm, n is the number of residues, m is the molecular mass in Daltons and c is the concentration of the protein in mg mL^{-1}.

3.5. Estimation of the Hydrodynamic Radius by SEC

The hydrodynamic radii (Stokes radii, R_S) of the proteins were estimated by analytical SEC using a HiLoad 16/600 Superdex 75 pg column (Cytiva, Marlborough, MA, USA). Buffer B was used as elution buffer. Typically, 250 µL of purified protein at 11 mg mL^{-1} were injected.

The Stokes radii of proteins eluted from the SEC column were deduced from a calibration curve obtained using globular proteins of known R_S (Conalbumin: 36.4 Å, Carbonic Anhydrase: 23 Å, RNAse A: 16.4 Å Aprotinin: 13.5 Å)

The R_S (in Å) of a natively folded (Rs^{NF}), fully unfolded state in urea (${R_S}^U$) and natively unfolded premolten globule (PMG) (${R_S}^{PMG}$) protein with a molecular mass (MM) (in Daltons) were calculated according to [44]:

$$\log ({R_S}^{NF}) = 0.357 \times (\log MM) - 0.204 \quad (3)$$

$$\log ({R_S}^{U}) = 0.521 \times (\log MM) - 0.649 \quad (4)$$

$$\log ({R_S}^{PMG}) = 0.403 \times (\log MM) - 0.239 \quad (5)$$

The R_S (in Å) of an IDP with N residues was also calculated according to [45] using the simple power-law model:

$${R_S}^{IDP} = R_0 N^{\nu} \quad (6)$$

where R_0 = 2.49 and ν = 0.509. The compaction index (CI) is expressed according to [46]:

$$CI = ({R_S}^U - {R_S}^{OBS})/({R_S}^U - {R_S}^{NF}) \quad (7)$$

This parameter, which allows comparison between proteins of different lengths, in principle varies between 0 and 1, with 0 indicating minimal compaction and 1 maximal compaction. In the case of the cysteine variant, as the SEC analysis was performed without DTT, a peak corresponding to a dimeric species was observed. However, since the latter was not well resolved, its corresponding R_S was not calculated.

3.6. Small-Angle X-ray Scattering (SAXS)

In order to ensure maximal monodispersity of the sample, SAXS studies were coupled to SEC. SEC-SAXS data were collected at SOLEIL (Gif-sur-Yvette, France), as described in Table 3. In both cases, the calibration was performed with water. Sample from each PNT3 variant at 5 mg mL^{-1} in buffer B containing 6M GDN was injected onto an AdvanceBio SEC 2.7 µm (Agilent) SEC column. Elution was carried out in buffer C (50 mM sodium phosphate buffer at pH 7.2). Data reduction and frames subtraction were done with the beamline software FOXTROT (available upon request from the SOLEIL staff). Gaussian decomposition was performed using the UltraScan solution modeler (US-SOMO) HPLC-SAXS module (https://somo.aucsolutions.com/, accessed on 6 September 2022) [47] or Chromixs (manual frames selection) [48] and the final deconvoluted scattering curves were submitted to the SHANUM program [49] to remove noisy, non-informative data at high angles.

Table 3. SEC-SAXS data acquisition parameters.

Instrument	SOLEIL Synchrotron (Gif-sur-Yvette, France) Beamline Swing
X-rays wavelength (Å)	1.033
Energy (keV)	12
Detector type	Dectris EIGER 4M
Sample-to-detector distance (m)	2.0
q-range	0.003 − 0.549 Å^{-1}
Temperature (°C)	20
Samples	
Concentration (mg mL^{-1})	5
Sample volume (µL)	50
Gel filtration column	AdvanceBio SEC 2.7 µm (Agilent)
Flow rate (mL min^{-1})	0.3
Buffer	50 mM sodium phosphate pH 7.2 (buffer C)

The data were analyzed using the ATSAS program package [49]. The radius of gyration (R_g) and I(0) were estimated at low angles ($q.R_g$ < 1.3) according to the Guinier approximation [50,51]:

$$Ln[I(q)] = Ln[I_0] - (q^2 R_g{}^2)/3 \quad (8)$$

The pairwise distance distribution functions P(r), from which the D_{max} and the R_g were estimated, were calculated with the program GNOM [52] and manually adjusted until a good CorMap p-value ($\alpha > 0.01$) was obtained [52].

The theoretical R_g value (in Å) expected for various conformational states was calculated using Flory's equation:

$$R_g = R_0 N^{\nu} \quad (9)$$

where N is the number of amino acid residues, R_0 a constant and ν a scaling factor. For IDPs, R_0 is 2.54 ± 0.01 and ν is 0.522 ± 0.01 [53], for chemically denatured (U) proteins R_0 is 1.927 ± 0.27 and ν is 0.598 ± 0.028 [53], and for natively folded (NF) proteins $R_0 = \sqrt{(3/5)} \times 4.75$ and $\nu = 0.29$ [54].

As in the case of the R_S, the *CI* allows comparing the degree of compaction of a given IDP, through comparison of the observed R_g to the reference values expected for a fully unfolded and a folded conformation of identical mass. The *CI* referred to the R_g can be calculated as follows [46]:

$$CI = (R_g{}^U - R_g{}^{OBS})/(R_g{}^U - R_g{}^{NF}) \quad (10)$$

where $R_g{}^{OBS}$ is the experimental value for a given protein, and $R_g{}^U$ and $R_g{}^{NF}$ are the reference values calculated for a fully unfolded (U) and natively folded (NF) form, as described above. Akin to the R_S-based *CI*, this index increases with increasing compaction.

The overall conformation and the flexibility of the proteins was assessed with the dimensionless Kratky plot ($(qR_g)^2$ I(q)/I_0 *vs* qR_g) and the Krakty-Debye plot (q^2I(q) *vs* q^2).

SEC-SAXS data have been deposited in the Small Angle Scattering Biological Data Bank (SASBDB) [55] under codes SASDQB7, SASDQC7, SASDQD7, SASDQE7 and SASDQF7 for the set of data of PNT3 *wt*, PNT3^{3A}, NiV PNT3, PNT3_C-terminal truncated and PNT3^{3A}_C-terminal truncated, respectively.

3.7. Congo Red Binding Assays

Quantitative measurement of Congo Red (Sigma Aldrich, Saint Louis, MO, USA) binding (CR shift assay) was carried out by using protein samples containing each PNT3 variant at 20 µM (in Buffer C) and 5 µM of CR in a final volume of 100 µL. The samples were then incubated at 37 °C for 4 or 7 days. The adsorption spectrum of the CR-containing samples was recorded using a PHERAstar FSX Microplate Reader (BMG LABTECH, Champignysur-Marne, France) in the 350–600 nm wavelength range. A solution of 5 μM CR in Buffer C without the protein was used as a control to normalize the analysis. Experiments were carried out in triplicate. Statistical analysis was performed using a Two-way ANOVA test implemented in the PRISM software.

3.8. Negative-Staining Transmission Electron Microscopy (ns-TEM)

All the variants, at a concentration of 200 μM, were prepared and analyzed at different times to monitor their evolution (0, 24, 96 h). Incubation was carried out at 37 °C in Buffer C. Prior to each measurement, the samples were diluted to reach a final concentration of 40 μM. EM grids (carbon coated copper grids, 300 mesh, Agar Scientific, UK) were exposed to plasma glow discharge for 20 s using GloQube (Quorum, UK) (Current 15 mA) in order to increase protein adhesion. Drops of 3.5 μL of the diluted protein solutions were deposited onto glow-discharged grids. After 1 min incubation with the sample, the grids were washed three times with 50 μL of buffer C, once in 35 μL 1% (*w*/*v*) Uranyl acetate solution (Laurylab, Brindas, France) and then stained for 1 min in the latter solution. Excess of uranyl was blotted and grids were left to dry for 1 h at RT. Images were collected on Tecnai 120 Spirit TEM microscope (FEI company, ThermoFisher, Illkirch-Graffenstaden France) operated at 120 kV using a Veleta 2K × 2K CCD camera (Olympus).

3.9. Kinetic Protein Aggregation Study by Taylor Dispersion Analysis (TDA)

TDA was performed as already described [41] using an Agilent 7100 (Waldbronn, Germany) capillary electrophoresis system with bare fused silica capillaries (Polymicro Technologies, USA) having 60 cm × 50 μm i.d. dimensions and a detection window at 51.5 cm. New capillaries were conditioned with the following flushes: 1 M NaOH for 30 min and ultrapure water for 30 min. Between each analysis, the capillaries were rinsed with Buffer C (2 min). Samples were injected hydrodynamically on the inlet end of the capillary (30 mbar, 6 s, injected volume is about 6.1 nL corresponding to less than 1% of the capillary volume to the detection point). Experiments were performed using a mobilization pressure of 100 mbar. The temperature of the capillary cartridge was set at 37 °C. The vial carrousel was thermostated using an external circulating water bath from Instrumat (Moirans, France). The solutes were monitored by UV absorbance at 198 nm. The mobile phase was Buffer C. (viscosity at 37 °C is 7.54×10^{-4} Pa.s.). Samples obtained after the desalting column were diluted to reach 700 μL at 200 μM solution and were immediately transferred to a vial and incubated at 37 °C in the capillary electrophoresis instrument's carrousel. The aggregation was conducted by injecting the sample every 1 h. The total number of TDA runs for each sample was about 120. The taylorgrams were recorded with Agilent Chemstation software and then exported to Microsoft Excel for subsequent data processing. Data were fitted to a first order exponential decay according to Equation (1). This resulted in a good quality fit as judged from the Reduced Chi-Sqr: 0.43723; R-Square (COD) = 0.93943, and Adj. R-Square = 0.93576. When necessary, the elution peaks were fitted with the sum of n Gaussian functions (in this work: $n \leq 3$) as already described [41,42].

4. Conclusions

This study constitutes a comprehensive analysis of the molecular basis of the fibrillation process of a small region (PNT3) within the N-terminal intrinsically disordered domain shared by the HeV P/V/W proteins. Biochemical and biophysical characterization of a set of HeV PNT3 variants bearing alanine substitutions in the amyloidogenic EYYY motif, along with the characterization of the corresponding PNT3 region from the cognate NiV, revealed that each of the three tyrosines in the motif are required for the elongation step of the fibrillation process. Remarkably, the present study also unveiled a role for the C-terminal domain of PNT3 in self-inhibition of fibrillation, possibly reminiscent of the α-synuclein fibrillation model, and of potential biological significance. Noteworthy, in light of the observation that amyloid-like fibrils form not only in vitro but also the cellular context, it is tempting to hypothesize that the amyloidogenicity of V/W proteins, which both encompass the PNT3 region, could be correlated with the pathogenic (and even encephalitogenic) properties of Henipaviruses. Therefore, the PNT3 variants that we have herein generated constitute valuable tools to further explore the functional impact of V/W fibrillation in transfected and infected cells. The present results therefore set the stage for further investigations aimed at illuminating the mechanisms underlying the disease as a preliminary step towards the rational design of antivirals.

Supplementary Materials: The following supporting information can be downloaded at: https://www.mdpi.com/article/10.3390/ijms24010399/s1.

Author Contributions: S.L. conceived, designed and supervised the study and acquired funding; C.B. generated all the bacterial expression constructs; H.B. purified all the proteins used in the work. F.G. recorded and analyzed the SAXS data, J.C. and H.C. performed the TDA analysis; D.P., G.P. and J.F.N. performed the TEM analyses; J.F.N. carried out all the other experiments. J.F.N., F.G., G.P., A.V.K. and S.L. analyzed and interpreted the data; J.F.N. generated the first draft of the manuscript. All the authors wrote and revised the manuscript. All authors have read and agreed to the published version of the manuscript.

Funding: This work was carried out with the financial support of the Agence Nationale de la Recherche (ANR), specific project Heniphase (ANR-21-CE11- 0012-01). It was also partly supported by the French Infrastructure for Integrated Structural Biology (FRISBI) (ANR-10-INSB- 0005) and by the CNRS. F.G. is supported by a post-doctoral fellowship from the FRM (Fondation pour la Recherche Médicale). G.P. is supported by a joint doctoral fellowship from the AID (Agence Innovation Défense) and Aix-Marseille University. J.F.N. is supported by a postdoctoral fellowship from the Infectiopôle Sud.

Institutional Review Board Statement: Not applicable.

Informed Consent Statement: Not applicable.

Data Availability Statement: The data present in the current study are available from the corresponding author on reasonable request.

Acknowledgments: We thank Aurélien Thureau (SOLEIL) and Anton Popov (ESRF) for their help in recording SEC-SAXS data. We thank both the ESRF and the SOLEIL synchrotrons for beamtime allocation. We are also grateful to Gerlind Sulzenbacher (AFMB lab) for efficiently managing the AFMB BAG. We thank all the AFMB technical and support staff (Denis Patrat, Patricia Clamecy, Béatrice Rolland, Paul Zamboni, Chantal Falaschi and Fabienne Amalfitano). We heartily thank Patrick Fourquet for mass spectrometry analyses done using the mass spectrometry facility of Marseille Proteomics (marseille-proteomique.univ-amu.fr), supported by IBISA (Infrastructures Biologie Santé et Agronomie), Plateforme Technologique Aix-Marseille, the Cancéropôle PACA, Région Sud-Alpes-Côte d'Azur, the Institut Paoli-Calmettes, the Centre de Recherche en Cancérologie de Marseille (CRCM), Fonds Européen de Développement Régional and Plan Cancer.

Conflicts of Interest: The authors declare no conflict of interest. The funders had no role in the design of the study; in the collection, analyses, or interpretation of data; in the writing of the manuscript, or in the decision to publish the results.

References

1. Eaton, B.T.; Mackenzie, J.S.; Wang, L.F. Henipaviruses. In *Fields Virology*, 5th ed.; Fields, B.N., Knipe, D.M., Howley, P.M., Eds.; Lippincott-Raven: Philadelphia, PA, USA, 2007; pp. 1587–1600.
2. Eaton, B.T.; Broder, C.C.; Middleton, D.; Wang, L.F. Hendra and Nipah viruses: Different and dangerous. *Nat. Rev. Microbiol.* **2006**, *4*, 23–35. [CrossRef]
3. Bloyet, L.M.; Welsch, J.; Enchery, F.; Mathieu, C.; de Breyne, S.; Horvat, B.; Grigorov, B.; Gerlier, D. HSP90 Chaperoning in Addition to Phosphoprotein Required for Folding but Not for Supporting Enzymatic Activities of Measles and Nipah Virus L Polymerases. *J. Virol.* **2016**, *90*, 6642–6656. [CrossRef]
4. Abdella, R.; Aggarwal, M.; Okura, T.; Lamb, R.A.; He, Y. Structure of a paramyxovirus polymerase complex reveals a unique methyltransferase-CTD conformation. *Proc. Natl. Acad. Sci. USA* **2020**, *117*, 4931–4941. [CrossRef]
5. Bloyet, L.M.; Schramm, A.; Lazert, C.; Raynal, B.; Hologne, M.; Walker, O.; Longhi, S.; Gerlier, D. Regulation of measles virus gene expression by P protein coiled-coil properties. *Sci. Adv.* **2019**, *5*, eaaw3702. [CrossRef]

6. Karlin, D.; Ferron, F.; Canard, B.; Longhi, S. Structural disorder and modular organization in Paramyxovirinae N and P. *J. Gen. Virol.* **2003**, *84*, 3239–3252. [CrossRef]
7. Habchi, J.; Mamelli, L.; Darbon, H.; Longhi, S. Structural Disorder within Henipavirus Nucleoprotein and Phosphoprotein: From Predictions to Experimental Assessment. *PLoS ONE* **2010**, *5*, e11684. [CrossRef]
8. Schiavina, M.; Salladini, E.; Murrali, M.G.; Tria, G.; Felli, I.C.; Pierattelli, R.; Longhi, S. Ensemble description of the intrinsically disordered N-terminal domain of the Nipah virus P/V protein from combined NMR and SAXS. *Sci. Rep.* **2020**, *10*, 19574. [CrossRef]
9. Wright, P.E.; Dyson, H.J. Intrinsically unstructured proteins: Re-assessing the protein structure-function paradigm. *J. Mol. Biol.* **1999**, *293*, 321–331. [CrossRef]
10. Uversky, V.N.; Gillespie, J.R.; Fink, A.L. Why are "natively unfolded" proteins unstructured under physiologic conditions? *Proteins* **2000**, *41*, 415–427. [CrossRef]
11. Tompa, P. Intrinsically disordered proteins: A 10-year recap. *Trends Biochem. Sci.* **2012**, *37*, 509–516. [CrossRef]
12. Habchi, J.; Tompa, P.; Longhi, S.; Uversky, V.N. Introducing Protein Intrinsic Disorder. *Chem. Rev.* **2014**, *114*, 6561–6588. [CrossRef]
13. Jensen, M.R.; Yabukarski, F.; Communie, G.; Condamine, E.; Mas, C.; Volchkova, V.; Tarbouriech, N.; Bourhis, J.M.; Volchkov, V.; Blackledge, M.; et al. Structural Description of the Nipah Virus Phosphoprotein and Its Interaction with STAT1. *Biophys. J.* **2020**, *118*, 2470–2488. [CrossRef]
14. Bruhn-Johannsen, J.F.; Barnett, K.; Bibby, J.; Thomas, J.; Keegan, R.; Rigden, D.; Bornholdt, Z.A.; Saphire, E.O. Crystal structure of the Nipah virus phosphoprotein tetramerization domain. *J. Virol.* **2014**, *88*, 758–762. [CrossRef]
15. Blocquel, D.; Beltrandi, M.; Erales, J.; Barbier, P.; Longhi, S. Biochemical and structural studies of the oligomerization domain of the Nipah virus phosphoprotein: Evidence for an elongated coiled-coil homotrimer. *Virology* **2013**, *446*, 162–172. [CrossRef]
16. Beltrandi, M.; Blocquel, D.; Erales, J.; Barbier, P.; Cavalli, A.; Longhi, S. Insights into the coiled-coil organization of the Hendra virus phosphoprotein from combined biochemical and SAXS studies. *Virology* **2015**, *477*, 42–55. [CrossRef]
17. Habchi, J.; Blangy, S.; Mamelli, L.; Ringkjobing Jensen, M.; Blackledge, M.; Darbon, H.; Oglesbee, M.; Shu, Y.; Longhi, S. Characterization of the interactions between the nucleoprotein and the phosphoprotein of Henipaviruses. *J. Biol. Chem.* **2011**, *286*, 13583–13602. [CrossRef]
18. Communie, G.; Habchi, J.; Yabukarski, F.; Blocquel, D.; Schneider, R.; Tarbouriech, N.; Papageorgiou, N.; Ruigrok, R.W.; Jamin, M.; Ringkjøbing-Jensen, M.; et al. Atomic resolution description of the interaction between the nucleoprotein and phosphoprotein of Hendra virus. *PLoS Pathog.* **2013**, *9*, e1003631. [CrossRef]
19. Douglas, J.; Drummond, A.J.; Kingston, R.L. Evolutionary history of cotranscriptional editing in the paramyxoviral phosphoprotein gene. *Virus Evol.* **2021**, *7*, veab028. [CrossRef]
20. Salladini, E.; Delauzun, V.; Longhi, S. The Henipavirus V protein is a prevalently unfolded protein with a zinc-finger domain involved in binding to DDB1. *Mol. Biosyst.* **2017**, *13*, 2254–2267. [CrossRef]
21. Pesce, G.; Gondelaud, F.; Ptchelkine, D.; Nilsson, J.F.; Bignon, C.; Cartalas, J.; Fourquet, P.; Longhi, S. Experimental Evidence of Intrinsic Disorder and Amyloid Formation by the Henipavirus W Proteins. *Int. J. Mol. Sci.* **2022**, *23*, 923. [CrossRef]
22. Fontana, J.M.; Bankamp, B.; Rota, P.A. Inhibition of interferon induction and signaling by paramyxoviruses. *Immunol. Rev.* **2008**, *225*, 46–67. [CrossRef]
23. Audsley, M.D.; Moseley, G.W. Paramyxovirus evasion of innate immunity: Diverse strategies for common targets. *World J. Virol.* **2013**, *2*, 57–70. [CrossRef]
24. Tsimbalyuk, S.; Cross, E.M.; Hoad, M.; Donnelly, C.M.; Roby, J.A.; Forwood, J.K. The Intrinsically Disordered W Protein Is Multifunctional during Henipavirus Infection, Disrupting Host Signalling Pathways and Nuclear Import. *Cells* **2020**, *9*, 1913. [CrossRef]
25. Salladini, E.; Gondelaud, F.; Nilsson, J.; Pesce, G.; Bignon, C.; Murrali, M.G.; Horvat, B.; Fabre, R.; Pierattelli, R.; Kajava, A.V.; et al. Identification of a region in the common amino-terminal domain of Hendra virus P, V and W proteins responsible for phase transition and amyloid formation. *Biomolecules* **2021**, *11*, 1324. [CrossRef]
26. Gondelaud, F.; Pesce, G.; Nilsson, J.F.; Bignon, C.; Ptchelkine, D.; Gerlier, D.; Mathieu, C.; Longhi, S. Functional benefit of structural disorder for the replication of measles, Nipah and Hendra viruses. *Essays Biochem.* **2022**, *in press*. [CrossRef]
27. Kajava, A.V.; Baxa, U.; Steven, A.C. Beta arcades: Recurring motifs in naturally occurring and disease-related amyloid fibrils. *FASEB J.* **2010**, *24*, 1311–1319. [CrossRef]
28. Ahmed, A.B.; Znassi, N.; Château, M.T.; Kajava, A.V. A structure-based approach to predict predisposition to amyloidosis. *Alzheimers Dement.* **2015**, *11*, 681–690. [CrossRef]
29. Swaminathan, R.; Ravi, V.K.; Kumar, S.; Kumar, M.V.; Chandra, N. Lysozyme: A model protein for amyloid research. *Adv. Protein Chem. Struct. Biol.* **2011**, *84*, 63–111. [CrossRef]
30. Tseng, W.H.; Chen, S.H.; Hiramatsu, H. pH-controlled stacking direction of the β-strands in peptide fibrils. *Sci. Rep.* **2020**, *10*, 22199. [CrossRef]
31. Pfefferkorn, C.M.; McGlinchey, R.P.; Lee, J.C. Effects of pH on aggregation kinetics of the repeat domain of a functional amyloid, Pmel17. *Proc. Natl. Acad. Sci. USA* **2010**, *107*, 21447–21452. [CrossRef]
32. Oeller, M.; Sormanni, P.; Vendruscolo, M. An open-source automated PEG precipitation assay to measure the relative solubility of proteins with low material requirement. *Sci. Rep.* **2021**, *11*, 21932. [CrossRef]
33. Uversky, V.N. Natively unfolded proteins: A point where biology waits for physics. *Protein Sci.* **2002**, *11*, 739–756. [CrossRef]

34. Schramm, A.; Bignon, C.; Brocca, S.; Grandori, R.; Santambrogio, C.; Longhi, S. An arsenal of methods for the experimental characterization of intrinsically disordered proteins - How to choose and combine them? *Arch. Biochem. Biophys.* **2019**, *676*, 108055. [CrossRef]
35. Anand, B.G.; Prajapati, K.P.; Shekhawat, D.S.; Kar, K. Tyrosine-Generated Nanostructures Initiate Amyloid Cross-Seeding in Proteins Leading to a Lethal Aggregation Trap. *Biochemistry* **2018**, *57*, 5202–5209. [CrossRef]
36. Hoyer, W.; Cherny, D.; Subramaniam, V.; Jovin, T.M. Impact of the acidic C-terminal region comprising amino acids 109-140 on alpha-synuclein aggregation in vitro. *Biochemistry* **2004**, *43*, 16233–16242. [CrossRef]
37. Sorrentino, Z.A.; Giasson, B.I. The emerging role of α-synuclein truncation in aggregation and disease. *J. Biol. Chem.* **2020**, *295*, 10224–10244. [CrossRef]
38. Holehouse, A.S.; Das, R.K.; Ahad, J.N.; Richardson, M.O.; Pappu, R.V. CIDER: Resources to Analyze Sequence-Ensemble Relationships of Intrinsically Disordered Proteins. *Biophys. J.* **2017**, *112*, 16–21. [CrossRef]
39. Gruet, A.; Dosnon, M.; Blocquel, D.; Brunel, J.; Gerlier, D.; Das, R.K.; Bonetti, D.; Gianni, S.; Fuxreiter, M.; Longhi, S.; et al. Fuzzy regions in an intrinsically disordered protein impair protein-protein interactions. *FEBS J.* **2016**, *283*, 576–594. [CrossRef]
40. Troilo, F.; Bonetti, D.; Bignon, C.; Longhi, S.; Gianni, S. Understanding Intramolecular Crosstalk in an Intrinsically Disordered Protein. *ACS Chem. Biol.* **2019**, *14*, 337–341. [CrossRef]
41. Deleanu, M.; Hernandez, J.F.; Cipelletti, L.; Biron, J.P.; Rossi, E.; Taverna, M.; Cottet, H.; Chamieh, J. Unraveling the Speciation of β-Amyloid Peptides during the Aggregation Process by Taylor Dispersion Analysis. *Anal. Chem.* **2021**, *93*, 6523–6533. [CrossRef]
42. Deleanu, M.; Deschaume, O.; Cipelletti, L.; Hernandez, J.F.; Bartic, C.; Cottet, H.; Chamieh, J. Taylor Dispersion Analysis and Atomic Force Microscopy Provide a Quantitative Insight into the Aggregation Kinetics of Aβ (1-40)/Aβ (1-42) Amyloid Peptide Mixtures. *ACS Chem. Neurosci.* **2022**, *13*, 786–795. [CrossRef] [PubMed]
43. Deleanu, M.; Hernandez, J.F.; Cottet, H.; Chamieh, J. Taylor dispersion analysis discloses the impairment of Aβ peptide aggregation by the presence of a fluorescent tag. *Electrophoresis* 2022, *in press*.
44. Uversky, V.N. What does it mean to be natively unfolded? *Eur. J. Biochem.* **2002**, *269*, 2–12. [CrossRef] [PubMed]
45. Marsh, J.A.; Forman-Kay, J.D. Sequence determinants of compaction in intrinsically disordered proteins. *Biophys. J.* **2010**, *98*, 2383–2390. [CrossRef]
46. Brocca, S.; Testa, L.; Sobott, F.; Samalikova, M.; Natalello, A.; Papaleo, E.; Lotti, M.; De Gioia, L.; Doglia, S.M.; Alberghina, L.; et al. Compaction properties of an intrinsically disordered protein: Sic1 and its kinase-inhibitor domain. *Biophys. J.* **2011**, *100*, 2243–2252. [CrossRef]
47. Brookes, E.; Rocco, M. Recent advances in the UltraScan SOlution MOdeller (US-SOMO) hydrodynamic and small-angle scattering data analysis and simulation suite. *Eur. Biophys. J.* **2018**, *47*, 855–864. [CrossRef]
48. Panjkovich, A.; Svergun, D.I. CHROMIXS: Automatic and interactive analysis of chromatography-coupled small-angle X-ray scattering data. *Bioinformatics* **2018**, *34*, 1944–1946. [CrossRef]
49. Manalastas-Cantos, K.; Konarev, P.V.; Hajizadeh, N.R.; Kikhney, A.G.; Petoukhov, M.V.; Molodenskiy, D.S.; Panjkovich, A.; Mertens, H.D.T.; Gruzinov, A.; Borges, C.; et al. ATSAS 3.0: Expanded functionality and new tools for small-angle scattering data analysis. *J. Appl. Cryst.* **2021**, *54*, 343–355. [CrossRef]
50. Guinier, A. La diffraction des rayons X aux tres petits angles; application a l'etude de phenomenes ultramicroscopiques. *Ann. Phys.* **1939**, *12*, 161–237. [CrossRef]
51. Guinier, A.; Fournet, F. *Small Angle Scattering of X-rays*; Wiley Interscience: New York, NY, USA, 1955. [CrossRef]
52. Svergun, D. Determination of the regularization parameters in indirect-trasform methods using perceptual criteria. *J. Appl. Cryst.* **1992**, *25*, 495–503. [CrossRef]
53. Bernado, P.; Blackledge, M. A self-consistent description of the conformational behavior of chemically denatured proteins from NMR and small angle scattering. *Biophys. J.* **2009**, *97*, 2839–2845. [CrossRef] [PubMed]
54. Wilkins, D.K.; Grimshaw, S.B.; Receveur, V.; Dobson, C.M.; Jones, J.A.; Smith, L.J. Hydrodynamic radii of native and denatured proteins measured by pulse field gradient NMR techniques. *Biochemistry* **1999**, *38*, 16424–16431. [CrossRef] [PubMed]
55. Valentini, E.; Kikhney, A.G.; Previtali, G.; Jeffries, C.M.; Svergun, D.I. SASBDB, a repository for biological small-angle scattering data. *Nucleic Acids Res.* **2015**, *43*, D357–D363. [CrossRef] [PubMed]

International Journal of
Molecular Sciences

MDPI

Article

Cytoophidia Maintain the Integrity of *Drosophila* Follicle Epithelium

Qiao-Qi Wang [1], Dong-Dong You [1] and Ji-Long Liu [1,2,*]

[1] School of Life Science and Technology, Shanghai Technology University, Shanghai 201210, China
[2] Department of Physiology, Anatomy and Genetics, University of Oxford, Oxford OX1 3PT, UK
* Correspondence: liujl3@shanghaitech.edu.cn or jilong.liu@dpag.ox.ac.uk

Abstract: CTP synthase (CTPS) forms a filamentous structure termed the cytoophidium in all three domains of life. The female reproductive system of *Drosophila* is an excellent model for studying the physiological function of cytoophidia. Here, we use $CTPS^{H355A}$, a point mutation that destroys the cytoophidium-forming ability of CTPS, to explore the in vivo function of cytoophidia. In $CTPS^{H355A}$ egg chambers, we observe the ingression and increased heterogeneity of follicle cells. In addition, we find that the cytoophidium-forming ability of CTPS, rather than the protein level, is the cause of the defects observed in $CTPS^{H355A}$ mutants. To sum up, our data indicate that cytoophidia play an important role in maintaining the integrity of follicle epithelium.

Keywords: CTP synthase; cytoophidium; *Drosophila*; epithelium; follicle cell; ingression

Citation: Wang, Q.-Q.; You, D.-D.; Liu, J.-L. Cytoophidia Maintain the Integrity of *Drosophila* Follicle Epithelium. *Int. J. Mol. Sci.* **2022**, *23*, 15282. https://doi.org/10.3390/ijms232315282

Academic Editors: Vladimir N. Uversky and Hans-Arno Müller

Received: 15 October 2022
Accepted: 25 November 2022
Published: 4 December 2022

1. Introduction

CTP synthase (CTPS) is a glutamate aminotransferase that catalyzes the transfer of amide nitrogen from glutamine to the C-4 position of UTP. CTP, the product of CTPS, is an important nucleotide and is a component of the synthesis of RNA, DNA, and sialoglycoprotein. It also acts as an energy coupler for some metabolic reactions, such as the synthesis of glycerophospholipids and glycosylated proteins [1,2].

In 2010, CTPS was found to form filamentous structures termed cytoophidia in *Drosophila* [3]. Subsequently, CTPS has been found to form filamentous structures in bacteria [4] and *S. cerevisiae* [5]. In the following years, the existence of cytoophidia was confirmed in human cells [6], *S. pombe* [7], *Arabidopsis thaliana* [8], and archaea [9], which indicates that cytoophidia are highly conserved in evolution.

Compartmentation is the basis for the function of organelles [10]. The classical cellular compartmentation in eukaryotic cells is achieved through membrane-bound organelles, such as the endoplasmic reticulum, mitochondrion, and Golgi apparatus [11]. Compartmentation establishes a physical boundary for the biological processes within cells, enabling cells to carry out different metabolic activities at the same time, generate specific microenvironments, regulate biological processes in time and space, and determine the specific location where biological processes should occur. The formation of cytoophidia realizes the regionalization of CTPS, and its location in cells may therefore have corresponding physiological significance.

Cells in the *Drosophila* ovary exhibit vigorous anabolic activity because they need nutrients for development. Cytoophidia are observed from region 2 of the germarium to stage 10A of oogenesis. Based on the widespread presence of cytoophidia in germline cells and follicle epithelial cells of *Drosophila* ovaries [12], and the characteristics of cytoophidium observed in germline cells and follicle epithelial cells at most stages of oogenesis, *Drosophila* ovarian follicle cells have become a classic model for studying cytoophidia.

Epidermal tissues form the boundaries of organs, where they perform a range of functions, including secretion, absorption, and protection. These tissues are usually composed of discrete cells, forming a single-cell thick sheet. Follicle epithelium is a simple epithelium. In the process of division, the cells of simple epithelium have a specific orientation of the spindle, so that both daughter cells are located in the epithelial plane. This is considered to be very important for maintaining the integrity of follicle epithelium and preventing hyperplasia [13–16].

An egg chamber consists of hundreds of follicle cells, and each follicle cell has multiple membrane domains including apical, basal, and lateral. The end adjacent to germline cells is defined as the apical side, and the end far away from germline cells is defined as the basal side [17]. Before mitosis, follicle cells will move toward the apical direction, which may be caused by the extrusion of neighbor cells. This movement results in the displacement of some cells from the tissue layer. Usually, the displaced cells need to be reintegrated to support tissue growth and maintain tissue architecture [18–20].

In a previous study, we find that cytoophidia are specifically distributed on the basolateral side of follicle cells, and this specific distribution is related to the polarity regulator of the cell membrane [21]. Therefore, we need to understand the function of cytoophidium which is specifically distributed in follicle cells.

In this study, we describe the effects of cytoophidium disassembly on follicle epithelium integrity. We are also concerned about whether these effects are directly related to the assembly of CTPS into cytoophidia, rather than to the level of CTPS protein. Our results indicate that cytoophidia play an important role in maintaining the integrity of follicle epithelium. In addition, we eliminate the influence of tissue-tissue interaction and find that cytoophidia can directly affect the integrity of follicle epithelium.

2. Results

2.1. CTPS Forms Cytoophidia in Drosophila Follicle Cells

Cells in *Drosophila* ovaries exhibit vigorous anabolic activity because they need nutrients for development. During *Drosophila* oogenesis, the follicle epithelium is a sheet of monolayer cells that encase germline cells. CTPS, as the synthase of CTP, plays an important role in the regulation of tissue growth and development. Cytoophidia exist in several different types of cells in the *Drosophila* ovary from region 2 of the germarium to stage 10A of oogenesis, including epithelial follicle cells (Figure 1A–C) and germline cells (Figure 1D–F).

2.2. Follicle Cells Undergo Ingression in $CTPS^{H355A}$ Egg Chambers

The amino acid histidine at the 355th position, or His355, lies at the tetramer-tetramer interface of CTPS [22]. If the H355 site is mutated, the cytoophidium cannot be formed. Previous studies showed that the H355 site is essential for its polymerization, but not enzymatic function [23,24]. Our laboratory has solved the structure of *Drosophila melanogaster* CTPS (dmCTPS) and found that the H355 site lies at the tetramer–tetramer interface and does not affect the catalytic site [25]. Therefore, we constructed an H355A point-mutated knock-in *Drosophila* strain to investigate whether the disassembly of cytoophidia would affect follicle cells. Former studies found that the H355A served as a dominant negative point mutation [26] (Figure S1).

In order to find out whether the abnormality is caused by the inability of CTPS to aggregate due to H355A point mutation or the addition of mCherry tag, our laboratory constructed another *Drosophila* strain with mCherry added to the C-terminus of CTPS based on w^{1118}. To determine whether the feature of cytoophidium localization was in fact introduced by protein fusion between CTPS and mCherry tag, we performed immunofluorescence microscopy and directly detected the CTPS protein of the w^{1118} fly and found no difference [27]. It is proven by the observation that the knock-in mCherry tag does not affect the polymerization of CTPS protein. The morphology of the *CTPS-mCherry*^{KI} *Drosophila(CTPS-mCh)* ovaries is consistent with that of the w^{1118}, which implies that the *CTPS-mCherry*^{KI} *Drosophila* can also be used as control in our experiment (Figure 2A–F).

Besides, our laboratory has used the *CTPS-mCherry*^{KI} *Drosophila* as control in previous studies [28,29].

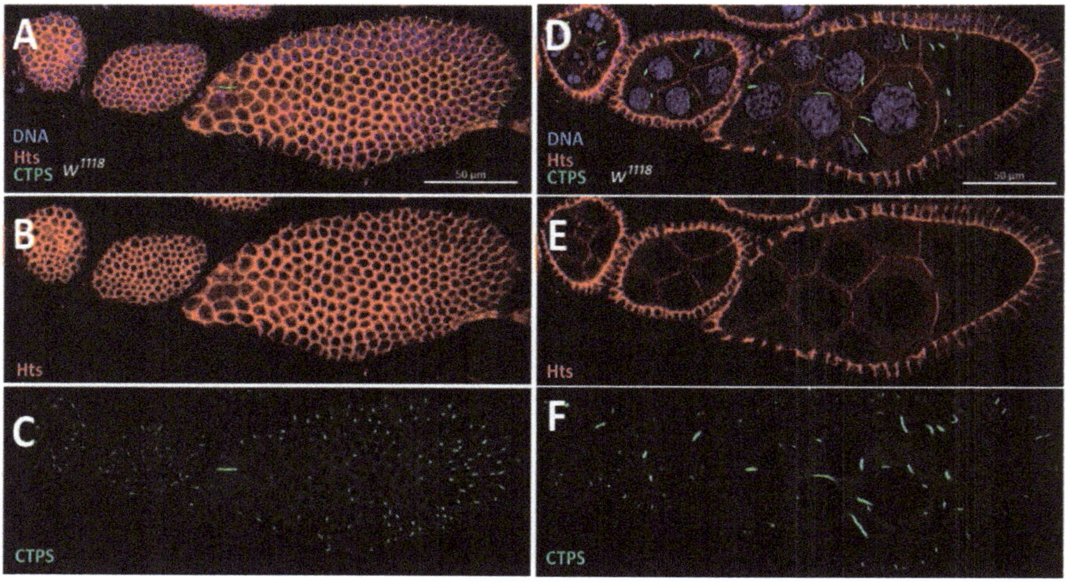

Figure 1. Follicle cells maintain monolayer during *Drosophila* oogenesis. (**A–C**) Surface view of a wild-type ovariole containing different stages of egg chambers. The ovariole is subjected to immunofluorescence analysis with antibodies against CTPS (green) and Hts (red, labelling cell membranes). DNA is labelled by Hoechst 33342 (blue). (**A**) Cytoophidia are distributed almost at different stages of each follicle cell. (**B**) The boundaies of follicle epithelia are displayed by a single projection of Hts staining. (**C**) CTPS staining shows the distribution of cytoophidia on the surface of egg chambers. (**D–F**) Side view of the same ovariole in (**A–C**). (**D**) Monolayer follicle cells envelop germline cells. (**E**) A single projection of Hts staining shows the monolayer structure of follicle epithelia. (**F**) CTPS staining shows the distribution of cytoophidia in follicle cells and germline cells. Scale bars, 50 µm.

When constructing the point-mutated *Drosophila* strain, we added a mCherry tag at the C-terminal of CTPS. Through confocal microscopy, we observed diffuse mCherry signal in *Drosophila* follicle cells, which confirmed that the CTPS could not form the cytoophidium after mutation at the H355 site (Figure 3A,F). In the egg chamber of wild-type flies, follicle cells are monolayer epidermal cells. We observed their morphological characteristics by immunofluorescence staining. The cell membrane was labeled with an antibody against Armadillo. We found that in the egg chamber of *CTPS*$^{H355A/H355A}$*-mCherry* knock-in homozygous fly (hereinafter referred to as *CTPS*^{H355A} strain), some follicle cells originally arranged in a monolayer migrated inward (ie. ingression), thus disrupting the monolayer arrangement. The ingression of follicle cells occurs not only in the early stages of oogenesis, such as stage 5 (Figure 3A–E), but also in the middle stages of oogenesis, such as stage 8 (Figure 3F–J).

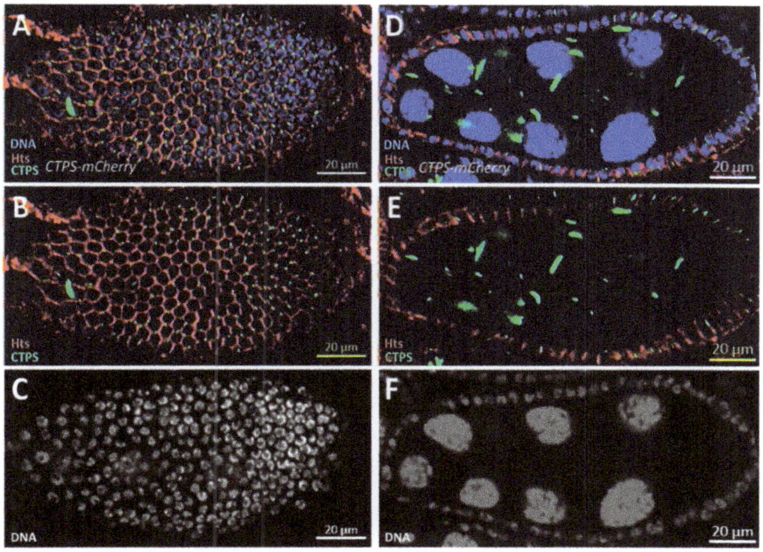

Figure 2. *CTPS-mCh* ovaries have same morphology as wild-type ovaries. (**A–C**) Surface view of a stage 8 CTPS-mCh egg chamber. The CTPS signal (green) shown is obtained using mCherry-tagged CTPS. Hts (red) staining marks cell membranes and Hoechst (blue) for DNA. (**A**) mCherry moiety doesn't affect CTPS assembly. (**B**) Merged panel of Hts and CTPS to display the cytoophidia location. (**C**) Single panel of the nucleus. (**D**) Lateral view of a stage 8 *CTPS-mCh* egg chamber. (**E**) Merged panel of Hts and CTPS to show the monolayer structure of follicle epithelia as well as cytoophidia distribution. (**F**) Single panel of nucleus to stress the single-layer follicle epithelia.

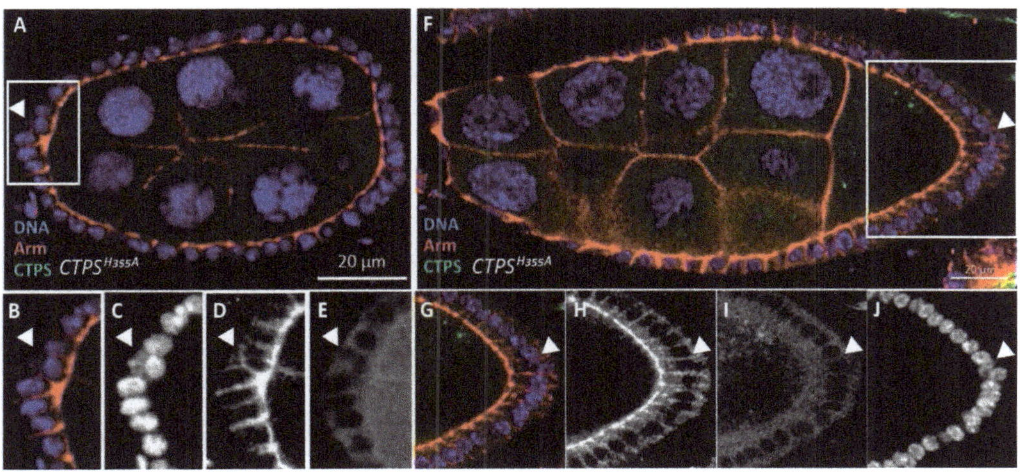

Figure 3. Follicle cell ingression in $CTPS^{H355A}$ mutant in early and middle stages of oogenesis. (**A**) Cross section of a $CTPS^{H355A}$ stage 6 egg chamber. The egg chamber is labeled with CTPS (green), Armadillo (red) for apical complex in follicle cells and Hoechst 33342 (blue) for DNA. The white rectangle emphasizes the ingression of a follicle cell. (**B–E**) Close-up images of the ingressive follicle cell, indicated by the yellow arrow. No cytoophidium is formed due to the H355A mutation in CTPS. (**F**) Cross section of a $CTPS^{H355A}$ egg chamber at stage 9. The white rectangle emphasizes the ingression of a follicle cell. (**G–J**) Close-up images of the ingressive follicle cell, indicated by the yellow arrow. Scale bars, 20 μm.

Our study mainly focused on stage 8 egg chambers. We demonstrated the ingression of follicle cells in stage 8 egg chambers through three-dimensional reconstruction (Figure 4A–C). Combined with the morphological changes of follicle cells observed on the surface of the egg chambers, we speculated that the integrity of follicle epithelia would be disturbed when CTPS could not assemble into cytoophidia. Through statistical analysis, we found that approximately 20% of egg chambers at stage 8 have follicle cells ingression by counting 20 stage 8 egg chambers of each genotype (Figure 4D). Our results indicate that the widely and specifically distributed cytoophidia play a role in maintaining the integrity of follicle epithelia.

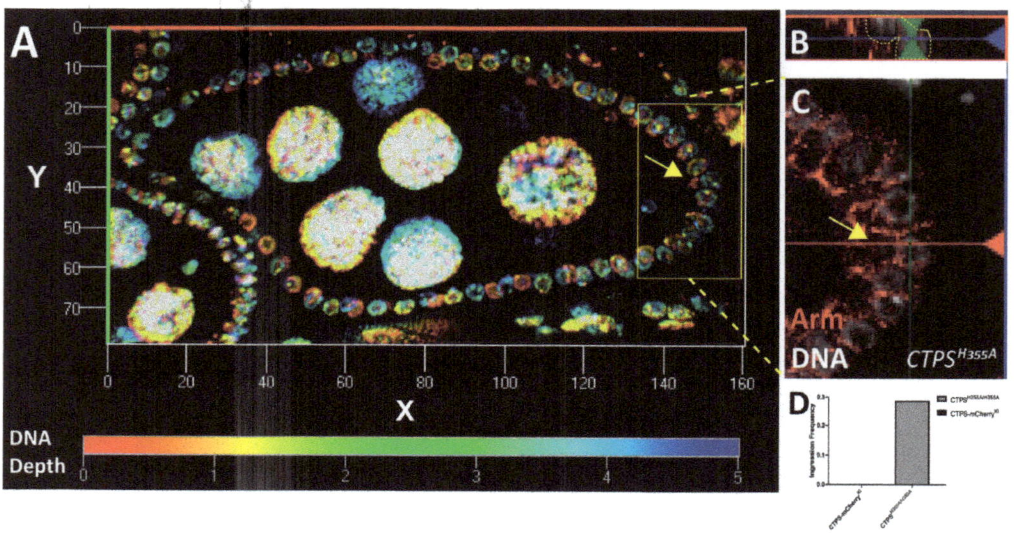

Figure 4. Three dimensional view of follicle cell ingression. (**A**) Cross section of a three-dimensional $CTPS^{H355A}$ egg chamber. A three-dimensional view of DNA stacked in layers. The interval between each layer is 0.5 μm, and a total of 12 layers are superimposed. The color from red to blue indicates the depth of DNA. (**B**) Side view of an ingressive nucleus. (**C**) On the xz plane, with the yellow dotted line marking the ingressive cell. (**D**) Quantification of the ingression frequency, 20 stage 8 egg chambers were counted per genotype.

2.3. *Ingressive Follicle Cells Display Abnormal DCAD2 Pattern*

In a previous study, we found that cytoophidia are specifically located on the lateral and basal sides of follicle cells [21]. The polarity regulators of follicle cells and adherens junctions have certain effects on the maintenance of cytoophidia. To explore whether the cell membrane components of ingressive follicle cells would be affected when cytoophidium fails to form, we labeled the basolateral regulator Dlg of follicle cells, adherens junctions DE-Cadherin DCAD2, and cell membrane protein Hts. After immunostaining, in the follicle epithelium labeled with Hts and Dlg, there was no significant difference between the cell membrane of ingressive follicle cells and that of normal follicle cells (Figure 5A–H).

Under the condition of DCAD2 labeling, we found that DCAD2 showed an abnormal pattern in the ingressive follicle cells compared with normally arranged follicle cells. In normal follicle cells, the end near the germline cells is defined as the apical side and the end near the muscle layer is defined as the basal side. From the cross-sectional view of the lateral side of the stage 8 egg chamber, the DCAD2 pattern should be adjacent to the germline cells. However, in the ingressive follicle cell, DCAD2 could be seen flipping in the direction rather than at the apical side (Figure 6).

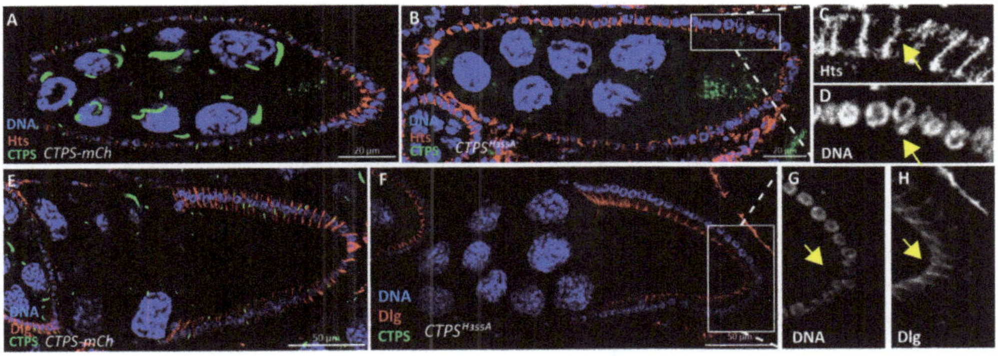

Figure 5. Ingression of follicle cells labeled with different membrane proteins. (**A**) Lateral view of a stage 8 egg chamber. Hts (red) labels the hulitaishao protein mainly presents at the lateral of follicle cell membranes. mCherry-tagged CTPS is shown green, and Hoechst (blue) for DNA. (**B**) A stage 8 egg chamber of *CTPS*^{H355A} *Drosophila*. mCherry-tagged *CTPS*^{H355A} is diffused. (**C**) The Hts pattern of an ingressive follicle cell. (**D**) Yellow arrow pointed out the ingressive cell nuclear. (**E**) Lateral view of a stage 10A egg chamber. Dlg (red) labels the discs large protein, which presents in the lateral and basal side of follicle cell. CTPS-mCherry cytoophidia can be observed, and Hoechst (blue) for DNA. (**F**) Lateral view of the *CTPS*^{H355A} egg chamber at stage 10A. (**G**) The ingressive cell nuclear is stressed by the yellow arrow. (**H**) The Dlg pattern of the follicle cell ingression.

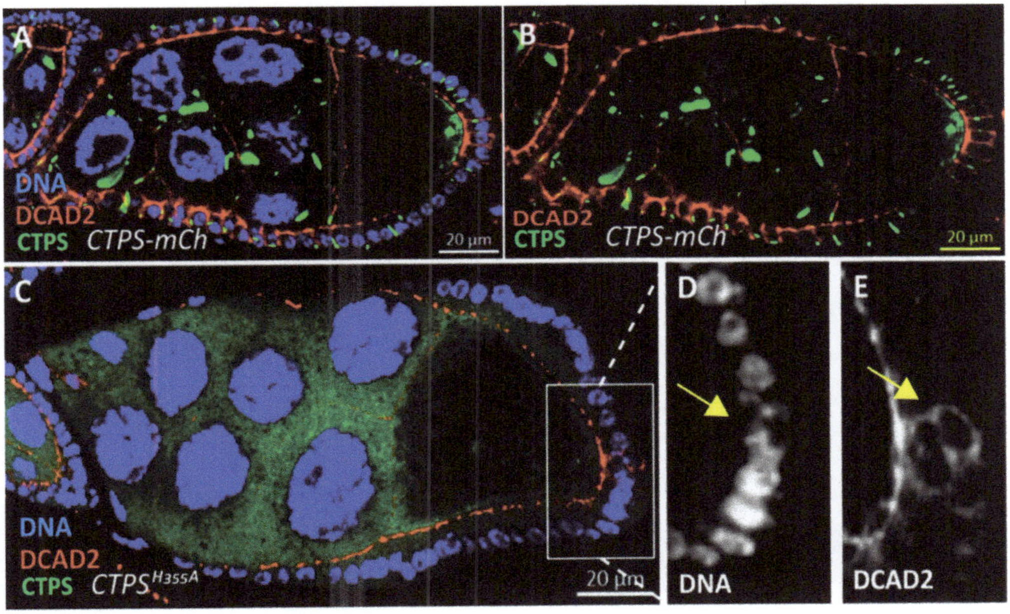

Figure 6. DCAD2 distribution is disturbed in *CTPS*^{H355A} follicle epithelia. (**A**) Cross section of Drosophila egg chamber expressing CTPS labelled by mCherry. The part of the follicle cell adjacent to the nurse cell is called apical, and the DE-Cadherin labeled by DCAD2 (red) is located at the apical of follicle cells. (**B**) DCAD2 together with CTPS as control to show the normal distribution of DCAD2. (**C**) Lateral view of a *CTPS*^{H355A}*–mCh* egg chamber with abnormal follicle cell. (**D**) Yellow arrow pointed to the follicle cell ingression. (**E**) Ingressive folllicle cell pointed to by yellow arrows show abnormal distribution of DCAD2.

2.4. $CTPS^{H355A}$ Follicle Cells Increase the Heterogeneity

The follicle epithelium of *Drosophila* consists of a monolayer of follicle cells, which surround the oocyte and 15 nurse cells. Follicle cells gradually differentiate into various subpopulations, which will undergo morphological changes. After stage 6, the follicle cells cease mitosis and are arranged in a hexagonal pattern, which means that under normal circumstances, each follicle cell contacts six adjacent cells, most of which are hexagonal and well arranged on the surface of egg chambers. However, in $CTPS^{H355A}$ mutant, the assembly of cytoophidia was disrupted and the number of non-hexagonal cells increased. We segmented the cell by the membrane staining of Hts protein, counting the different shapes cell by cell. In $CTPS^{H355A}$ mutant, we observed many pentagonal follicle cells, and the heptagonal cells increased by about 10% (Figure 7A–D).

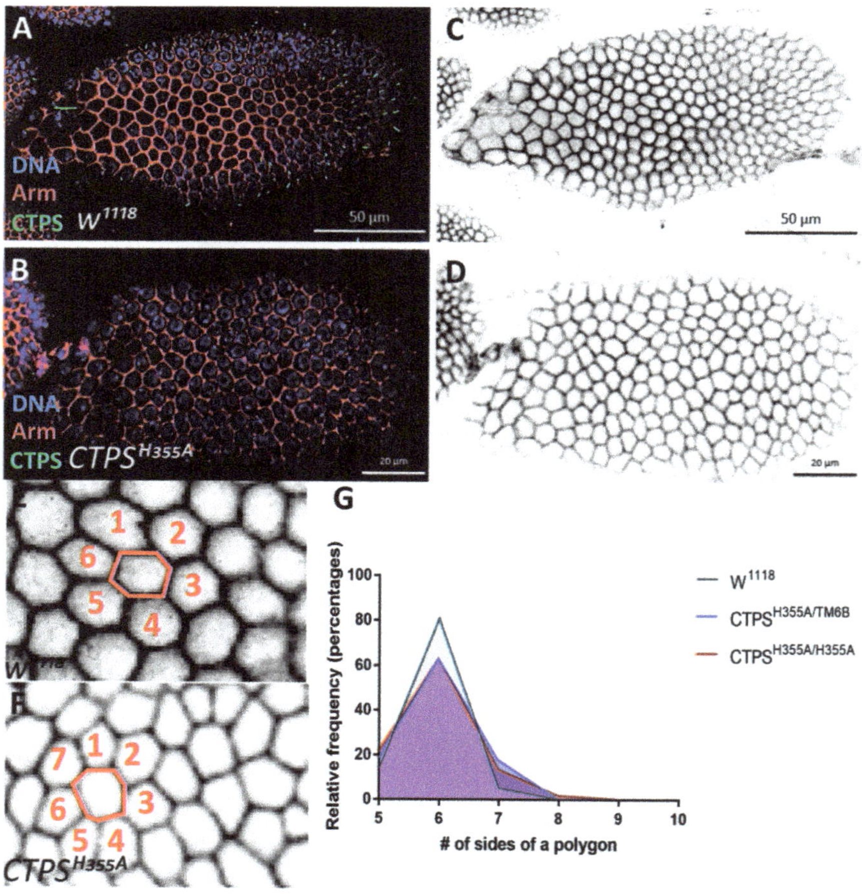

Figure 7. Morphological comparision of follicles between wild-type and $CTPS^{H355A}$. (**A–F**) Surface view of stage 8 egg chambers. Membrane was labeled by Armadillo (red), and CTPS is shown in green. (**A**,**C**) w^{1118} egg chamber. Scale bars, 50 µm. (**B**,**D**) $CTPS^{H355A}$ egg chamber. Scale bars, 20 µm. (**E**) Zoom-in view of (**C**). (**F**) Zoom-in view of (**D**). Redlines in E and F outline a central follicle cell with neighboring follicle cells in numbers. Note that the number of neighboring cells reflects the number of sides of the central polygonal follicle cells. (**G**) Quantitative analysis of the morphological difference between the wild-type control and $CTPS^{H355A}$ follicle cells (6 egg chambers were quantified per genotype, biological repeats = 3).

The quantification further confirmed that compared with w^{1118}, there was difference in the number of heptagonal follicle cells in stage 8 egg chambers of the $CTPS^{H355A}$ mutant. Moreover, the number of hexagonal follicle cells in $CTPS^{H355A}$ egg chambers decreased by about 15% at stage 8, while the number of polygonal cells increased by about 10% (Figure 7E–G). Considering that the more sides the polygon, the closer it is to the round circle, we speculate that when cytoophidia cannot be formed, the cell membrane will be affected, and the tight arrangement of epithelial follicle cells will not be maintained. Morphology changes indicate that cytoophidia located at the basolateral side may play a role in maintaining the integrity of follicle epithelium.

2.5. Follicle Epithelia Reduce Compactness in $CTPS^{H355A}$ Mutant

In the wild-type flies, follicle epithelial cells at stages 4–9 of oogenesis are tightly packed, and most of the hexagonal follicle cells enclose germ cells. In the case that CTPS could not be assembled into cytoophidia, we observed that closely arranged epithelial follicle cells became relatively loose, and follicle cells of similar size in the wild-type became relatively very large or very small, which was not conducive to compact arrangement (Figure 8A–L).

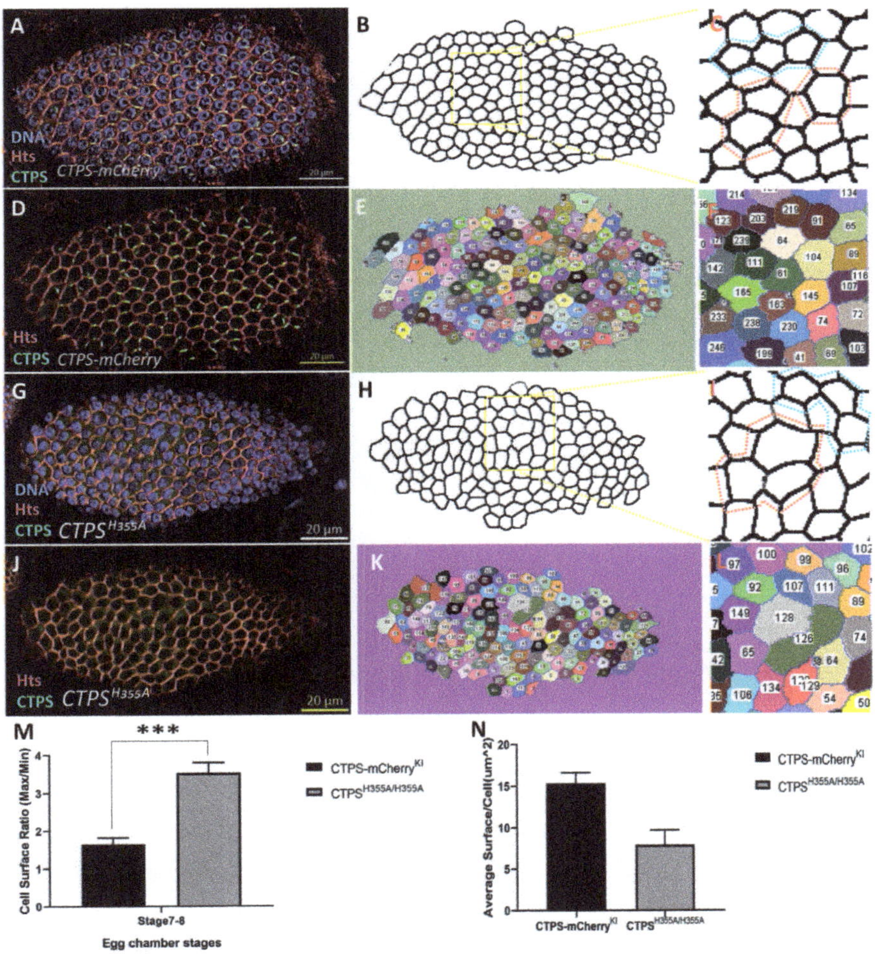

Figure 8. Quantitative morphological analysis of follicle cells. (**A**) Surface view of a stage 8 wild-type egg chamber. CTPS-mCherry (green), Hts (red) and DNA (blue, labeled with Hoechst 33342).

(**B**) Morphology of cell membrane. The yellow rectangle highlights an area of the follicle epithelium. (**C**) Enlarged part of the framed region in B. Larger follicular cells are circled in red, and smaller follicular cells are circled in blue. (**D**) CTPS-mCherry (green) and Hts (red) of the same egg chamber shown in A. (**E**) The area of each cell was measured after dividing each cell along the cell membrane. (**F**) corresponds to (**D**). (**G**) Surface view of a *CTPS^{H355A}* stage 8 egg chamber. CTPS^{H355A}-mcherry knock-in (green), Hts (red, labels cell membrane), and DNA (blue, labeled with Hoechst 33342). (**H**) Morphology of the cell membrane. A yellow rectangle highlights an area of the follicle epithelium. (**I**) Enlarged part of the framed region in (**H**). Larger follicle cells are circled in red, and smaller follicle cells are circled in blue. (**J**) CTPS^{H355A}-mCherry (green) and Hts (red) of the same egg chamber shown in (E). Surface view of the cell membrane with CTPS. (**K**) The area of each cell was measured after dividing each cell along the cell membrane. (**L**) corresponds to (**K**). Scale bars, 20 μm. (**M**) The ratio of the average area of the three largest cells to the average area of the three smallest cells in an egg chamber. N = 6, ***, $p < 0.0001$. Mann-Whitney U test. (5 stage 7–8 egg chambers/genotypes, 3 biological replicates) (**N**) Average follicle cell surface. N = 3.

In order to further clarify the observed phenomenon, we segmented the follicle cell surface based on the cell membrane and calculated the basal area of each follicle cell on an egg chamber through software. The average area ratio of the group with the largest area of three adjacent follicle cells and the group with the smallest area of three adjacent follicle cells was used as an indicator of follicle cell heterogeneity. The higher the ratio, the higher the heterogeneity of surface follicle cell. The quantitative analysis showed that the average area of follicle cells at stage 8 *CTPS^{H355A}* was smaller than that of the wild-type, but the heterogeneity was much higher than that of the wild-type (Figure 8M,N).

2.6. Follicle Cell Ingression Occurs in Egg Chamber Overexpressing *CTPS^{H355A}*

In previous studies, our laboratory found that the formation of cytoophidia can prolong the half-life of CTPS protein in mammalian cells. Therefore, we want to know whether H355A point mutation affects CTPS protein level in *Drosophila* ovaries. Western blot results confirmed that the level of CTPS protein in *Drosophila* ovaries after *CTPS^{H355A}* mutation was lower than that in the *wild-type* (Figure 9H,I). Thus, we want to investigate whether the phenotypes observed in the *CTPS^{H355A}* strain are caused by the decrease of CTPS protein level.

To eliminate the influence of protein level, we used the *Actin-Gal4* driver to overexpress *CTPS^{H355A}* in *Drosophila* ovaries. We constructed *Actin-Gal4-driven Drosophila* strains overexpressing *CTPS^{H355A}* (*Actin > UAS CTPS^{H355A}-mCherry-OE*) or *wild-type CTPS* (*Actin > UAS CTPS-mCherry-OE*). Western blot confirmed that there was no significant difference in CTPS level between the *Actin > UAS CTPS-mCherry-OE* heterozygous strain and the *Actin > UAS CTPS^{H355A}-mCherry-OE* homozygous strain (Figure 9J,K).

We found that the distribution of cytoophidia in the basolateral side of follicle cells could be clearly observed in *Actin > UAS CTPS-mCherry-OE* heterozygous egg chambers (Figure 9A,B). Almost every follicle cell had one or two cytoophidia, and follicle cells were arranged in a single layer. In *Actin > UAS CTPS^{H355A}-mCherry-OE* homozygous flies, the diffused distribution of CTPS^{H355A} could be observed, and the ingressive follicle cells appeared as well (Figure 9C–G). These results indicate that loss of the cytoophidium-forming ability of CTPS, rather than its protein level, is the primary cause of follicle cell ingression in the *CTPS^{H355A}* mutant.

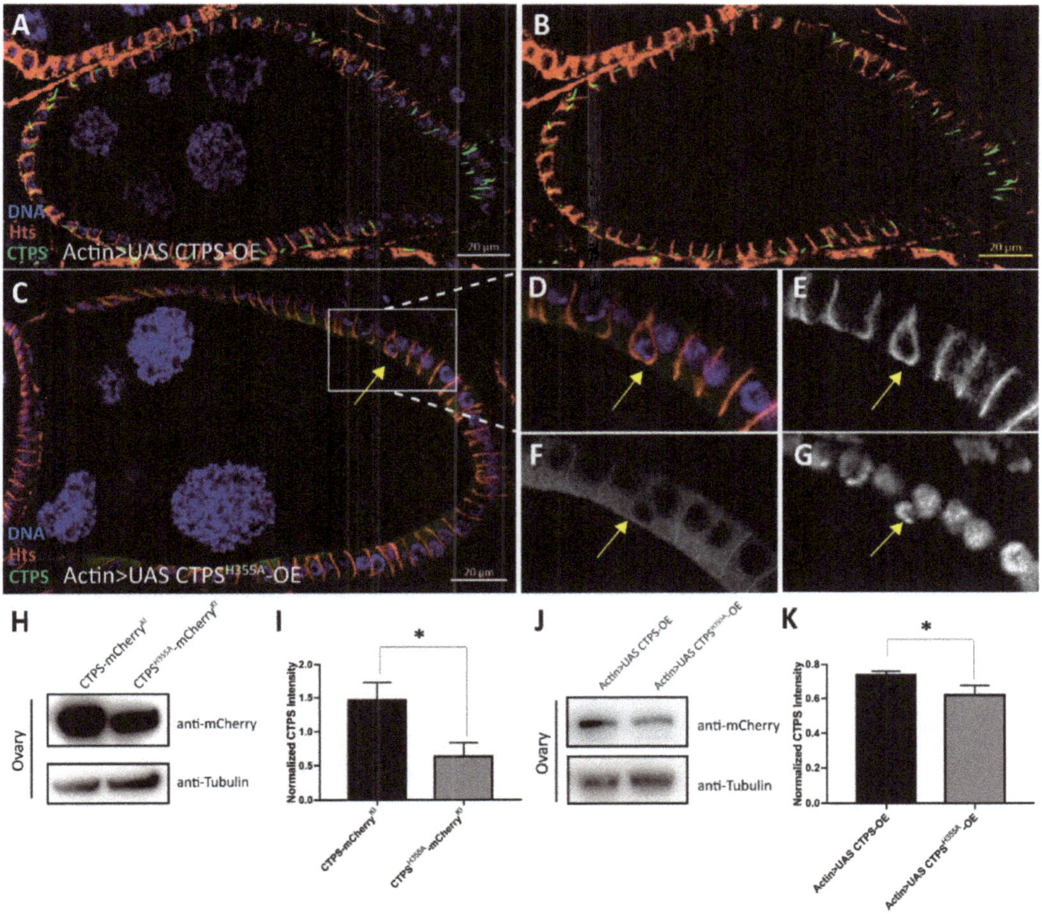

Figure 9. Follicle cell ingression occurs in egg chamber overexpressing $CTPS^{H355A}$. (**A**,**B**) Side view of stage 8 egg chamber overexpressing *CTPS*. (**C**) Cross section of stage 8 egg chamber overexpressing $CTPS^{H355A}$. The ingression is framed by a rectangle. (**D–G**) Zoom-in images of the ingression follicle cells in (**C**). Scale bars, 20. (**H**) Western blot is detected with antibodies against mCherry and tubulin on the ovarian lysates of *CTPS-mCh* and $CTPS^{H355A}$*-mCh* mutants. Scale bars, 20 μm. (**I**) Quantitative analysis of the CTPS protein level of samples represented in (**H**), the mean and standard deviation. (**J**) Western blot of the ovarian lysates of *Actin > UAS CTPS-OE* and *Actin > UAS* $CTPS^{H355A}$*-OE* mutants, detected with antibodies against mCherry and tubulin. (**K**) Quantitative analysis of the CTPS protein level of samples represented in (**J**), the mean and standard deviation. *, $p < 0.05$; Mann-Whitney U test.

2.7. Overexpression of $CTPS^{H355A}$ Increases the Heterogeneity of Follicle Cells

Similarly, we wanted to examine whether the heterogeneity of follicle cells was affected by the level of CTPS protein. According to our study, there was a long and curly cytoophidium in each follicle cell on the surface of *Actin > UAS CTPS-mCherry-OE* heterozygous egg chamber. Compared with the wild-type egg chambers, where cytoophidia are mostly rod-shaped and distributed along the cell membrane, the elongated cytoophidia were still distributed along the cell membrane after the overexpression of CTPS (Figure 10A–C).

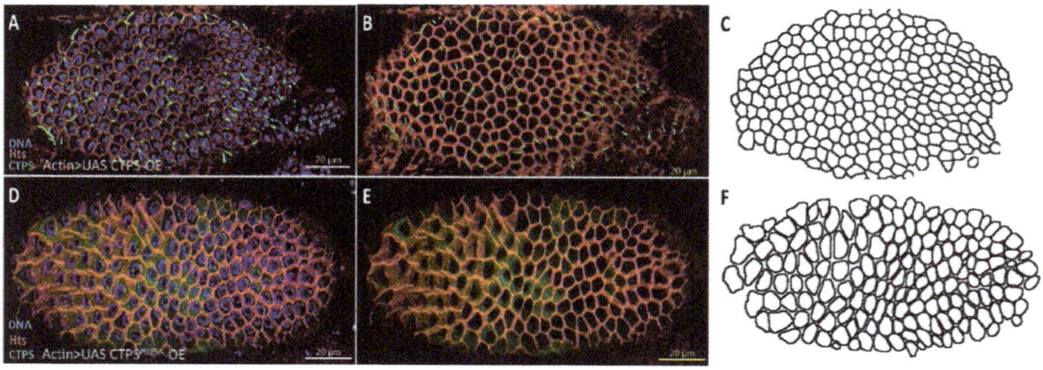

Figure 10. The integrity of follicle epithelium is compromised when overexpressing *CTPS*^{H355A}. (**A–C**) Surface view of a stage 8 egg chamber with follicle cells overexpressing *CTPS*. Large cytoophidia are detectable in almost all follicle cells. (**D–F**) Surface view of a stage 8 egg chamber with follicle cells overexpressing *CTPS*^{H355A}. Note that the heterogenous sizes of follicle cells and increased gaps between neighbouring follicle cells. CTPS-mCherry (green), Hts (red) and DNA (blue, labelled with Hoechst 33342). Scale bars, 20 μm.

The diffused distribution of CTPS was confirmed on the surface of the egg chamber of *Actin > UAS CTPS*^{H355A}*-mCherry-OE* homozygous fly. The changes of morphology and cell size showed that the heterogeneity of follicle cells was enhanced because there was no cytoophidium on the cell membrane. It seemed that these follicle cells could not even be arranged tightly (Figure 10D–F).

2.8. Follicle Cell-Specific Overexpression of CTPS^{H355A} *Impairs the Integrity of Follicle Epithelium*

After excluding the influence of CTPS protein level, we wanted to further eliminate the effect of the inter-tissue interaction caused by the ubiquitous expression of *Actin-Gal4*. To this end, we constructed a strain overexpressing CTPS^{H355A} using the same *UAS CTPS*^{H355A}*-mCherry-OE* and *UAS CTPS-mCherry-OE* strains together with *Tj-Gal4* specifically expressed in follicle cells. As a control, *wild-type CTPS* was overexpressed specifically in follicle cells using the *Tj-Gal4* driver. Our results confirmed that the integrity of follicle epithelium was impaired when *CTPS*^{H355A} was overexpressed specifically in follicle cells (Figure 11A–N).

2.9. Space between Muscle Sheath and Egg Chamber Increases in CTPS^{H355A}

IF is a member of the integrin complex and widely exists in the muscle layer that encloses the ovarioles. We found that in the wild-type ovary, the muscle sheath tightly wrapped the ovarioles and drove their movement (Figure 12A,B), which was conducive to common life activities such as oogenesis. In *CTPS*^{H355A} ovaries, the space between the muscle sheath and the egg chamber was significantly increased, and the egg chamber almost collapsed from the muscle sheath (Figure 12C–E), which might affect normal physiological activities.

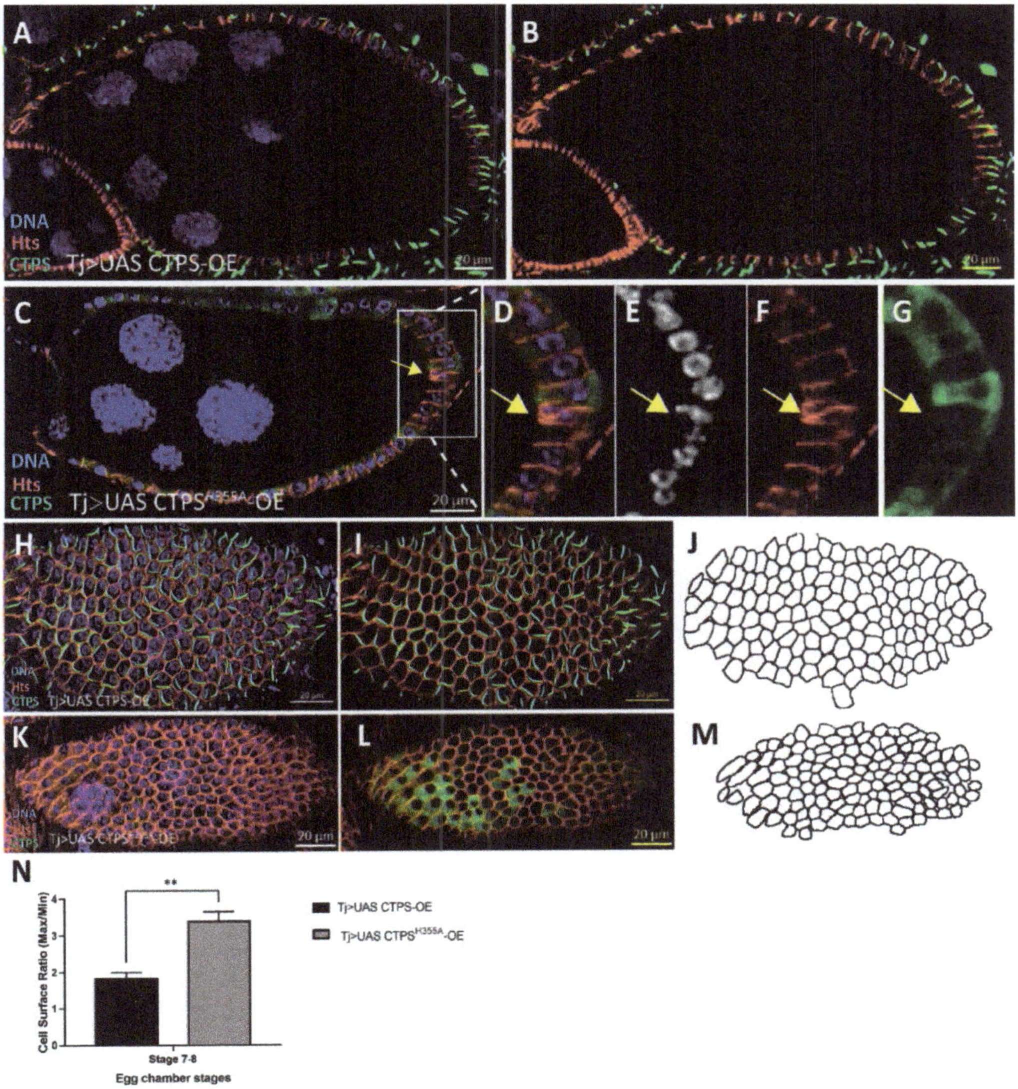

Figure 11. Follicle cell ingression occurs with follicle cell-specific overexpresson of *CTPS*^{H355A}. (**A**,**B**) Side view of a stage 8 egg chamber overexpressing CTPS in follicle epithelium. (**C**) Cross section of stage 8 egg chamber after overexpression of CTPS^{H355A} in follicle epithelium. The ingression is framed by a rectangle. (**D–G**) Zoom-in of the ingressive follicle cell in (**C**). (**H–J**) Surface view of a stage 8 egg chamber with follicle cell-specific overexpression of CTPS. Large cytoophidia are detectable in almost all follicle cells. Note that most cytoophidia are distributed on or near the cortex of follicle cells. (**K–M**) Surface view of a stage 8 egg chamber with follicle cell-specific overexpression of CTPS^{H355A}. No cytoophidium is detectable. CTPS-mCherry (green), Hts (red) and DNA (blue, labelled with Hoechst 33342). Scale bars, 20 μm. (**N**) Quantitative analysis of the ratio of three largest cells versus three smallest cells (5 images/genotypes, 3 biological replicates). Mann-Whitney U test, **, $p = 0.002$.

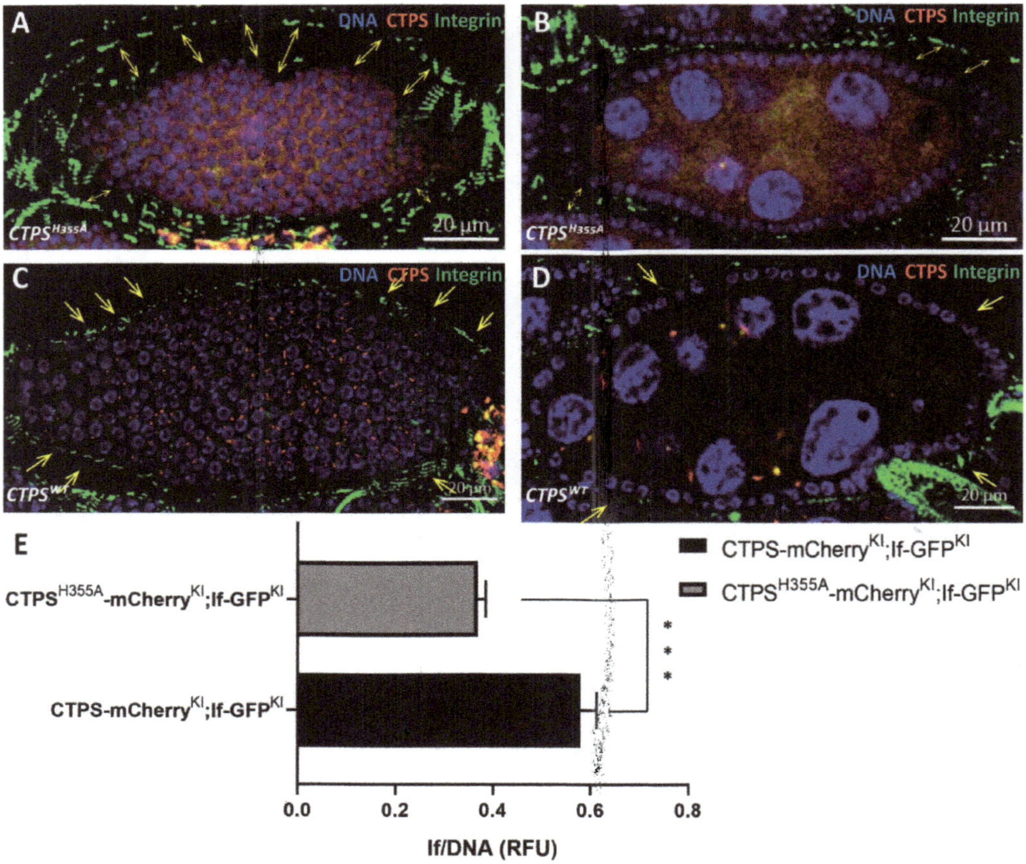

Figure 12. Compared with wild-type control, the space increases between muscle sheath and egg chambers in ***$CTPS^{H355A}$*****.** (**A**,**B**) A stage 8 egg chamber of the $CTPS^{H355A}$ mutant. Integrin is labeled with knock-in GFP (green). $CTPS^{H355A}$ is labeled with knock-in mCherry (red). Yellow arrows point to gap between egg chamber and the muscle sheath. (**C**,**D**) A stage 8 wild-type egg chamber. Integrin is labeled by knock-in GFP. CTPS is labeled by knock-in mCherry. Scale bars, 20 µm. (**E**) The ratio of GFP intensity of the integrin to DNA from (**A**–**D**). The value is normalized to the control (5 images/genotypes, 3 biological replicates). Mann-Whitney U test. ***, $p = 0.0005$.

3. Discussion

To explore the physiological function of cytoophidia in *Drosophila* follicle cells, we analyze the changes in follicle cells in CTPS mutant when cytoophidia cannot be formed. Our results indicate that the integrity of follicle epithelium is compromised when CTPS lose its cytoophidium-forming capability.

In this study, we generate transgenic flies with a point mutation in CTPS. Mutations do not affect enzymatic activity but lead to the disassembly of cytoophidia. In the mutant flies, the integrity of follicle epithelia is impaired with two related phenotypes: (1) ingression of follicle cells and (2) heterogeneous follicle cells.

We have previously discovered that cytoophidia are specifically distributed on the basolateral side of follicle cells [21]. Moreover, when the polarity of follicle cells is disrupted, cytoophidia will become unstable, especially due to the disruption of apical regulators. In this study, the apical polarity of follicle cells is indeed affected by the absence of cytoophidia.

Our data indicate that the cytoophidium, as a kind of membraneless organelle, maintains its specific subcellular localization in biological processes.

In these experiments, we also notice that cytoophidia play a role in maintaining the integrity of follicle epithelium. We speculate that cytoophidia located at the basolateral side of follicle cells may play a role in supporting follicle cells. In the absence of cytoophidia, the mechanical tension of the follicle cell membrane will be reduced, making it more difficult to maintain the cell morphology. Therefore, follicle cells are more likely to be drawn into polygons and expanded by surrounding cells or squeezed and reduced by surrounding cells. Similarly, due to the weakening of membrane mechanical tension, the follicle cells migrating inward after mitosis cannot be reintegrated into the follicular monolayer, resulting in the ingression. Our laboratory also found that in the male reproductive system of *Drosophila*, when CTPS cannot form cytoophidia, the main cells on the surface of the accessory gland may be difficult to maintain their cell shape, and two horizontally arranged nuclei appear to be vertically arranged. This further support our hypothesis [30].

When cytoophidia are disassembled, the observed separation of egg chamber and muscle sheath may also be due to the disappearance of the supporting force of cytoophidia. When cytoophidia cannot be formed, the internal supporting force of each follicle cell is weakened, leading to the collapse of the entire egg chamber. Considering that the follicle epithelium will develop into the eggshell of a fertilized egg in the later stages [31], it is possible that its shell hardness and the hatchability of the fertilized egg will also be affected accordingly.

However, we could not simply rescue the phenotypes found in the $CTPS^{H355A}$*-mCh* mutant by expressing CTPS-mCherry protein. Our previous studies in mammals [26] and *Drosophila* [27] confirmed that the $CTPS^{H355A}$ point mutation is dominant-negative, which is to say that as long as the $CTPS^{H355A}$ protein exists, the CTPS protein would not be able to assemble into cytoophidium [26,27]. Because the H355A point mutation of CTPS would disrupt the assembly of the cytoophidia dominant negatively, we have analyzed the $CTPS^{H355A/TM6B}$*-mCh* egg chambers and found that $CTPS^{H355A/TM6B}$ also have defects in follicle epithelial integrity mentioned above (Supplementary Figure S1). These results further validate our hypothesis that the cytoophidium structure plays a certain role in maintaining epithelial integrity, and the dominant negative CTPS point mutation confirmed that it is crucial for the assembly of cytoophidium.

Since the first discovery of cytoophidia in our laboratory in 2010, great progress has been made in the research on the existence of cytoophidia in different species and different types of cells. However, knowledge concerning the function of this new type of organelle widely existing in organisms is still in the initial stage. Therefore, our work has potential reference value for understanding the role of cytoophidia in *Drosophila* follicle cells. Our results indicate that forming cytoophidia is crucial to epithelial integrity.

4. Materials and Methods

4.1. Fly Stocks

All stocks were maintained at 25 °C on standard cornmeal. Both w^{1118} and C-terminal mChe-4V5 tagged CTPS knock-in flies out of w^{1118} produced in our laboratory were used as wild-type controls unless stated otherwise. The stocks used were: (1) $CTPS^{H335A}$ *mutated with mChe-4V5 tagged CTPS Knock-in fly*, (2) *Actin-Gal4/Cyo* (A gift from Guanjun Gao's lab, ubiquitous expression under strong promoter, a chromosome II insertion balanced over Curly of Oster [32]), (3) *Tj-Gal4* (A gift from Kun Dou's lab [33,34]), (4) *Sp/Cyo; Sb/Tm6B* (Institute of Biochemistry and Cell Biology, Chinese Academy of Sciences, Drosophila Resources and Technology Platform), (5) *UAS CTPS-mCherry-OE, and UAS* $CTPS^{H355A}$*-mCherry-OE* [27].

4.2. Fly Genetics

Transgenic flies expressing full-length CTPS (isoform C, LD25005) under the UAS promoter (*UAS-CTPS*) were generated in our lab (Back cross 5 generations before use) [35]. To generate the *Tj* > *UAS CTPS-OE*, we crossed the homozygous *UAS CTPS-OE Drosophila* and *Tj-Gal4 Drosophila* with the double balancer *Drosophila* for one generation, and the target generation was inbred for two generations to get the homozygous *Tj-Gal4; UAS CTPS-mCherry-OE Drosophila*. Same as the Tj-Gal4; *UAS* $CTPS^{H355A}$*-mCherry-OE; Actin-Gal4; UAS CTPS-mCherry-OE* and *Actin-Gal4; UAS* $CTPS^{H355A}$*-mCherry-OE* strains.

4.3. Generation of Transgenic Flies

For polymerase chain reaction (PCR), PUASTattb plasmids were used as the template, and phanta Maxa Super-Fidelity DNA Polymerase (Vazyme, #P505) as the polymerase. Sequences for primers were as below:

H355A-F: GAGCAAGTACGCCAAGGAGTGGCAGAAGCTATGCGATAGCCAT;

H355A-R: TGCCACTCCTTGGCGTACTTGCTCGGCTCAGAATGCAAAGTTT

After obtaining the required plasmids, the $CTPS^{H355A}$ *Drosophila* strain was constructed by microinjection.

4.4. Immunohistochemistry

Ovaries from flies were dissected in Grace's Insect Medium (Gibco) and then fixed in 4% formaldehyde (Sigma) diluted in PBS for 10 min before immunofluorescence staining. The samples were then washed twice using PST (0.5% horse serum + 0.3% Triton × 100 in PBS). For membrane staining, samples were incubated with primary antibodies at room temperature overnight, and then washed using PST. Secondary antibodies were used to incubate the samples at room temperature for another night.

Primary antibodies used in this study were rabbit anti-CTPS (1:1000; y-88, sc-134457, Santa Cruz BioTech Ltd., Santa Cruz, CA, USA), mouse anti-Discs Large (1:500, Developmental Studies Hybridoma Bank, Iowa City, IA, USA), mouse anti-D-E Cadherin (1:500, Developmental Studies Hybridoma Bank), mouse anti-HTS (1:1000, Developmental Studies Hybridoma Bank, Cat. No. AB_528070), mouse anti-Armadillo (1:500, Developmental Studies Hybridoma Bank). Secondary antibodies used in this study were anti-mouse, rabbit, or goat antibodies that were labeled with Alexa Fluor® 488 (Molecular Probes), or with Cy5 (Jackson ImmunoResearch Laboratories, West Grove, PA, USA). Hoechst 33342 was used to label DNA.

4.5. Microscopy and Image Analysis

All images were obtained under laser-scanning confocal microscopy (Zeiss 880). Image processing was performed using Zeiss Zen. ImageJ was used to analyze the area and number of follicle cells.

We used the ImageJ SCF to segment the follicle cells by the membrane, then use ImageJ cell counter to calculate different shapes of cells to get the number of polygonal cells. We used ImageJ to measure the area of each cell. For each statistical quantification, we collected the surface images using Zeiss 880 with the interval as 0.5 μm for z-stack, 5 stage 8 egg chambers were quantified per genotype, biological repeats = 3. Mann-Whitney U test was conducted for comparison.

4.6. Western Blotting

Female adult ovaries of *Drosophila* were collected with gathered into lysis buffer RIPA (Meilunbio, Dalian, China) with protease inhibitor cocktail (Bimake, Shanghai, China) for Western blotting, and then ground with 1 mm Zirconia beads in Sonicator (Shanghai Jing Xin, Shanghai, China). The sample would then lysis on ice for up to 30 min. Samples were centrifuged for 10 min at 10,000 g at 4 °C. The 6× protein loading buffer was pipetted into the supernatants and boiled at 99 °C for 15 min to obtain protein. Then, the protein sample was run through 10% SDS-PAGE gels and transferred to PVDF membranes. At room temperature, membrane was incubated with 5% w/v nonfat dry milk dissolved by 1× TBST for 1 h of blocking. Then, the membrane was incubated with primary antibodies in 5% w/v nonfat milk at 4 °C and gently shaken overnight.

The following primary antibodies were used in this study: anti-mCherry Tag Monoclonal antibodies (Cat. No. A02080, Abbkine, Beijing China), mouse anti-a-Tubulin antibodies (Cat. No. T6199, Sigma). The membranes were washed three times for 5 min per time with shaking, then incubated with secondary antibodies (anti-mouse IgG, HRP-linked antibody, Cell Signaling, Danvers, MA, USA) diluted in 5% w/v nonfat milk at room temperature for 1 h. An Amersham Imager 600 (General Electric, Boston, MA, USA) and Pierce ECL Reagent Kit (Cat. No. 32106, Thermo Fisher, Waltham, MA, USA) were adopted for the chemiluminescence immunoassay. Protein levels were quantified on ImageJ (National Institutes of Health, Bethesda, MD, USA) and normalized to tubulin. At least three biological replicates were quantified.

4.7. Data Analysis

Images collected by confocal microscopy were processed using Adobe Illustrator and ImageJ. Cell segmentation based on the cell membrane was achieved using CellPose and SCF methods. Quantitative analysis was processed by Excel and GraphPad. The Mann–Whitney U test was conducted to get the p-value.

Supplementary Materials: The following supporting information can be downloaded at: https://www.mdpi.com/article/10.3390/ijms232315282/s1, Figure S1. *CTPS*$^{H355A/TM6B}$ mutants have no significant difference from *CTPS*^{H355A} mutant.

Author Contributions: Q.-Q.W. and J.-L.L. conceived the studies. Q.-Q.W. and J.-L.L. performed the experiments. D.-D.Y. assisted fly genetics. Q.-Q.W. drafted the manuscript. J.-L.L. revised the manuscript. All authors have read and agreed to the published version of the manuscript.

Funding: This work was funded by the Ministry of Science and Technology of the People's Republic of China (Grant No. 2021YFA0804700), the National Natural Science Foundation of China (No. 31771490), the Shanghai Science and Technology Commission (20JC1410500), and the UK Medical Research Council (Grant No. MC_UU_12021/3 and MC_U137788471). This research was supported by Shanghai Frontiers Science Center for Biomacromolecules and Precision Medicine at ShanghaiTech University.

Institutional Review Board Statement: Not applicable.

Informed Consent Statement: Not applicable.

Data Availability Statement: Not applicable.

Acknowledgments: We thank the Molecular Imaging Core Facility (MICF) at School of Life Science and Technology, ShanghaiTech University for providing technical support.

Conflicts of Interest: The authors have no relevant financial or non-financial interest to disclose.

References

1. Barry, R.M.; Bitbol, A.-F.; Lorestani, A.; Charles, E.J.; Habrian, C.H.; Hansen, J.M.; Li, H.-J.; Baldwin, E.P.; Wingreen, N.S.; Kollman, J.M.; et al. Large-scale filament formation inhibits the activity of CTP synthetase. *Elife* **2014**, *3*, e03638. [CrossRef] [PubMed]
2. Lieberman, I. Enzymatic Amination of Uridine Triphosphate to Cytidine Triphosphate. *J. Biol. Chem.* **1956**, *222*, 765–775. [CrossRef] [PubMed]
3. Liu, J.L. Intracellular compartmentation of CTP synthase in Drosophila. *J. Genet. Genom.* **2010**, *37*, 281–296. [CrossRef] [PubMed]
4. Ingerson-Mahar, M.; Briegel, A.; Werner, J.N.; Jensen, G.J.; Gitai, Z. The metabolic enzyme CTP synthase forms cytoskeletal filaments. *Nat. Cell Biol.* **2010**, *12*, 739–746. [CrossRef] [PubMed]
5. Noree, C.; Sato, B.K.; Broyer, R.M.; Wilhelm, J.E. Identification of novel filament-forming proteins in Saccharomyces cerevisiae and Drosophila melanogaster. *J. Cell Biol.* **2010**, *190*, 541–551. [CrossRef] [PubMed]
6. Chen, K.; Zhang, J.; Tastan, Y.; Deussen, Z.A.; Siswick, M.Y.-Y.; Liu, J.-L. Glutamine analogs promote cytoophidium assembly in human and Drosophila cells. *J. Genet. Genom.* **2011**, *38*, 391–402. [CrossRef]
7. Zhang, J.; Hulme, L.; Liu, J.L. Asymmetric inheritance of cytoophidia in *Schizosaccharomyces pombe*. *Biol. Open* **2014**, *3*, 1092–1097. [CrossRef] [PubMed]

8. Chang, C.C.; Jeng, Y.-M.; Peng, M.; Keppeke, G.D.; Sung, L.-Y.; Liu, J.-L. CTP synthase forms the cytoophidium in human hepatocellular carcinoma. *Exp. Cell Res.* **2017**, *361*, 292–299. [CrossRef]
9. Zhou, S.; Xiang, H.; Liu, J.L. CTP synthase forms cytoophidia in archaea. *J. Genet. Genom.* **2020**, *47*, 213–223. [CrossRef]
10. Uzman, A. Molecular Cell Biology, Sixth Edition. *Biochem. Mol. Biol. Educ.* **2010**, *38*, 60–61. [CrossRef]
11. Ovadi, J.; Saks, V. On the origin of intracellular compartmentation and organized metabolic systems. *Mol. Cell. Biochem.* **2004**, *256–257*, 5–12. [CrossRef] [PubMed]
12. Aughey, G.N.; Grice, S.J.; Liu, J.L. The Interplay between Myc and CTP Synthase in Drosophila. *PLoS Genet.* **2016**, *12*, e1005867. [CrossRef] [PubMed]
13. Pease, J.C.; Tirnauer, J.S. Mitotic spindle misorientation in cancer – out of alignment and into the fire. *J. Cell Sci.* **2011**, *124 Pt 7*, 1007–1016. [CrossRef]
14. McCaffrey, L.M.; Macara, I.G. Epithelial organization, cell polarity and tumorigenesis. *Trends Cell Biol.* **2011**, *21*, 727–735. [CrossRef] [PubMed]
15. Bergstralh, D.; Lovegrove, H.E.; Johnston, D.S. Lateral adhesion drives reintegration of misplaced cells into epithelial monolayers. *Nat. Cell Biol.* **2015**, *17*, 1497–1503. [CrossRef]
16. Sauer, F.C. The interkinetic migration of embryonic epithelial nuclei. *J. Morphol.* **1936**, *60*, 1–11. [CrossRef]
17. Bilder, D.; Li, M.; Perrimon, N. Cooperative Regulation of Cell Polarity and Growth by *Drosophila* Tumor Suppressors. *Science* **2000**, *289*, 113–116. [CrossRef]
18. Cammarota, C.; Finegan, T.; Wilson, T.J.; Yang, S.; Bergstralh, D.T. An Axon-Pathfinding Mechanism Preserves Epithelial Tissue Integrity. *Curr. Biol.* **2020**, *30*, 5049–5057.e3. [CrossRef]
19. Spear, P.C.; Erickson, C.A. Interkinetic nuclear migration: A mysterious process in search of a function. *Dev. Growth Differ.* **2012**, *54*, 306–316. [CrossRef] [PubMed]
20. Sorce, B.; Escobedo, C.; Toyoda, Y.; Stewart, M.P.; Cattin, C.J.; Newton, R.; Banerjee, I.; Stettler, A.; Roska, B.; Eaton, S.; et al. Mitotic cells contract actomyosin cortex and generate pressure to round against or escape epithelial confinement. *Nat. Commun.* **2015**, *6*, 8872. [CrossRef]
21. Wang, Q.Q.; Zhao, P.A.; Tastan, Y.; Liu, J.L. Polarised maintenance of cytoophidia in Drosophila follicle epithelia. *Exp. Cell Res.* **2021**, *402*, 112564. [CrossRef] [PubMed]
22. Zhou, X.; Guo, C.-J.; Hu, H.-H.; Zhong, J.; Sun, Q.; Liu, D.; Zhou, S.; Chang, C.C.; Liu, J.-L. Drosophila CTP synthase can form distinct substrate- and product-bound filaments. *J. Genet. Genom.* **2019**, *46*, 537–545. [CrossRef] [PubMed]
23. Lynch, E.M.; Hicks, D.R.; Shepherd, M.; Endrizzi, J.A.; Maker, A.; Hansen, J.M.; Barry, R.M.; Gitai, Z.; Baldwin, E.P.; Kollman, J.M. Human CTP synthase filament structure reveals the active enzyme conformation. *Nat. Struct. Mol. Biol.* **2017**, *24*, 507–514. [CrossRef] [PubMed]
24. Sun, Z.; Liu, J.L. mTOR-S6K1 pathway mediates cytoophidium assembly. *J. Genet. Genom.* **2019**, *46*, 65–74. [CrossRef]
25. Zhou, X.; Guo, C.-J.; Chang, C.-C.; Zhong, J.; Hu, H.-H.; Lu, G.-M.; Liu, J.-L. Structural basis for ligand binding modes of CTP synthase. *Proc. Natl. Acad. Sci. USA* **2021**, *118*, e2026621118. [CrossRef]
26. Sun, Z.; Liu, J.L. Forming cytoophidia prolongs the half-life of CTP synthase. *Cell Discov.* **2019**, *5*, 32. [CrossRef]
27. Liu, J.; Zhang, Y.; Zhou, Y.; Wang, Q.Q.; Ding, K.; Zhao, S.; Lu, P.; Liu, J.L. Cytoophidia coupling adipose architecture and metabolism. *Cell. Mol. Life Sci.* **2022**, *79*, 534. [CrossRef]
28. Wu, Z.; Liu, J.L. CTP synthase does not form cytoophidia in Drosophila interfollicular stalks. *Exp. Cell Res.* **2022**, *418*, 113250. [CrossRef]
29. Zhang, Y.; Liu, J.; Liu, J.L. The atlas of cytoophidia in *Drosophila* larvae. *J. Genet. Genom.* **2020**, *47*, 321–331. [CrossRef]
30. You, D.D.; Zhou, X.L.; Wang, Q.Q.; Liu, J.L. Cytoophidia safeguard binucleation of *Drosophila* male accessory gland cells. *Exp. Cell Res.* **2022**, *422*, 113433. [CrossRef]
31. Wu, X.; Tanwar, P.S.; Raftery, L.A. Drosophila follicle cells: Morphogenesis in an eggshell. *Semin. Cell Dev. Biol.* **2008**, *19*, 271–282. [CrossRef] [PubMed]
32. Chintapalli, V.R.; Kato, A.; Henderson, L.; Hirata, T.; Woods, D.J.; Overend, G.; Davies, S.A.; Romero, M.F.; Dow, J.A.T. Transport proteins NHA1 and NHA2 are essential for survival, but have distinct transport modalities. *Proc. Natl. Acad. Sci. USA* **2015**, *112*, 11720–11725. [CrossRef] [PubMed]
33. Schotthofer, S.K.; Bohrmann, J. Analysing bioelectrical phenomena in the Drosophila ovary with genetic tools: Tissue-specific expression of sensors for membrane potential and intracellular pH, and RNAi-knockdown of mechanisms involved in ion exchange. *BMC Dev. Biol.* **2020**, *20*, 15. [CrossRef] [PubMed]
34. Sahai-Hernandez, P.; Nystul, T.G. A dynamic population of stromal cells contributes to the follicle stem cell niche in the Drosophila ovary. *Development* **2013**, *140*, 4490–4498. [CrossRef] [PubMed]
35. Azzam, G.; Liu, J.L. Only One Isoform of Drosophila melanogaster CTP Synthase Forms the Cytoophidium. *PLoS Genet.* **2013**, *9*, e1003256. [CrossRef]

International Journal of *Molecular Sciences*

Article

Super-Resolution Imaging Reveals Dynamic Reticular Cytoophidia

Yi-Fan Fang [1], Yi-Lan Li [1], Xiao-Ming Li [1] and Ji-Long Liu [1,2,*]

1 School of Life Science and Technology, Shanghai Tech University, Shanghai 201210, China
2 Department of Physiology, Anatomy and Genetics, University of Oxford, Oxford OX1 3PT, UK
* Correspondence: liujl3@shanghaitech.edu.cn or jilong.liu@dpag.ox.ac.uk

Abstract: CTP synthase (CTPS) can form filamentous structures termed cytoophidia in cells in all three domains of life. In order to study the mesoscale structure of cytoophidia, we perform fluorescence recovery after photobleaching (FRAP) and stimulated emission depletion (STED) microscopy in human cells. By using an EGFP dimeric tag as a tool to explore the physical properties of cytoophidia, we find that cytoophidia are dynamic and reticular. The reticular structure of CTPS cytoophidia may provide space for other components, such as IMPDH. In addition, we observe CTPS granules with tentacles.

Keywords: CTP synthase; cytoophidium; fluorescence recovery after photobleaching (FRAP); stimulated emission depletion (STED)

Citation: Fang, Y.-F.; Li, Y.-L.; Li, X.-M.; Liu, J.-L. Super-Resolution Imaging Reveals Dynamic Reticular Cytoophidia. *Int. J. Mol. Sci.* **2022**, *23*, 11698. https://doi.org/10.3390/ijms231911698

Academic Editor: Vladimir N. Uversky

Received: 24 August 2022
Accepted: 27 September 2022
Published: 2 October 2022

1. Introduction

In addition to organelles with membranes, proteins with important functions in the cell can also be compartmented into membraneless organelles. CTP synthase (CTPS), a metabolic enzyme for de novo synthesis of CTP, was found to form filament-like compartments in cells called cytoophidia [1]. Describing their shape vividly, the word "cytoophidia" means "cellular snakes" in Greek. Cytoophidia were found in many species in all three domains of life, which means cytoophidia are conserved in evolution [1–19].

A glutamine analog, 6-diazo-5-oxo-L-norleucine (DON), promotes cytoophidia formation in *Drosophila* and human cells [5]. DON binds CTPS with covalent bonds [20]. Glutamine deprivation promotes cytoophidium formation in mammalian cells [21]. IMPDH can form cytoophidia [22] and both CTPS and IMPDH are related to glutamine and NH_3 metabolism. It was reported that there is an interaction between CTPS and IMPDH [23]. The function of cytoophidia may be closely related to glutamine and NH_3 metabolism.

In metabolic regulation, the activity of CTPS is inhibited via filament formation [24,25]. The half-life of CTPS is prolonged when forming cytoophidia [26]. Given the high-level metabolism in cancer cells, cytoophidium formation is highly related to oncogenes. Myc is required for cytoophidia assembly, and cytoophidia formation is regulated by Myc expression levels [27]. Ack kinase regulates cytoophidium morphology and CTPS activity [28]. Cytoophidium assembly was found to be regulated by the mTOR-S6K1 pathway [29]. Cytoophidia were also found in human hepatocellular carcinoma [10].

CTPS can be assembled into thin filaments in vitro, and the structures of CTPS filaments at near-atomic resolution have been solved by cryo-EM [20,25,30]. However, nanometer-scale CTPS filaments are different from the micron-scale cytoophidia observed via confocal microscopy. How CTPS filaments assemble into big micron-scale cytoophidia is still unclear. The physical properties of cytoophidia at the mesoscale remain to be explored.

To study cytoophidium properties in human cell lines, we performed fluorescence recovery after photobleaching (FRAP) microscopy to study the dynamic characteristics and stimulated emission depletion (STED) microscopy to study the super-resolution structure. By measuring the intensity and recovery speed of bleached ROIs, we were able to quantify the relative dynamic characteristics of cytoophidia under different treatments. STED allows fluorescence imaging to achieve a resolution of 50 to 70 nm [31,32].

2. Results

2.1. Assembly of CTPS Filaments into Cytoophidia

The cytoophidium is a compartment of metabolic enzymes, such as filamentous CTPS, which can be observed via confocal microscopy [1–3]. In vitro experiments showed that CTPS can also form filaments built from tetramer units [20,25,30]. In human cells, CTPS can be assembled into cytoophidia under DON treatment or glutamine deprivation. In this study, we did not distinguish hCTPS1 and hCTPS2. We found that after DON treatment, hCTPS1 can form granules in 293T cells with hCTPS1 over-expression (Figure 1A,B). When we observed living cells, CTPS granules existed in a small population of cells. CTPS granules can exist in the same cells as cytoophidia (Figure 1A,B).

How CTPS filaments are arranged in cytoophidia remains unclear. To conceive the arrangement model from CTPS filaments to cytoophidia, we constructed hCTPS1-overexpression vectors with different fluorescence proteins and different mutations (Supplemental Table S1). In order to show and evaluate the effect of exogenous protein overexpression in the experiment, we compared the protein levels of overexpressed hCTPS1 and endogenous hCTPS1/2 (Figure 1D) and tested the transfection efficiencies (Figure 1E,F). In transfection-positive cells, exogenous hCTPS1 expression was approximately twice as high as that of endogenous hCTPS1/2 (Figure 1G).

EGFP is a weak dimer while $EGFP^{A206K}$ is a monomer [33]. Overexpression of *hCTPS1-EGFP* forms cytoophidium-like condensates in 293T cells (Figure 1H). The force of forming dimer between EGFP pulls hCTPS1 together, and hCTPS is assembled into filaments. The hCTPS filaments may be compressed together by a simple force of EGFP dimerization (Figure 1C). The hCTPS1-EGFP group was a control for cytoophidium induction. CTPS1 with the H355A mutation disassembles cytoophidia. We found that overexpressed $\mathit{hCTPS1}^{H355A}$*-EGFP* could not form cytoophidia in 293T cells (Figure 1H).

2.2. Dynamic Equilibria of Cytoophidia

To test the dynamic characteristics of cytoophidia, we performed FRAP on four groups of hCTPS1 cytoophidia (Figure 2A,B). The intensity of the FRAP ROI was normalized as $Normalized\ Intensity = \frac{I(non-bleach\ ROI)_{Pre-bleach,\ 0} - I(background)_0}{I(non-bleach\ ROI)_n - I(background)_n} \times [I(bleach\ ROI)_n - I(background)_n]$ [34]. The bleached ROIs on cytoophidia induced by 20 μg/mL for 8 h (low concentration and short time) before imaging recovered very quickly (Figure 2C). However, ROIs in cells with *hCTPS1* overexpression treated with DON in 100 μg/mL (higher concentration) recovered fluorescence slowly and ended at a lower intensity (Figure 2D; Supplemental Figure S1A).

By extending the DON treatment time to 25 h, ROIs were able to recover as quickly as possible and in a relatively short time (Figure 2E; Supplemental Figure S1B). The fluorescence intensity restored by bleaching the ROIs of cytoophidium-like condensates of hCTPS-EGFP cells was very low (Figure 2F; Supplemental Figure S1C). There was a significant difference in dynamics between hCTPS1-EGFP cytoophidium-like condensates and DON-induced hCTPS cytoophidia at low concentrations and over short time periods (Figure 2G).

This shows that DON-induced cytoophidia have very different dynamic characteristics from hCTPS-EGFP cytoophidium-like condensates. DON-induced cytoophidia seem not to be assembled by simple forces, like the hCTPS-EGFP cytoophidium-like condensates.

The bleached ROIs gradually recovered throughout, rather than from any particular side. Neither of the two ROIs moved to either side, nor were they far away or close to each other (Figure 2C). Our results showed that bleached hCTPS1 molecules in cytoophidia could exchange with free hCTPS1 molecules in the cytosol (Figure 2H).

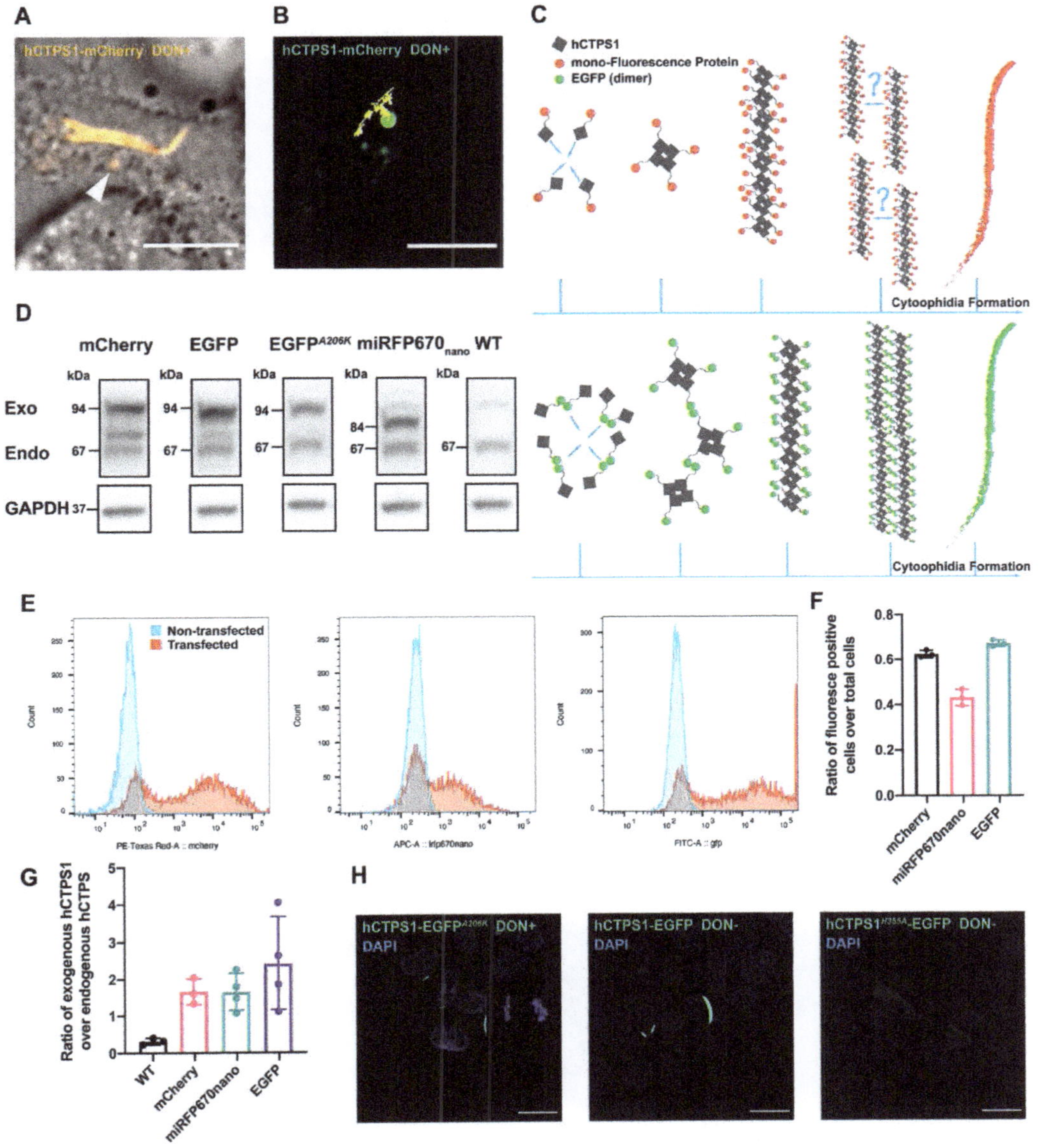

Figure 1. Assembly of CTPS filaments into cytoophidia. (**A**) hCTPS1 forms granules in the same cell that coexist with hCTPS1 cytoophidia. The arrowhead points to a granule. (**B**) The trajectory of hCTPS1 granules is a random walk. For the DON treatment, 20 µg/mL DON in PBS solution was added to fresh DMEM medium 8 to 25 h before live-cell imaging. (**C**) hCTPS1-EGFP forms cytoophidium-like condensates by simple force between EGFP. The arrangement from hCTPS1 filaments to hCTPS cytoophidia is the problem that needed to be solved. (**D**) The quantities of transfected over-expressed hCTPS in 293T cells were measured. (**E**,**F**) The transfection efficiencies were quantified. (**G**) Estimated ratio of exogenous hCTPS1 to endogenous hCTPS. (**H**) hCTPS-EGFPA206K cytoophidia can be induced by DON treatment. hCTPS1-EGFP can form cytoophidium-like condensates, which are wider and larger than hCTPS cytoophidia. hCTPS1H355A-EGFP cannot form cytoophidium-like condensates. For the DON treatment, 20 µg/mL DON (PBS solution) was added to fresh DMEM medium 8 h before fixation. Scale bars, 10 µm (**A**,**B**) and 20 µm (**H**).

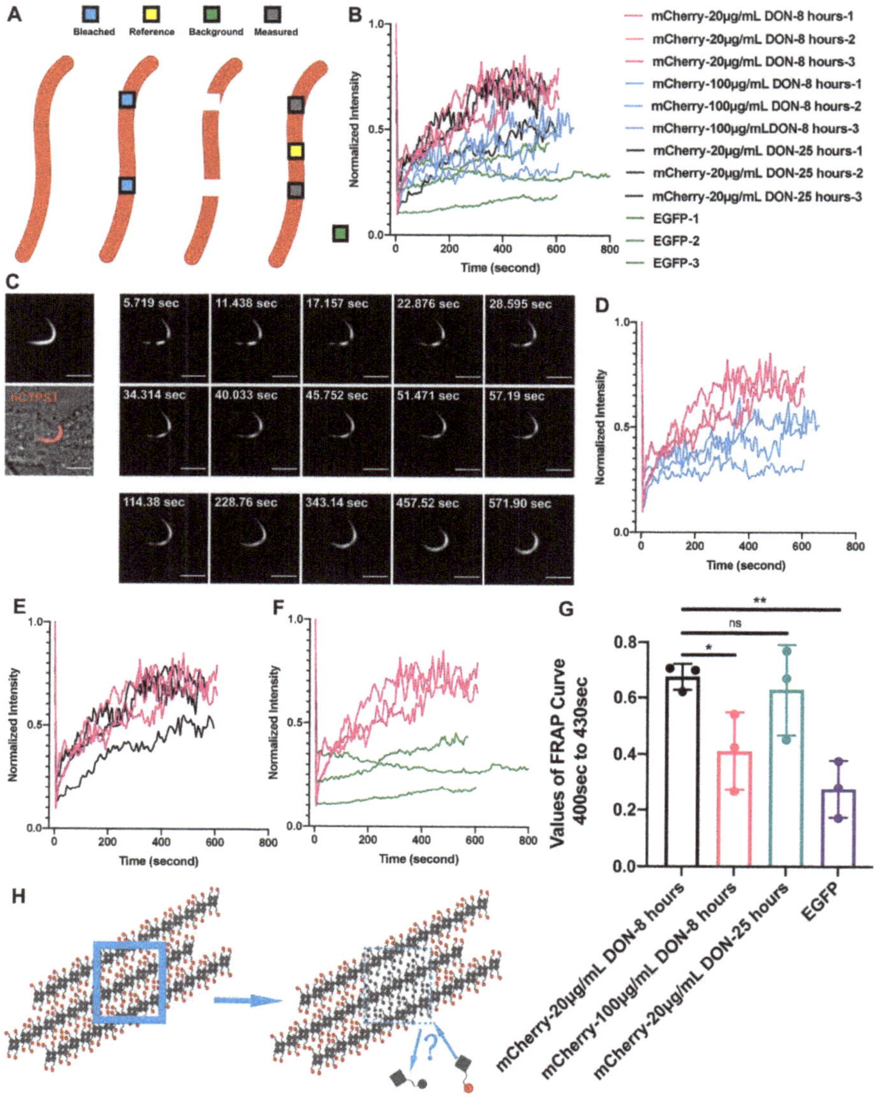

Figure 2. Dynamic equilibria of cytoophidia. (**A**) ROIs were used for bleaching, measurement and normalization of data. (**B**) Normalized intensity curves of FRAP results in different groups were merged. (**C**) Live-cell images of FRAP in hCTPS1-mCherry cytoophidia induced by 20 µg/mL DON for 8 h. (**D**) Comparison of FRAP curves for hCTPS1-mCherry between 20 µg/mL DON for 8 h and 100 µg/mL DON for 8 h. (**E**) Comparison of FRAP curves for hCTPS1-mCherry with 20 µg/mL DON for 8 h (pink curves) and 20 µg/mL DON for 25 h (black curves). (**F**) Comparison of FRAP curves for hCTPS1-mCherry cytoophidia (pink curves) induced by DON and hCTPS1-EGFP cytoophidium-like condensates (green curves). (**G**) Analysis of the level and speed of fluorescence recovery. *, p-value < 0.05; **, p-value < 0.01; ns, no significant difference. (**H**) The model of the structure of compact or condensed hCTPS cytoophidia does not fit the results of the FRAP images. A new model of structures is needed to explain the recovery from bleached fluorescence in the cytoophidia. The intensity of the FRAP ROI was normalized as $Normalized\ Intensity = \frac{I(non-bleach\ ROI)_{Pre-bleach,0} - I(background)_0}{I(non-bleach\ ROI)_n - I(background)_n} \times [I(bleach\ ROI)_n - I(background)_n]$. Scale bars, 10 µm (**C**).

2.3. The Reticular Structure of the hCTPS1 Cytoophidium and Its Localization with hIMPDH2

In order to build a model to fit the dynamic-equilibrium characteristics of cytoophidia, we obtained super-resolution structures of hCTPS cytoophidia using stimulated emission depletion microscopy (STED). The images under conventional confocal microscopy could not show the structure inside cytoophidia, while the STED images revealed the super-resolution structure with a resolution of 50 to 70 nm (Figures 3A and S3A,B), which implied a possible mechanism of highly dynamic cytoophidia under FRAP. We estimated the resolution by measuring the distance between two distinguishable nearby particles and it was 50 to 70 nm (Figure S3A,B).

STED revealed a heterogeneous structure for hCTPS cytoophidia. Some parts were relatively more condensed, while other parts were looser. More importantly, it seemed that there were many tiny filaments inside, which in different orientations formed reticular structures (Figure 3A; Supplemental Figure S2A).

However, the super-resolution results were homogeneous inside hCTPS1-EGFP cytoophidium-like condensates (Figure 3A). The structures of the hCTPS cytoophidia and those of the hCTPS1-EGFP cytoophidia were totally different. No condensed and loose parts or reticular structures knitted with tiny filaments could be observed in the cytoophidium-like condensates.

When performing super-resolution imaging, there might be some interfering factors against reliability of the super-resolution imaging results, such as the efficiencies of antibodies, optical properties of the fluorescent labeling, the steric hindrance of fluorescence proteins, and the influence of sample preparation. To eliminate the effects of antibodies and Cy5, we used 293T cells with hCTPS1-miRFP670nano for the live-cell imaging. We performed immunofluorescence staining on DON-treated 293T cells to eliminate the effects of potential steric hindrance of fluorescence protein tags and overexpression. Both results showed reticular structures (Figure 3A). To avoid the difference in optical properties between Cy5 and EGFP, we also performed immunofluorescence staining with Cy5 on hCTPS1-EGFP-overexpressed 293T cells, and the signal obtained was from Cy5 (Figure 3A).

In addition, we performed immunofluorescence staining on SW480 cells cultured in glutamine-free medium, which showed a reticular structure (Figure 3A). Glutamine is an NH_3 donor in metabolic reactions. This means that the reticular structure of hCTPS is not only a phenomenon induced by DON but also a common structure of metabolic enzymes when cells are under metabolic stress. Without changing the super-resolution structural results, deconvolution of the STED images could improve their resolution and signal-to-noise ratios (Figure 3A; Supplemental Figure S2A).

These tiny filaments appeared as subunits of hCTPS cytoophidia (Figure 3A,B). In vitro experiments showed that CTPS can be assembled into filaments [20,25,30]. Based on the in vitro and in vivo results, we envisioned a model to illustrate the reticular structure of hCTPS cytoophidia (Figure 3B). Inside cytoophidia, subunit filaments are weaved into a reticulation. The model can make FRAP results clearer. The dynamic equilibrium of assembly and disassembly occurs in the tiny filaments of hCTPS, rather than the assembly and disassembly of the whole cytoophidium. FRAP procedures performed on untreated and dispersive hCTPS1 signals resulted in fast recovery, which meant that we were unable to capture the images after bleaching, which were similar to the images before bleaching (Supplemental Figure S2B). Due to the limitation of STED resolution, it is unclear whether the subunit filament is one CTPS filament, a bundle of CTPS filaments or some other form of CTPS.

After obtaining the super-resolution reticular structures, we wanted to determine the reason for and function of this reticular structure. There might be some unknown molecules in the space between hCTPS filaments (Figure 3B). It was reported that IMPDH2 interacted with CTPS1 cytoophidia under DON treatment [23]. The cytoophidia of hCTPS1 and hCTPS2 were located together in 293T cells (Supplemental Figure S2C). IMPDH and CTPS are both part of the glutamine and NH_3 metabolic pathways (Supplemental Figure S2D). We overexpressed *hCTPS1-EGFP*^{A206K}, a monomer version of EGFP, and labeled IMPDH2 with Cy5 by immunofluorescence staining. Under DON treatment, IMPDH2 and hCTPS1 were localized spatially adjacent to each other (Figure 3C). IMPDH2 and hCTPS1 were not exclusive but were positioned mutually in each Z stack (Figure 3D). Therefore, IMPDH2 could be one of the molecules located between hCTPS1 filaments.

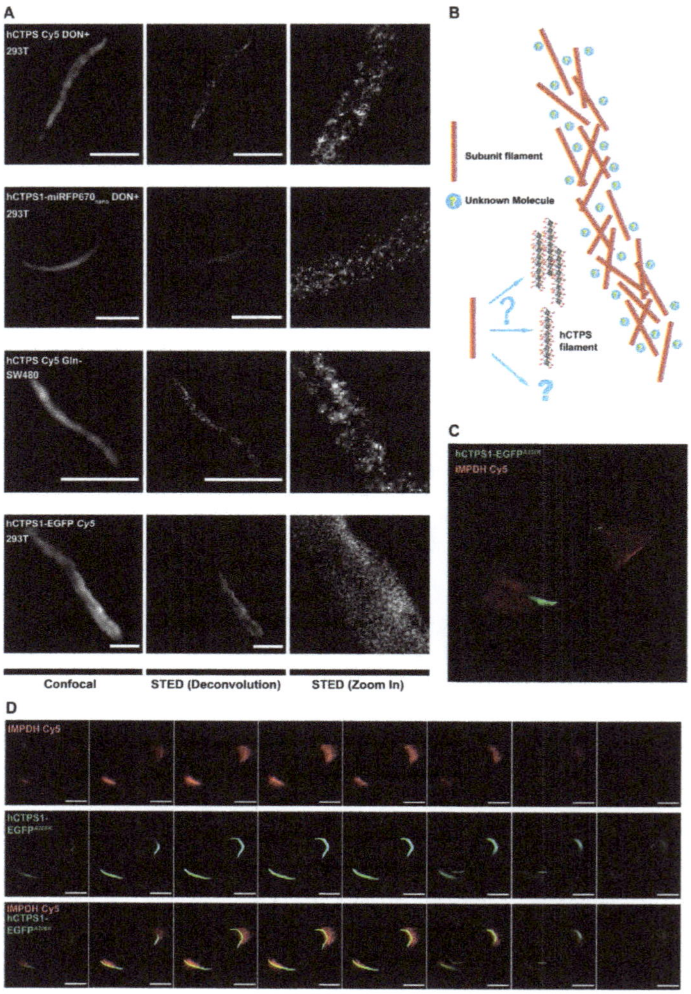

Figure 3. The reticular structure of the hCTPS1 cytoophidium and its localization with hIMPDH2. (**A**) Confocal STED with deconvolution and zoomed-in images of cytoophidia from the following groups: (1) hCTPS1/2 cytoophidia (Cy5 antibody-stained) induced by DON in fixed 293T cells, (2) hCTPS1-miRFP670nano cytoophidia induced by DON in live 293T cells, (3) hCTPS1/2 (Cy5-stained) cytoophidia induced by glutamine deprivation in fixed SW480 cells and (4) hCTPS1/2 (Cy5-stained)- cytoophidium-like condensates in live 293T (hCTPS1-EGFP-overexpression) cells. For the SW480 culture, DMEM without glutamine replaced DMEM 8 h before fixation. For the DON treatment, DON (PBS solution) was added to fresh DMEM medium 8 to 25 h before live-cell imaging or 8 h before being fixed. Scale bars, 3 μm. (**B**) The model of the arrangement of hCTPS filaments into cytoophidia. (**C**) In 293T cells, hCTPS1-EGFPA206K was localized adjacently with DON-induced IMPDH2 (Cy5-stained). (**D**) In each slice along the Z stacks, hCTPS1-EGFPA206K and IMPDH2 (Cy5-stained) were localized to each other. Scale bars, 10 μm.

2.4. CTPS Granules with Tentacles

We performed live-cell imaging via confocal microscopy in 293T cells overexpressing *hCTPS1-mCherry* and found that hCTPS1 could not only form cytoophidia but also formed DON-induced granules (Figure 1A). The movement of most granules was a random walk, like that of ordinary granules in cells (Figure 1B). Fortunately, when we observed live cells treated with DON, we found a filiform structure connecting hCTPS1 granules (Figure 4A). We name these filiform structures connecting granules "tentacles", and the main part of the granules is called the "granule body". The tentacle slowly extended out of the granule body and retracted as quickly as a rubber band after reaching its longest length (Figure 4B).

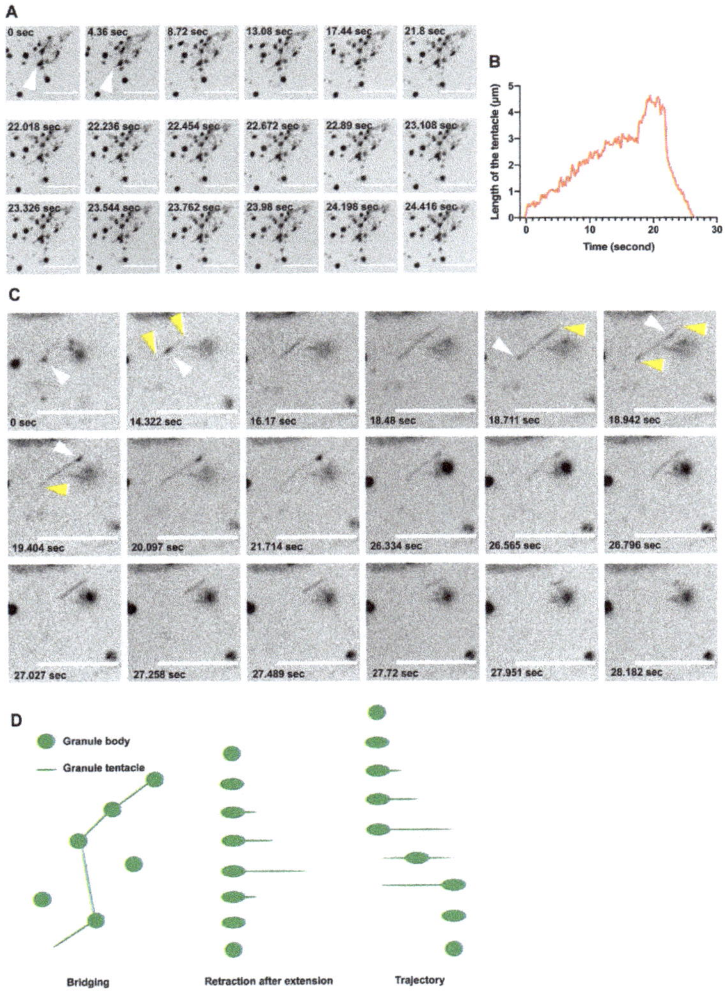

Figure 4. CTPS granules with tentacles. (**A**) Tentacles connect hCTPS1 granules. Tentacles extend and retract. Arrowheads indicate the tentacle. (**B**) Tentacles extend slowly and retract rapidly after reaching the maximum length. (**C**) hCTPS1 granules move from one side to the other along the tentacles. Yellow arrowheads indicate the tentacles; white arrowheads indicate the granules. (**D**) hCTPS1 granular tentacles have three different behaviors and characteristics, bridging, retraction after extension and the trajectory of hCTPS1 granule movement. For the DON treatment, DON (PBS solution) was added to fresh DMEM medium 8 to 25 h before live-cell imaging. Scale bars, 10 μm (**A**,**C**).

We wanted to know the function of the granular tentacles. We found that tentacles were different from small granule bodies. When the tentacle stretched out, the granule body moved from one side of the tentacle to the other side along the tentacle, and then the tentacle retracted into the granule body in the new location (Figure 4C). Granules with tentacles move with clear direction along the tentacles rather than in a random walk. The movement of tentacled granules was different from that of non-tentacled granules. Granular tentacles are tiny structures that bridge granules, move granules and retract after extension (Figure 4D).

3. Discussion

Taking advantage of multiple fluorescence tags, we study the physical characteristics of hCTPS1-containing compartments via fluorescence microscopy. We perform FRAP and STED analyses to reveal the dynamic and reticular structure of cytoophidia. In addition, we observe that hCTPS1 forms granules with tentacles.

3.1. Cytoophidia Are Not Condensates

Protein compartments, similar to droplets, can be assembled by physical forces in cells [35]. Since the discovery of CTPS-forming cytoophidia, the exact phases of cytoophidia and the arrangements of CTPS in cytoophidia are still unclear. Cytoophidia are presumed to be static bundles of filaments [4](Liu, 2011) or in a liquid phase, just like LLPS. However, when we performed live-cell imaging on CTPS cytoophidia, we found that CTPS can not only form long filamentous structures, that is, cytoophidia, but also form granules in the same cell (Figure 1A). This means that, as compartments of CTPS, cytoophidia may not be static, concentrated and rigid structures. For another hypothesis, the puzzling question is why this compartment is not a spherical droplet if it is in the liquid phase, such as LLPS.

According to previous studies, the residue 355H of CTPS (CTPS-355H) is the key site for the formation of this filamentous structure. CTPS-355H lying at the tetramer–tetramer interface plays a critical role in CTPS polymerization. In in vitro experiments, the CTPS tetramer assembly mechanism of cytoophidia is more like that of actin filaments than droplets in cells assembled by physical force [36].

To study the role of CTPS-355H, we used both dimeric EGFP and monomeric EGFP^{A206K} tags [32]. We generated *hCTPS1^{H355A}* mutations. hCTPS1-EGFP^{A206K} can form cytoophidia with DON treatment. Without DON treatment, hCTPS1-EGFP^{A206K} cannot form cytoophidia, suggesting that EGFP^{A206K} does not promote CTPS assembly. mCherry and miRFP670nano are also monomeric tags, just like EGFP^{A206K}. However, hCTPS1-EGFP can form filament-shaped condensates without DON treatment (Figure S2E). Since EGFP has a force to form dimer-like "sticky" features, and hCTPS1-EGFP molecules stick to each other in filament-shaped condensates, we refer to these filament-shaped hCTPS1-EGFP structures as "cytoophidium-like condensates" (to be distinguished from the term "cytoophidia") (Figure 1C).

Are cytoophidia just condensates? If they are, the key cytoophidium-forming site CTPS-355H may provide a directional force of assembly, which should be important for filamentous condensate formation. For hCTPS1^{H355A}-EGFP, the force of assembly is provided by dimeric EGFP, since CTPS-355H has been mutated to CTPS^{H355A}. If either CTPS-355H or EGFP can provide force for condensate formation, we would expect that both hCTPS1^{H355A}-EGFP and hCTPS1-EGFP^{A206K} can form condensates. Our results show that hCTPS1^{H355A}-EGFP cannot form cytoophidium-like condensates, suggesting that CTPS-H355 is an essential site of connection rather than just there to provide a directional force (Figure 1H).

Therefore, our data argue against the idea that cytoophidia are condensates. Two factors appear to be required for assembling CTPS into cytoophidia. First, CTPS molecules are brought together by some forces of assembly. Second, CTPS tetramers need to be connected via CTPS-355H.

3.2. Cytoophidia Are Dynamic

In order to solve the problem of the physical phase of cytoophidia, we carried out FRAP assays to measure the dynamic features. We used hCTPS1^{H355A}-mCherry as the control for complete diffusion, which recovered its intensity quickly after bleaching, so that we could not capture the difference before and after bleaching or measure its dynamic value. We used hCTPS1-EGFP cytoophidium-like condensates as static controls. The intensity recovered in cytoophidia was achieved significantly faster than in hCTPS1-EGFP cytoophidium-like condensates (Figure 2F,G). This means that the cytoophidium is not a condensate but a highly dynamic structure.

A low concentration of DON can induce cytoophidia, mimicking the stress of severe glutamine deprivation. We also measured the effects of time and concentration of DON treatment to determine whether the dynamic results were valid only in particular circumstances. Compared with these curves, the treatment time of DON had little effect on the kinetics (Figure 2E,G), while the concentration of DON had a significant effect on the kinetics of cytoophidia (Figure 2D,G). Due to the covalent bond between DON and CTPS, excessive DON may destroy the conformation of CTPS, thus damaging cells.

3.3. Cytoophidia Are Reticular

The cytoophidium is on a micrometric scale. A previous assumption was that the cytoophidium might be a bundle of CTPS filaments, similar to actin filaments. When a bleached ROI gradually recovered its intensity, no treadmill phenomenon was observed, which means that the assembly mechanism of cytoophidia is different from that of actin filaments. Moreover, recovery did not come from either side of the bleached ROI. The intensity of the entire ROI recovered steadily at the same speed.

If the cytoophidium is a bundle of CTPS filaments, how can the bleached ROI recover its intensity without a treadmill phenomenon (Figure 2H)? To solve this problem, we need to resolve the fine structure of cytoophidia. We performed super-resolution STED imaging which revealed the reticular structure of cytoophidia. In order to minimize artificial influences on the imaging results, we used miRFP670nano as a fluorescence tag to conduct STED imaging directly on live-cell samples, which gave a clear reticular structure (Figure 3A). We called it a reticular structure because small subunit filaments were interconnected, crossed and woven into the reticulation (Figure 3B).

At present, it is unclear whether the subunit filament is a CTPS filament, a bundle of CTPS filaments or some other form of CTPS. To maximize the resolution of the STED imaging (Figure 3B), all fluorescent tags were infrared-emitting. miRFP670nano is much smaller than other infrared fluorescent proteins, and it can minimize the impact of tag-space obstruction. We performed immunofluorescence staining to confirm the results and avoid the effects of spatial obstruction and overexpression. Cytoophidia with immunofluorescence dye Cy5 on hCTPS also showed the reticular structure.

We also performed immunofluorescence staining on SW480 cells. Cytoophidia in SW480 cells can be induced by glutamine deprivation and present reticular structures, which means reticular cytoophidia are common in different cell types and under glutamine metabolic stress conditions but cannot be treated artificially. Even DON is used on cells to mimic the stress of glutamine metabolic stress.

We used hCTPS1-EGFP as the structurally static control to represent the condensate assembled by physical forces. hCTPS1-EGFP cytoophidium-like condensates do not show the reticular characteristics. The cytoophidium-like condensates appear homogeneous, and their internal parts look the same. There are neither subunit filaments nor can space be observed inside the cytoophidium-like condensates. In summary, the structure of the cytoophidium is reticular, and the arrangements of CTPS assembled on cytoophidia are different from those of condensates or actin filaments.

The cytoophidium reticulation provides a structural basis for the localization of other enzymes, such as IMPDH (Figure 3C,D). Both CTPS and IMPDH are associated with glutamine and NH_3 metabolism. In reticular cytoophidia, there may be dynamic interaction between CTPS, IMPDH and their substrates and the microenvironment. The reticular structure can also provide elasticity for the bending or twisting of cytoophidia (Figure S1B). It was reported that cytoophidia are related to the regulation of IMPDH activity [37].

The reticular structure provides a structural basis for the FRAP results. No treadmill phenomenon has been observed. How can CTPS molecules in a free state replace those composed of the bleached ROIs? Based on the assumption of the reticular structure, the treadmill may occur in the subunit filaments rather than the whole cytoophidium. The assembly and disassembly of CTPS on subunit filaments may have a dynamic equilibrium, thus changing the CTPS molecules after bleaching. This may be a potential explanation for the dynamic equilibrium of large-scale FRAP.

Due to the limitations of STED and confocal devices, we could not achieve super-resolution, high-speed live-cell imaging and low phototoxicity for the cells. In order to verify the speculations, more advanced microscope technology is required. Dynamic equilibria of assembly and disassembly in subunit filaments may contribute to the metabolic regulation of reactions in the microenvironment.

In *Caulobacter crescentus*, a small amount of CTPS forms a bundle, while a large amount of CTPS forms a splayed structure [2]. A large amount of hCTPS transforms the morphologies of CTPS bundles into complex structures. Cytoophidia in *Drosophila* female germ cells also exhibit reticular characteristics [1]. Cytoophidia are regulated by the level of molecular crowding in a cell [38]. The formation and maintenance of the reticular cytoophidia may be related to molecular crowding.

3.4. CTPS Can Form Granules with Tentacles

While using live-cell imaging to capture CTPS-containing structures, we found interesting CTPS granules with tentacles. CTPS granules move in a random-walk mode, but the granules with tentacles move in a clear direction. We assume that these two entities are different forms of compartments with similar shapes. We use the term "tentacle" because the structure slowly stretches out of a granule and quickly retracts, just like the tentacle of an octopus or snail (Figure 4B). The tentacle may be extended to find something that can be connected. The tentacles, like bridges, connect different granules (Figure 4A).

If the granules are very small, the tentacles play a role in directional movement (Figure 4C). The granules extend the tentacles to the maximal length, and the granules move quickly from one side to the other along the tentacles, just like a slingshot. The only function of tentacles we know of is related to the directional movement of granules. We still do not know the function of tentacles as bridges. Moreover, it is unclear whether the granules with tentacles have membranes, their movements being similar to the movements of mitochondria and vesicles from the Golgi.

Interestingly, CTPS granules, tentacled granules and cytoophidia appeared under the same conditions, i.e., treatment with DON. They can even exist in the same cell, with potential for interactions and transformations. The tentacles are also capable of directional movements, just like cytoophidia. However, due to their tiny size, the tentacles may not share the same reticular structure as cytoophidia.

The structure of the tentacle may be similar to the subunit CTPS filament to obtain directional characteristics, or it may be in a liquid state in the membrane vesicle. The intensities of granules with tentacles are far lower than those of cytoophidia, and the tentacles exist in fewer cells than cytoophidia do. Due to their low intensities and small volumes, it is difficult to analyze the properties and fine structures of the tentacles. Compared with cytoophidia reacting to metabolism, it is not clear whether granules with tentacles are related to glutamine metabolism or the role of DON.

4. Materials and Methods

4.1. Cell Culture

The 293T and SW480 cells were cultured in DMEM (SH30022.01, Hyclone; Cytiva; 100 Results Way, Marlborough, MA USA 01752) supplemented with 10% FBS (04–001;

Biological Industries; Kibbutz Beit-Haemek, 25115, Israel) in a humidified atmosphere containing 5% CO_2 at 37 °C. All the commercial cell lines used in this article were purchased from the Shanghai Institutes for Biological Sciences, Chinese Academy of Sciences (Shanghai, China). They were originally purchased from ATCC. DON was dissolved in PBS and was added to the culture medium as described in individual experiments. DMEM without glutamine (C11960500BT, Gibco; Thermo Fisher Scientific; 168 Third Avenue, Waltham, MA, USA 02451) replaced DMEM 8 h before imaging.

4.2. Constructs and Transfection

The pLV-hCTPS1-EGFP over-expression vector was kindly provided by Dr. Zhe Sun from ShanghaiTech University. mCherry, miRFP670nano replaced EGFP and $EGFP^{A206K}$ was mutated back into EGFP using PCR and a Gibson Assembly System (NEB). Cell transfection was performed with PEI reagent (24765-1, Polysciences; Polysciences, Inc.; 400 Valley Road, Warrington, PA 18976, USA), according to the instructions provided by the manufacturer. The sequences of oligonucleotides used in this study are listed in Supplemental Table S2.

4.3. Immunoblotting

Cells were harvested in lysis buffer (containing 20 mM Tris, 150 mM NaCl and 1% Triton X-100; P0013J, Beyotime; Beyotime Biotechnology; Building 30, Songjiang Science and Technology Entrepreneurship Center, 1500 Lane, Xinfei Road, Songjiang District, Shanghai, China 201611). Undissolved cell fractions were separated by centrifugation at 12,000 rpm for 10 min at 4 °C, and the supernatants were boiled in SDS-PAGE loading buffer for 10 min. Proteins in total cell lysates were separated by SDS-PAGE and transferred to PVDF membranes. Membranes were blocked in 5% nonfat milk and incubated with the appropriate primary antibodies. Protein bands were visualized using horseradish peroxidase (HRP)-conjugated secondary antibodies with ECL reagent (34577, Thermo Fisher Scientific; 168 Third Avenue Waltham, MA, USA 02451).

4.4. Immunofluorescence

Cells were fixed with 4% paraformaldehyde added into media for 25 min. Then, the fixed cells were washed in 1xPBS 3 times. Samples were incubated with appropriate primary antibodies (rabbit anti-IMPDH2, Proteintech 12948-1-AP; rabbit anti-CTPS, Proteintech 15914-1-AP; Proteintech Group, Inc.; 5500 Pearl Street, Suite 400 Rosemont, IL 60018, USA) overnight at 4 °C and washed in PBS 3 times. Samples were incubated with Cy5-conjugated secondary antibodies (donkey anti-rabbit Cy5-conjugated antibody, Jackson 711-175-152; Jackson ImmunoResearch Inc.; 872 West Baltimore Pike, West Grove, PA 19390, USA) at room temperature for 1 h (in the dark) and washed with PBS 3 times after incubation. The mountant used for STED imaging (Figures 3A,C,D and S2A) was Prolong™ Diamond Antifade (Invitrogen, P36965; Thermo Fisher Scientific; 168 Third Avenue, Waltham, MA, USA 02451). The mountant used for confocal imaging (Figure 1H) was HardSet Mounting Medium with DAPI (VECTASHIELD, H-1500; Vector Laboratories, Inc.; 6737 Mowry Ave, Newark, CA 94560, USA)

4.5. Microscopy

Images (Figures 1A,B, 2C, 4A,C, S1A,B,C and S2B) were acquired under 100× objectives with a confocal microscope (Nikon CSU-W1 SoRa). Confocal images and super-resolution images (Figures 3A,C,D and S2A) were acquired under 100× objectives with an STED confocal microscope (Leica TCS SP8 STED 3X). Images (Figure S2C) were acquired under 63× objective with a Lattice SIM microscope (Zeiss Elyra 7) in wide-field mode. Confocal images (Figure 1H) were acquired under 63× objective with a confocal microscope (Zeiss LSM 980 Airyscan2).

4.6. Live Imaging

The 293T cells transfected with hCTPS1-mCherry and hCTPS1-miRFP670nano constructs were cultured on glass-bottom culture dishes (C8-1.5H-N, Cellvis; Vitro Scientific; Mountain View, CA 94039, USA) with medium and maintained at 37 °C when the live imaging was performed.

4.7. Image Analysis

Fluorescence images was analyzed with the software IMAGEJ (NIH, Bethesda, MD, USA). The ROIs of bleached regions for intensity measurement shown in Figure 2 were selected and measured manually with IMAGEJ. The FRAP curve unpaired *t*-test was analyzed with Graph Prism 8.4.0 (GraphPad Software, LLC; San Diego, CA, USA). The deconvolution of STED images shown in Figure 3 was analyzed by the lighting algorithm of the Leica LAS X. The 3D model shown in Figure 3 was analyzed with the Leica LAS X. The lengths of the tentacles shown in Figure 4 were measured manually with IMAGEJ. Quantity data were collected with Microsoft Excel.

4.8. Fluorescence-Activated Cell Sorting (FACS) Analysis

The flow cell analyzer used was an LSRFortessa X20 (BD). Flow cell data were analyzed with FlowJo 10.4 (FlowJo, LLC; Ashland, OR, USA) software. Data were collected with Microsoft Excel 16.61 (Microsoft Corporation; Redmond, WA, USA).

5. Conclusions

To sum up, the main purpose of this study is to understand the structure and arrangement of CTPS in CTPS filaments with near-atomic-resolution and micron-scale cytoophidia observed under confocal microscopy. We use dimeric EGFP tags as controls to provide aggregation viscosity and identified the connecting role of the CTPS-355H site. FRAP analysis shows that the cytoophidium is highly dynamic, while STED analysis reveals the reticular structure of cytoophidia.

According to the comparison with CTPS-EGFP cytoophidium-like condensates, the dynamic and reticular characteristics of cytoophidia are different from those of condensates (Figure 3B). Moreover, we find that the compartments of CTPS not only exist in the snake-shaped cytoophidia but also in the granules. CTPS granules move in different ways depending on whether they have tentacles. To understand the functions of CTPS granules with tentacles, further studies are required.

Supplementary Materials: The following supporting information can be downloaded at: https://www.mdpi.com/article/10.3390/ijms231911698/s1.

Author Contributions: Y.-F.F. and J.-L.L. conceived the project. Y.-F.F. designed experiments. Y.-F.F., Y.-L.L. and X.-M.L. performed the experiments. Y.-F.F. analyzed the data. Y.-F.F. wrote the original manuscript. Y.-F.F., Y.-L.L., X.-M.L. and J.-L.L. revised the manuscript. All authors have read and agreed to the published version of the manuscript.

Funding: This research was funded by the Ministry of Science and Technology of China (grant number 2021YFA0804700), the National Natural Science Foundation of China (grant number 31771490), the Shanghai Science and Technology Commission (grant number 20JC1410500) and the UK Medical Research Council (grant numbers MC_UU_12021/3 and MC_U137788471), and the APC was funded by ShanghaiTech University.

Institutional Review Board Statement: Not applicable.

Informed Consent Statement: Not applicable.

Data Availability Statement: Not applicable.

Acknowledgments: We thank the Molecular Imaging Core Facility (MICF) at the School of Life Science and Technology, ShanghaiTech University, for providing technical support. This work was supported by grants from the Ministry of Science and Technology of China (no. 2021YFA0804700), the National Natural Science Foundation of China (no. 31771490), the Shanghai Science and Technology Commission (no. 20JC1410500) and the UK Medical Research Council (nos. MC_UU_12021/3 and MC_U137788471) (to J.L.L).

Conflicts of Interest: The authors declare no conflict of interest. The funders had no role in the design of the study; in the collection, analyses, or interpretation of data; in the writing of the manuscript; or in the decision to publish the results.

References

1. Liu, J.-L. Intracellular compartmentation of CTP synthase in Drosophila. *J. Genet. Genom.* **2010**, *37*, 281–296. [CrossRef]
2. Ingerson-Mahar, M.; Briegel, A.; Werner, J.N.; Jensen, G.J.; Gitai, Z. The metabolic enzyme CTP synthase forms cytoskeletal filaments. *Nat. Cell Biol.* **2010**, *12*, 739–746. [CrossRef] [PubMed]
3. Noree, C.; Sato, B.K.; Broyer, R.M.; Wilhelm, J.E. Identification of novel filament-forming proteins in Sac-charomyces cerevisiae and Drosophila melanogaster. *J. Cell Biol.* **2010**, *190*, 541–551. [CrossRef] [PubMed]
4. Liu, J.-L. The enigmatic cytoophidium: Compartmentation of CTP synthase via filament formation. *BioEssays* **2011**, *33*, 159–164. [CrossRef] [PubMed]
5. Chen, K.; Zhang, J.; Tastan, O.Y.; Deussen, Z.A.; Siswick, M.Y.; Liu, J.L. Glutamine analogs promote cy-toophidium assembly in human and Drosophila cells. *J. Genet. Genom.* **2011**, *38*, 391–402. [CrossRef] [PubMed]
6. Carcamo, W.C.; Satoh, M.; Kasahara, H.; Terada, N.; Hamazaki, T.; Chan, J.Y.; Yao, B.; Tamayo, S.; Covini, G.; von Muhlen, C.A.; et al. Induction of cytoplasmic rods and rings structures by inhibition of the CTP and GTP synthetic pathway in mammalian cells. *PLoS One* **2011**, *6*, e29690. [CrossRef] [PubMed]
7. Zhang, J.; Hulme, L.; Liu, J.-L. Asymmetric inheritance of cytoophidia in Schizosaccharomyces pombe. *Biol. Open* **2014**, *3*, 1092–1097. [CrossRef]
8. Shen, Q.J.; Kassim, H.; Huang, Y.; Li, H.; Zhang, J.; Li, G.; Wang, P.Y.; Yan, J.; Ye, F.; Liu, J.L. Filamentation of Metabolic Enzymes in Saccharomyces cerevisiae. *J. Genet. Genom.* **2016**, *43*, 393–404. [CrossRef] [PubMed]
9. Liu, J.-L. The Cytoophidium and Its Kind: Filamentation and Compartmentation of Metabolic Enzymes. *Annu. Rev. Cell Dev. Biol.* **2016**, *32*, 349–372. [CrossRef]
10. Chang, C.-C.; Jeng, Y.-M.; Peng, M.; Keppeke, G.D.; Sung, L.-Y.; Liu, J.-L. CTP synthase forms the cytoophidium in human hepatocellular carcinoma. *Exp. Cell Res.* **2017**, *361*, 292–299. [CrossRef] [PubMed]
11. Li, H.; Ye, F.; Ren, J.Y.; Wang, P.Y.; Du, L.L.; Liu, J.L. Active transport of cytoophidia in Schizosaccharomyces pombe. *FASEB J.* **2018**, *32*, 5891–5898. [CrossRef] [PubMed]
12. Daumann, M.; Hickl, D.; Zimmer, D.; DeTar, R.A.; Kunz, H.H.; Möhlmann, T. Characterization of filament-forming CTP synthases from Arabidopsis thaliana. *Plant J.* **2018**, *96*, 316–328. [CrossRef] [PubMed]
13. Zhang, S.; Ding, K.; Shen, Q.J.; Zhao, S.; Liu, J.L. Filamentation of asparagine synthetase in Saccharomyces cerevisiae. *PLoS Genet.* **2018**, *14*, e1007737. [CrossRef] [PubMed]
14. Zhou, S.; Xiang, H.; Liu, J.-L. CTP synthase forms cytoophidia in archaea. *J. Genet. Genom.* **2020**, *47*, 213–223. [CrossRef]
15. Chang, C.C.; Keppeke, G.D.; Sung, L.Y.; Liu, J.L. CTPS forms the cytoophidium in zebrafish. *Exp. Cell Res.* **2021**, *405*, 112684. [CrossRef]
16. Zhang, Y.; Liu, J.; Liu, J.-L. The atlas of cytoophidia in Drosophila larvae. *J. Genet. Genom.* **2020**, *47*, 321–331. [CrossRef]
17. Wang, Q.Q.; Zhao, P.A.; Tastan, O.Y.; Liu, J.L. Polarised maintenance of cytoophidia in Drosophila follicle epithelia. *Exp. Cell Res.* **2021**, *402*, 112564. [CrossRef]
18. Li, Y.L.; Liu, J.L. Hypoosmolality impedes cytoophidium integrity during nitrogen starvation. *Yeast* **2021**, *38*, 276–289. [CrossRef]
19. Liu, J.; Zhang, Y.; Zhou, Y.; Wang, Q.Q.; Ding, K.; Zhao, S.; Lu, P.; Liu, J.L. Cytoophidia coupling adipose architecture and metabolism. *Cell Mol. Life Sci.* **2022**, *79*, 534. [CrossRef]
20. Zhou, X.; Guo, C.-J.; Chang, C.-C.; Zhong, J.; Hu, H.-H.; Lu, G.-M.; Liu, J.-L. Structural basis for ligand binding modes of CTP synthase. *Proc. Natl. Acad. Sci. USA* **2021**, *118*, e2026621118. [CrossRef]
21. Calise, S.J.; Carcamo, W.C.; Krueger, C.; Yin, J.D.; Purich, D.L.; Chan, E.K.L. Glutamine deprivation initiates reversible assembly of mammalian rods and rings. *Cell. Mol. Life Sci.* **2014**, *71*, 2963–2973. [CrossRef] [PubMed]
22. Ji, Y.; Gu, J.; Makhov, A.M.; Griffith, J.D.; Mitchell, B.S. Regulation of the Interaction of Inosine Monophosphate Dehydrogenase with Mycophenolic Acid by GTP. *J. Biol. Chem.* **2006**, *281*, 206–212. [CrossRef] [PubMed]
23. Chang, C.C.; Keppeke, G.D.; Sung, L.Y.; Liu, J.L. Interfilament interaction between IMPDH and CTPS cy-toophidia. *FEBS J.* **2018**, *285*, 3753–3768. [CrossRef] [PubMed]
24. Barry, R.M.; Bitbol, A.-F.; Lorestani, A.; Charles, E.J.; Habrian, C.H.; Hansen, J.M.; Li, H.-J.; Baldwin, E.P.; Wingreen, N.S.; Kollman, J.M.; et al. Large-scale filament formation inhibits the activity of CTP synthetase. *eLife* **2014**, *3*, e03638. [CrossRef] [PubMed]

25. Lynch, E.; Hicks, D.R.; Shepherd, M.; Endrizzi, J.A.; Maker, A.; Hansen, J.M.; Barry, R.M.; Gitai, Z.; Baldwin, E.P.; Kollman, J.M. Human CTP synthase filament structure reveals the active enzyme conformation. *Nat. Struct. Mol. Biol.* **2017**, *24*, 507–514. [CrossRef] [PubMed]
26. Sun, Z.; Liu, J.-L. Forming cytoophidia prolongs the half-life of CTP synthase. *Cell Discov.* **2019**, *5*, 32. [CrossRef]
27. Aughey, G.N.; Grice, S.J.; Shen, Q.J.; Xu, Y.; Chang, C.C.; Azzam, G.; Wang, P.Y.; Freeman-Mills, L.; Pai, L.M.; Sung, L.Y.; et al. Nucleotide synthesis is regulated by cytoophidium formation during neurodevelopment and adaptive metabolism. *Biol. Open* **2014**, *3*, 1045–1056. [CrossRef] [PubMed]
28. Strochlic, T.I.; Stavrides, K.P.; Thomas, S.V.; Nicolas, E.; O'Reilly, A.M.; Peterson, J.R. Ack kinase regulates CTP synthase filaments during Drosophila oogenesis. *EMBO Rep.* **2014**, *15*, 1184–1191. [CrossRef] [PubMed]
29. Sun, Z.; Liu, J.-L. mTOR-S6K1 pathway mediates cytoophidium assembly. *J. Genet. Genom.* **2019**, *46*, 65–74. [CrossRef] [PubMed]
30. Zhou, X.; Guo, C.-J.; Hu, H.-H.; Zhong, J.; Sun, Q.; Liu, D.; Zhou, S.; Chang, C.C.; Liu, J.-L. Drosophila CTP synthase can form distinct substrate- and product-bound filaments. *J. Genet. Genom.* **2019**, *46*, 537–545. [CrossRef] [PubMed]
31. Jans, D.C.; Wurm, C.A.; Riedel, D.; Wenzel, D.; Stagge, F.; Deckers, M.; Rehling, P.; Jakobs, S. STED su-per-resolution microscopy reveals an array of MINOS clusters along human mitochondria. *Proc. Natl. Acad. Sci. USA* **2013**, *110*, 8936–8941. [CrossRef] [PubMed]
32. Sezgin, E.; Schneider, F.; Zilles, V.; Urbancic, I.; Garcia, E.; Waithe, D.; Klymchenko, A.S.; Eggeling, C. Polar-ity-Sensitive Probes for Superresolution Stimulated Emission Depletion Microscopy. *Biophys. J.* **2017**, *113*, 1321–1330. [CrossRef] [PubMed]
33. Shaner, N.; A Steinbach, P.; Tsien, R.Y. A guide to choosing fluorescent proteins. *Nat. Methods* **2005**, *2*, 905–909. [CrossRef] [PubMed]
34. Warrington, S.J.; Strutt, H.; Strutt, D. Use of Fluorescence Recovery After Photobleaching (FRAP)Fluorescence Recovery after Photobleaching (FRAP) to Measure In Vivo Dynamics of Cell Junction–Associated Polarity Pro-teins. In *Cell Polarity Signaling: Methods and Protocols*; Chang, C., Wang, J., Eds.; Springer: New York, NY, USA, 2022; pp. 1–30.
35. Brangwynne, C.P.; Eckmann, C.R.; Courson, D.S.; Rybarska, A.; Hoege, C.; Gharakhani, J.; Julicher, F.; Hyman, A.A. Germline P Granules Are Liquid Droplets That Localize by Controlled Dissolution/Condensation. *Science* **2009**, *324*, 1729–1732. [CrossRef] [PubMed]
36. Banani, S.F.; Lee, H.O.; Hyman, A.A.; Rosen, M.K. Biomolecular condensates: Organizers of cellular bio-chemistry. *Nat. Rev. Mol. Cell Biol.* **2017**, *18*, 285–298. [CrossRef]
37. Chang, C.-C.; Lin, W.-C.; Pai, L.-M.; Lee, H.-S.; Wu, S.-C.; Ding, S.-T.; Liu, J.-L.; Sung, L.-Y. Cytoophidium assembly reflects upregulation of IMPDH activity. *J. Cell Sci.* **2015**, *128*, 3550–3555. [CrossRef] [PubMed]
38. Chang, C.-C.; Peng, M.; Zhong, J.; Zhang, Z.; Keppeke, G.D.; Sung, L.-Y.; Liu, J.-L. Molecular crowding facilitates bundling of IMPDH polymers and cytoophidium formation. *Cell. Mol. Life Sci.* **2022**, *79*, 420. [CrossRef] [PubMed]

International Journal of
Molecular Sciences

MDPI

Review

Phase Separation-Mediated Chromatin Organization and Dynamics: From Imaging-Based Quantitative Characterizations to Functional Implications

Woei Shyuan Ng [1,2], Hendrik Sielaff [1,2] and Ziqing Winston Zhao [1,2,3,*]

1 Department of Chemistry, Faculty of Science, National University of Singapore, Singapore 119543, Singapore; w.shyuan@u.nus.edu (W.S.N.); hesielaf@nus.edu.sg (H.S.)
2 Centre for BioImaging Sciences (CBIS), Faculty of Science, National University of Singapore, Singapore 117557, Singapore
3 Mechanobiology Institute (MBI), National University of Singapore, Singapore 117411, Singapore
* Correspondence: zhaozw@nus.edu.sg

Abstract: As an effective and versatile strategy to compartmentalize cellular components without the need for lipid membranes, phase separation has been found to underpin a wide range of intranuclear processes, particularly those involving chromatin. Many of the unique physico-chemical properties of chromatin-based phase condensates are harnessed by the cell to accomplish complex regulatory functions in a spatially and temporally controlled manner. Here, we survey key recent findings on the mechanistic roles of phase separation in regulating the organization and dynamics of chromatin-based molecular processes across length scales, packing states and intranuclear functions, with a particular emphasis on quantitative characterizations of these condensates enabled by advanced imaging-based approaches. By illuminating the complex interplay between chromatin and various chromatin-interacting molecular species mediated by phase separation, this review sheds light on an emerging multi-scale, multi-modal and multi-faceted landscape that hierarchically regulates the genome within the highly crowded and dynamic nuclear space. Moreover, deficiencies in existing studies also highlight the need for mechanism-specific criteria and multi-parametric approaches for the characterization of chromatin-based phase separation using complementary techniques and call for greater efforts to correlate the quantitative features of these condensates with their functional consequences in close-to-native cellular contexts.

Keywords: phase separation; chromatin organization; nuclear condensate; intrinsically disordered region; transcription; DNA damage repair; super-enhancer; quantitative imaging

Citation: Ng, W.S.; Sielaff, H.; Zhao, Z.W. Phase Separation-Mediated Chromatin Organization and Dynamics: From Imaging-Based Quantitative Characterizations to Functional Implications. *Int. J. Mol. Sci.* **2022**, *23*, 8039. https://doi.org/10.3390/ijms23148039

Academic Editor: Vladimir N. Uversky

Received: 20 June 2022
Accepted: 19 July 2022
Published: 21 July 2022

1. Introduction

The cell nucleus is known to be a highly crowded environment in which a myriad of biochemical reactions take place simultaneously. Hence, compartmentalization of intranuclear components and processes is an essential and effective strategy to achieve precise spatio-temporal coordination of such complex dynamics. The nucleolus and Cajal bodies, which were discovered over a century ago [1–3], are among the most conspicuous and structurally stable membraneless compartments observed within the nucleus. The constituents of these compartments were later found to be highly dynamic rather than static protein aggregates [4–6], although the underlying physical nature of these compartments was not clearly understood. Since evidence of a liquid-like state was demonstrated for P granules in germ cells of *Caenorhabditis elegans* [7], a growing number of membraneless nuclear bodies/structures, including paraspeckles [8,9], nuclear speckles [10], promyelocytic leukemia (PML) bodies [11] and DNA damage repair foci [12], have been revisited through the lens of phase separation, which has greatly expanded and re-shaped our understanding of the importance of intranuclear compartmentalization. As a unifying conceptual framework accounting for the formation and unique physico-chemical properties of such membraneless compartments, phase separation has emerged as a general mechanism that underpins a wide range of intracellular processes both inside and outside of the nucleus and involves a variety of biomolecular species [13–17]. In particular, adding on to the many types of phase separation phenomena discovered earlier that involve RNAs and RNA-binding proteins [9,18–26] or are implicated in processes related to RNA metabolism [27–29], more recent studies have uncovered the involvement of phase separation in regulating DNA- or chromatin-based molecular transactions. Here, we survey key recent findings on this growing body of phase separation-mediated phenomena specifically related to chromatin-based intranuclear processes, as revealed primarily through various quantitative imaging methods, and illustrate the critical functional roles of phase separation in regulating the organization and dynamics of these processes. More importantly, by illuminating the complex interplay between chromatin and various chromatin-interacting molecular players mediated by phase separation, this review sheds important light on an emerging multi-scale, multi-modal and multi-faceted landscape that hierarchically organizes the eukaryotic genome within the highly crowded and dynamic nuclear space.

2. Intranuclear Phase Separation: Physico-Chemical Properties and Molecular Driving Forces

Just as oil tends to "demix" with water, chemically and structurally distinct biomolecules that exist as a homogenously mixed solution within the cell can similarly separate themselves into distinct and stably co-existing phases, each enriched with a distinct composition and/or concentration of biomolecules, resulting in liquid-like droplets known as biomolecular condensates (or simply condensates) [14]. The existence of the liquid–liquid phase separation (LLPS) of proteins was first observed and correlated to the physics of phase transition using lysozyme as the model system [30], which subsequently paved the way for understanding LLPS as the physico-chemical underpinning of certain pathological states (e.g., cataracts) [31,32]. However, despite the demonstration of its disease implications, it was only until recently that LLPS re-emerged as a new framework for conceptualizing membraneless intracellular organelles, hence encouraging biologists to revisit many of its previously under-explored properties.

The formation of biomolecular condensates can be best understood from the perspective of the thermodynamics and kinetics underlying polymer demixing in solution, a concept firmly rooted in soft matter physics. Put simply, biomolecules can be driven to phase separate by the balance between two counteracting thermodynamic properties: entropy (which favors the well-mixed state) and enthalpy (in the form of attractive interactions between them). Beyond a particular concentration threshold, at which point interactions between the biomolecules exceed their interactions with the solvent (i.e., the cytoplasm or nucleoplasm of a cell) as a consequence of molecular enrichment, the biomolecules become less and less soluble and thus separate into phases with different concentrations but the same chemical potential to minimize the overall free energy of the system. At the same time, perturbations such as alterations in biomolecular structure or affinity and environmental changes that shift the equilibrium of the system can lead to changes in the material and/or physico-chemical properties of the condensates. Such behaviors have key functional consequences in various biological contexts, where condensates enriched with certain biomolecular species can assemble at specific intracellular locations to perform specialized tasks and readily disassemble in a regulated manner.

While biomolecular condensates are diverse in their molecular make-ups, intracellular locations and functions, they often share a similar set of physico-chemical properties in terms of morphology, dynamics and assembly/disassembly behaviors. To begin with, phase condensates often exhibit the characteristics of liquid-like droplets (e.g., spherical in shape, tendency to coalesce and low surface tension) and can exist stably while being able to dynamically alter their compositions in response to environmental conditions via molecular exchange with the surrounding cellular milieu [33,34]. Secondly, molecular enrichment within such condensates is often driven by preferential interactions between proteins, RNAs and DNA (Figure 1A), particularly multivalent interactions that can be achieved via repetitive modules [14,35]. These modules harbor multiple elements for intra- or inter-molecular interactions, in line with the classic polymer physics descriptions of multivalent molecules in a mixture. Associated with multivalency is a molecular feature known as an intrinsically disordered region (IDR), a type of protein domain with low structural complexity that is often enriched with specific amino acid residues, repetitive motifs or patches of alternating charges. IDRs are commonly implicated in LLPS, in which the formation and selective partitioning of condensates is attributed to transient and weak interactions between IDR-containing biomolecules, including π–π stacking, π–cation interaction, Van der Waals forces, hydrogen bonding and electrostatic and hydrophobic interactions [36]. Modular proteins can also act as scaffolds when recruiting clients that harbor IDRs, which in turn form a multi-modal interaction network to enhance the avidity of weak interactions in the condensed phase [37,38]. In addition to IDRs, oligomerization domains have also recently been shown to enhance the LLPS of protein domains and can potentially serve as an alternative molecular signature associated with LLPS [39].

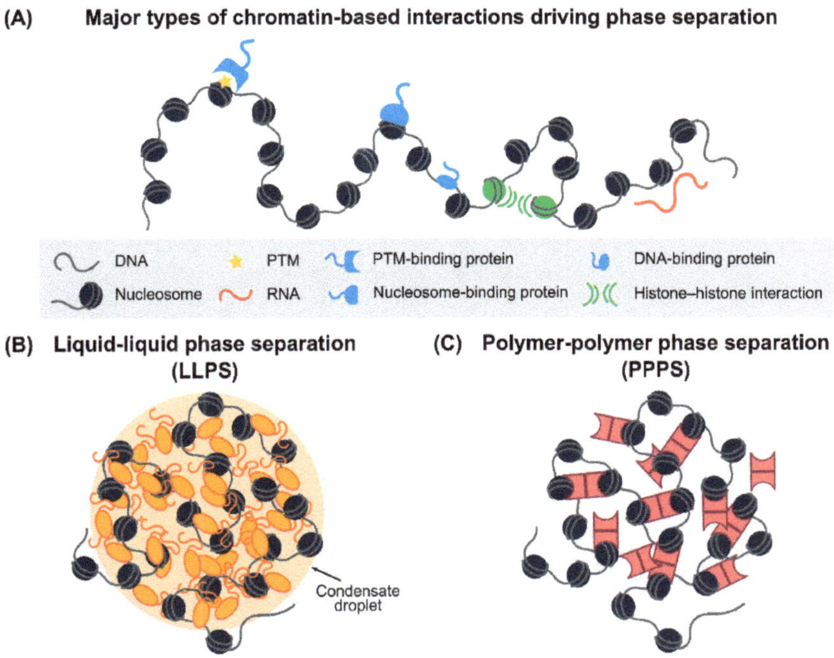

Figure 1. Schematic representation of chromatin-based phase separation. (**A**) Major types of chromatin–protein and chromatin–RNA interactions that can drive chromatin-based phase separation, including direct binding of proteins or RNAs to DNA/nucleosomes or to post-translational modifications (PTMs), as well as inter-nucleosome or -histone tail interactions. (**B**) Liquid–liquid phase separation as promoted by weak and multivalent interactions between chromatin and chromatin-associated factors (orange). (**C**) Polymer–polymer phase separation takes place through the oligomerization of multiple bridging proteins (pink) that draw different regions of the chromatin scaffold together via nonspecific interactions.

In the context of chromatin (Figure 1B), LLPS can drive the formation of chromatin-associated liquid-like droplets via electrostatic attractions between charged residues, dipoles or aromatic groups. In addition to multivalency, site-specific phase condensation can also be promoted and tuned using DNA, RNA and free nucleotides [19,40–42]. In particular, repetitive DNA sequences and the epigenetic states of chromatin can modulate the nucleation and dynamics of intranuclear condensates, contributing to chromatin compaction and other chromatin-based processes. For instance, CpG islands (i.e., CG-rich DNA sequence elements) can recruit the Polycomb repressive complex 2 (PRC2) for the maintenance of the stability of repressed genome at these sites [43], which has been subsequently found to be involved in phase separation (see below for details). LLPS of repetitive telomeric DNA sequences is also implicated in the induction of alternative lengthening of telomeres (ALT) [44], as well as in promoting ALT-dependent telomere maintenance [45].

Alternatively, polymer–polymer phase separation (PPPS), also known as bridging-induced phase separation, can take place through the oligomerization of multiple modular or bridging proteins that link different regions of the chromatin scaffold together via non-specific interactions (Figure 1C). The molecular compositions inside and outside of the condensate formed by PPPS are the same and do not impact the size of the condensate formed, as opposed to LLPS in which changes in the concentration of multivalent binders can affect the size of the condensate. In addition, LLPS droplets have been predicted to be able to persist after the removal of chromatin scaffolds, whereas PPPS condensates rely on chromatin scaffolds for their formation [46]. PPPS was first conceived theoretically using polymer physics models and demonstrated via simulations. For example, in the "strings and binders switch" model, diffusible binding factors establish interactions between binding sites on nonrandom chromatin conformations, leading to stable chromatin architectures [47,48]. On the other hand, PPPS can also be driven by entropic bridging-induced attractions through local DNA distortions induced by bridging proteins that bridge distant DNA regions together; the associated entropic penalties can be minimized by clustering these distorted elements, which results in a local increase in DNA concentration to attract more bridging molecules into the condensate [49–51]. Recently, PPPS has been shown to underlie the formation of DNA–cohesin clusters in vivo [52], pointing to the potential applicability of this previously under-explored mechanism of phase separation in various DNA–protein complexes.

Since the theoretical framework [13,53–55] and the various computational models [56] for understanding chromatin-based phase separation have been expertly reviewed elsewhere, we focus here instead on their quantitative characterizations via imaging-based approaches, as well as their functional implications in organizing and regulating intranuclear structures and processes. Even though the physical processes that underlie LLPS and PPPS can be separated well in theory and simulations, distinguishing between them is often hampered in practice by experimental limitations, and most of the studies reviewed here do not make a specific distinction between these two mechanisms.

3. Quantitative Imaging Techniques for Probing Chromatin-Based Phase Condensates

Over the years, a variety of technical approaches have been employed to characterize chromatin-based phase separation from different fronts, including in vitro biochemical reconstitution, optical imaging (both in cellulo and in vivo) and genomic methodologies (e.g., Hi-C, ChIP-seq and ATAC-seq), as well as theoretical/computational modeling. Among these, optical imaging-based approaches (in both fixed and live samples) arguably provide the most direct and comprehensive capabilities for the in situ quantification of these phase condensates across a wide range of spatial and temporal scales, as has been demonstrated for other intranuclear processes [57]. Despite their respective capabilities, advantages and limitations (Table 1), most of these techniques rely on the use of fluorescent proteins or dyes (via, e.g., SNAP, CLIP and Halo-tags [58–60]) for the labeling and visualization of condensate components inside the cell. In addition to the more conventional imaging configurations (such as wide-field and confocal), many of these techniques also employ total internal reflection fluorescence (TIRF) or light-sheet illuminations in order to leverage their superior optical sectioning capabilities and therefore achieve enhanced sensitivity.

In the time domain, a powerful technique for quantifying the dynamics of chromatin-based phase condensates is fluorescence correlation spectroscopy (FCS), which monitors the fluctuations in fluorescence intensity produced by molecules as they diffuse across a small confocal observation volume, followed by autocorrelation analysis of these time traces and model fitting to extract quantitative parameters (Figure 2A) [61,62]. Combining FCS with photoactivatable fluorescent proteins (paFCS) enables us to fine-tune the level of fluorescent molecules detected, hence making it suitable for probing high-background intracellular environments, such as the nucleus [63]. Other related fluctuation-based techniques include polarization-sensitive FCS [64], number and brightness (N&B) analysis [65,66] and imaging FCS and raster image correlation spectroscopy (RICS) [67,68], each of which is suitable for quantifying a particular aspect of condensate dynamics. A complementary technique to FCS is single-particle tracking (SPT), which leverages the ability to detect the fluorescence signal of individual biomolecules to precisely localize their positions and track their dynamics over time (Figure 2B) [69]. The sensitivity of SPT, especially when measuring inside the highly crowded cell nucleus, can be enhanced through integration with various light-sheet-based illumination schemes [70–73], which selectively excite only a thin section of the nucleus to cut down the out-of-focus background that could easily overwhelm the signal of a single biomolecule. Finally, photobleaching-based techniques, such as fluorescence recovery after photobleaching (FRAP) and fluorescence loss in photobleaching (FLIP), probe intranuclear dynamics by photobleaching the fluorescent molecules in a specific region of the nucleus and then monitoring either the recovery of fluorescence as bleached molecules in the region get replenished after a single photobleaching (FRAP) (Figure 2C) or the propagation of fluorescence loss through the nucleus after repeated photobleaching (FLIP) [74,75].

In the spatial realm, super-resolution microscopy (SRM) has been widely used to characterize the spatial features of chromatin-based phase condensates at resolutions an order of magnitude below those afforded by conventional imaging techniques (such as confocal microscopy). Among the various approaches for breaking the diffraction limit, single-molecule localization-based methods, such as photoactivated localization microscopy (PALM) and stochastic optical reconstruction microscopy (STORM), leverage the labeling of a cellular structure with photoswitchable or photoactivable fluorophores, a sparse subset of which can be randomly activated, individually resolved and localized with nanometer precision. Iterating the process multiple times with a different subset of fluorophores activated each time allows a super-resolution image to be reconstructed from the collective localizations of all fluorophore molecules in the target structure (Figure 2D) [76–78]. Alternatively, methods based on spatially patterned illumination, such as structured illumination microscopy (SIM) and stimulated emission depletion (STED) microscopy, make use of sub-diffraction-limit spatial features introduced into the excitation light either to generate Moiré patterns from cellular structures that can be used to reconstruct a super-resolution image (SIM) [79] or to suppress fluorescence emission from fluorophores located off the center of the excitation region and effectively shrink the point spread function (STED) [80]. Similar to SPT, these super-resolution techniques can also be combined with various implementations of light-sheet illumination [73,81–83], especially when resolving highly dense intranuclear structures, such as those involving chromatin. In addition, DNA or RNA fluorescence in situ hybridization (FISH) enables us to spatially correlate chromatin-based phase condensates with their genomic locations or transcriptional outputs, although no dynamic information can be obtained due to the need for cell fixation.

Table 1. Commonly used quantitative imaging techniques for the characterization of chromatin-based phase condensates and their respective capabilities, advantages and limitations.

Technique	Condensate Parameters Measurable	Spatial/Temporal Resolutions	Sample Types Compatible	Pros and Cons
FCS (and associated variants)	Diffusion coefficient Concentration Residence time for binding (e.g., to DNA) Local viscosity (polarization-sensitive FCS) Oligomerization state (N&B analysis) Spatial context of condensate dynamics (RICS and imaging FCS)	*Spatial:* Diffraction-limited *Temporal:* Microseconds	Live cells/organisms (e.g., embryos)	*Pros:* Wide coverage of temporal dynamics (from microseconds to seconds) Low photodamage/photobleaching to/of live samples due to low illumination power used *Cons:* Poor signal quality could result from high molecular concentrations commonly found in condensates Difficult to probe condensates smaller than diffraction limit
SPT	Diffusion coefficient Residence time for binding (e.g., to DNA) Spatial context of condensate dynamics	*Spatial:* Diffraction-limited (with nm localization precision) *Temporal:* Milliseconds	Live cells/organisms (e.g., embryos)	*Pros:* Direction visualization of condensate dynamics Less reliant on calibrations/corrections commonly required for other techniques *Cons:* Tracking duration can be limited by photobleaching (especially when using fluorescent proteins) Signal quality for single molecules can be reduced in high-background/dense intracellular environments
FRAP/ FLIP	Mobility (as measured by characteristic half-time for fluorescence recovery or loss) Local viscosity (indirectly derived)	*Spatial:* Diffraction-limited *Temporal:* Seconds	Live cells/organisms (e.g., embryos)	*Pros:* More suitable for probing dynamics at longer timescales (seconds to minutes or longer) *Cons:* Requires complex data analysis/modeling Not suitable for probing fast and transient dynamics Ensemble nature masks intrinsic heterogeneities among individual biomolecules
SRM (e.g., PALM/STORM, SIM, STED)	Spatial/morphological features (e.g., size, area, aspect ratio); Intranuclear distribution and density Molecular stoichiometry Colocalization between components	*Spatial:* 10 s of nm or better *Temporal:* Up to seconds (for live samples)	Fixed or live cells/ tissues	*Pros:* Superior spatial resolution Possible to perform molecular counting *Cons:* Limited imaging speed/temporal resolution due to the need to accumulate sufficient localizations (PALM/STORM) Computationally demanding image reconstruction (SIM) Requires complex optical instrumentation and high laser power to achieve fluorescence depletion (STED)

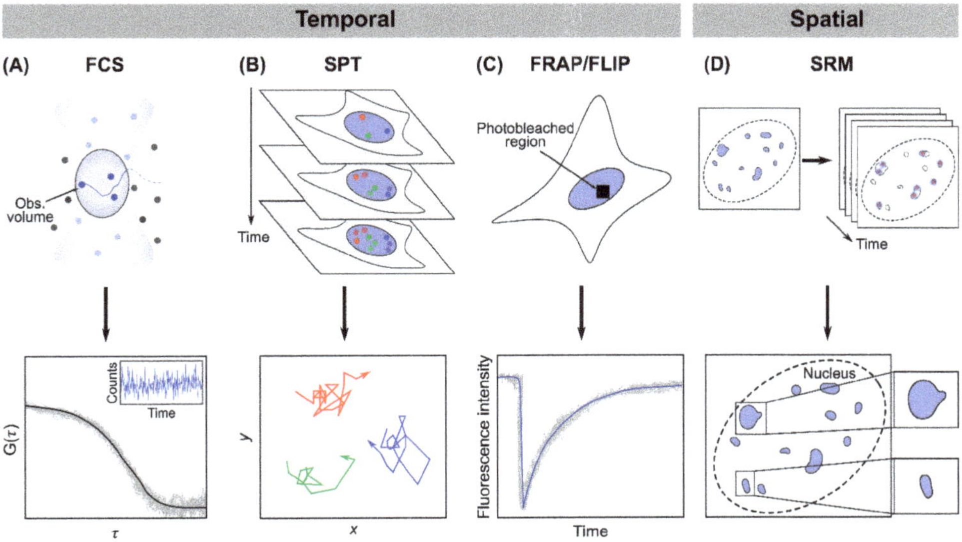

Figure 2. Principles of major types of quantitative imaging techniques commonly used for characterizing chromatin-based phase condensates. (**A**) FCS quantifies intranuclear dynamics by monitoring the fluorescence intensity fluctuations as biomolecules move in and out of a small observation volume; a typical intensity trace (inset) and the autocorrelation function curve calculated from it are shown. (**B**) SPT detects individual fluorescent biomolecules inside the nucleus and tracks their movements over time; a few typical single-particle trajectories (colored red, green and blue) are depicted. (**C**) Photobleaching-based techniques, such as FRAP and FLIP, where a small region of the cell nucleus is selectively photobleached; a typical FRAP curve is shown. (**D**) The SRM technique PALM/STORM labels an intranuclear structure with photoswitchable fluorophores, activates a random subset of the fluorophores each time and localizes their individual positions with ultra-high spatial precision; iterating the process multiple times then reconstructs a super-resolution image of the structure.

While these imaging techniques are by no means specific to probing only intranuclear phase condensates (as opposed to condensates at other intracellular locations), they are nevertheless among the most widely adopted methods in previous studies of chromatin-based phase separation and are often used in combination with each other or in conjunction with other complementary approaches (such as in vitro biochemistry measurements of the concentration range associated with phase separation for a particular condensate). The extensive application of these quantitative methods has not only shed light on previously hidden physico-chemical parameters of a variety of chromatin-based phase condensates (as summarized in Table 2), but also in many cases led to illuminating insights into the mechanistic and functional implications of these phenomena, as will be discussed in detail in the following section.

Table 2. Various phase separation-mediated chromatin structures and processes, the molecular players involved in them, as well as their quantitative characterizations using different imaging methods.

Chromatin-Based Structure	Molecular Species Involved	Mechanistic Role(s) of Phase Separation	Imaging Methods Used	Quantitative Parameters In Vivo		In Vitro Validation /Concentration Range for LLPS	Refs.
				Spatial	Temporal		
High-order chromatin domains	Chromatin	Nucleosome arrays can phase separate under physiological conditions; BRD4 induces LLPS of acetylated chromatin	Live-cell imaging, FRAP, IF	-	-	Yes 50–750 nM (nucleosome)	[41]
	Histone H1	H1 phase separation facilitated by ATP partitions large segments of DNA or polynucleosomes	Live-cell imaging, FRET, FCS, FRAP, IF	-	-	Yes 0.1–0.3 μM	[40]
Constitutive heterochromatin	HP1a/HP1α	LLPS of HP1a/HP1α drives formation of heterochromatin domains	FRAP, LLSM, RICS	-	Diffusion coeff. $D \sim 1.09\ \mu m^2\ s^{-1}$	Yes 0.05–1 mg/mL	[84]
		Phosphorylation of HP1α NTD promotes its LLPS by forming higher-order oligomers	Live-cell imaging, FRAP	-	FRAP half-time: $\tau_{1/2}$ ~75 s *	Yes 100–200 μM (phosphorylated)	[85,86]
	HP1β	Multivalent interactions between CDs in HP1β complexes with nucleosomes drive LLPS of heterochromatin	Live-cell imaging, FRAP, IF	Size: ~0.5–1 μm *	FRAP half-time: $\tau_{1/2}$ ~50–75 s *	Yes 0.8–50 μM (CD)	[86]
	MeCP2	MeCP2 condensates selectively partition HP1α and enhances the separation between heterochromatin and euchromatin	Live-cell imaging, FRAP, IF	Volume: ~1–5 μm^3 *	FRAP half-time: $\tau_{1/2}$ ~10 s *	Yes 2–10 μM	[87]
		MeCP2 competes with histone H1 to form mutually exclusive chromatin condensates	FRAP, IF, EM	Size: ~0.1–0.3 μm *	-	Yes 1.25–10 μM	[88]
	53BP1	53BP1 undergoes LLPS with HP1α to maintain heterochromatin and prevent DNA damage and genomic instability	Live-cell imaging, FRAP, IF	Area (median): 1.243 μm^2	FRAP half-time: $\tau_{1/2}$ ~10–20 s *	Yes ~10 μM	[89]
Facultative heterochromatin	CBX2/PRC1	CBX2 drives LLPS of PRC1; CBX2-PRC1 condensates compact chromatin by concentrating DNA and nucleosomes via direct binding	Live-cell imaging, FRAP, IF	Area: 0.1–0.2 μm^2 *	FRAP half-time: $\tau_{1/2}$ ~35 s *	Yes 0.8–12.5 μM	[90,91]

Table 2. *Cont.*

Chromatin-Based Structure	Molecular Species Involved	Mechanistic Role(s) of Phase Separation	Imaging Methods Used	Quantitative Parameters In Vivo: Spatial	Quantitative Parameters In Vivo: Temporal	In Vitro Validation /Concentration Range for LLPS	Refs.
Chromatin loops	CTCF	CTCF-mediated chromatin loops act as a topological framework for the formation of phase-separated transcriptional condensates at SEs mediated by Pol II	FISH, FRAP, PALM, STORM	-	Lifetimes: ~10 s (transient) >100 s (stable) FRAP half-time: $\tau_{1/2}$ ~20 s * (Pol II clusters)	No	[92]
	Cohesin	Cohesin induces phase separation of DNA–cohesin–homocomplex clusters	AFM, FRAP	Size: 1.14 µm (In vitro)	FRAP half-time: $\tau_{1/2}$ ~126 s	Yes 1–1000 nM	[52]
DNA damage repair (DDR) hubs	RAD52	Rad52 condensates coupled with nuclear microtubule filaments drive nucleoplasmic flow and DNA repair center formation	Live-cell imaging, FLIP	Area (mean): 0.1–1.2 µm^2 *	FLIP half-time: $\tau_{1/2}$ ~5 s *	Yes 5–20 µM	[93]
	FET-family TFs (FUS/EWS/ TAF15)	FET family TFs form DDR hubs through LLPS on PAR-seeded DNA damage sites in early DDR response and exclude 53BP1	Live-cell imaging, EM, IF, FRAP, OT, SIM	Size: 0.5–2.5 µm * Aspect ratio: ~1	FRAP half-time: $\tau_{1/2}$ ~0.2–1 s	Yes 0.1–500 µM	[12,18]
				Other parameters: viscosity: 10–100 mPa·s			
		FUS drives LLPS of DDR hubs to recruit downstream DDR factors and reorganize γH2AX nano-foci in an FUS-dependent manner	Live-cell imaging, IF, SIM	-	Recruitment time: ~40 s	No	[94]
	53BP1	53BP1 nucleates at DNA damage sites and undergoes LLPS to organize damaged chromatin into larger repair compartments and shield it from nucleolytic processing	Live-cell imaging, FRAP, IF, STORM	Size: 0.6–2.8 µm *	FRAP half-time: $\tau_{1/2}$ ~2–20 s *	Yes	[95,96]
				Other parameters: viscosity: 2.5 Pa·s; surface tension: γ ~0.5 µN m^{-1}			
Transcription-related hubs	MED1/BRD4	MED1 and BRD4 form phase condensates to concentrate transcriptional machineries at SE-regulated genes to activate their transcription, promoted by short RNAs and low RNA levels via positive feedback loops	Live-cell imaging, FISH, FRAP, IF, PALM	Size: 0.2–1.3 µm * Aspect ratio: ~1.1 (In vitro) *	FRAP half-time: $\tau_{1/2}$ ~4 s Diffusion coeff.: D = 0.14–0.37 µm^2 s^{-1} Cluster lifetime: ~3–25 s	Yes 0.2–20 µM	[42,97]
	OCT4	OCT4 can phase separate with MED1 or be incorporated into MED1 condensates	Live-cell imaging, FISH, FRAP, IF, PALM	Size: ~0.3 µm *	-	Yes 10–40 µM	[98]

Table 2. *Cont.*

Chromatin-Based Structure	Molecular Species Involved	Mechanistic Role(s) of Phase Separation	Imaging Methods Used	Quantitative Parameters In Vivo		In Vitro Validation /Concentration Range for LLPS	Refs.
				Spatial	Temporal		
Transcription-related hubs	Pol II/MED1	Pol II and MED1 form clusters of different sizes and lifetimes; large and stable clusters exhibit phase condensate properties and associate with chromatin at SEs in a transcription-dependent manner	Live-cell imaging, FRAP, PALM, LLSM	Cluster size: ~0.1 μm (small) >0.3 μm (large)	Cluster lifetime: ~12 s (transient) >100 s (stable) FRAP half-time: $\tau_{1/2}$ ~10 s Sub-diffusivity: $\alpha \sim 0.4$	No	[99]
	YAP	YAP redistributes into the nucleus upon hyper-osmotic stress and forms a phase condensate to reorganize chromatin and enrich TFs for transcription of YAP target genes	Live-cell imaging, FRAP, IF, PALM	Size: 0.2–1.6 μm *	FRAP half-time: $\tau_{1/2}$ ~1 s *	Yes 40 μM	[100]
	TAZ	TAZ condensates compartmentalize transcription machineries to promote TAZ-specific gene expression and shield themselves against upstream regulators	Live-cell imaging, FRAP, IF, SIM	Size: 0.3–1.2 μm *	FRAP half-time: $\tau_{1/2}$ ~0.5–6 s Diffusion coeff.: $D = 0.11\ \mu m^2\ s^{-1}$	Yes 10–80 μM	[101]
	FET-family TFs (FUS/EWS/TAF15)	FET family TFs form condensates at SEs via both homotypic and heterotypic interactions; EWS/FLI1 form transactivation hubs via LLPS to target GGAA microsatellites at SE loci for oncogene activation/expression; TAF15 condensates nucleated by nascent Pol II CTD form transcriptional initiation hubs to activate transcription but exclude phosphorylated Pol II CTD	Live-cell imaging, FCS, FISH, FRAP, IF, LLSM, SPT	Size: 0.2–2 μm *	Recovery time: 7–10 s Residence time in cluster: 5–20 s Diffusion coeff.: D ~2 $\mu m^2\ s^{-1}$ (nucleolus) or ~0.8 $\mu m^2\ s^{-1}$ (nucleoplasm)	No	[102–104]
				Other parameters: critical concentration for LLPS: ~8 μM (cytoplasm) or ~2.6 μM (nucleus) (TAF15); 1–2 μM (cytoplasm and nucleus) (EWS and FUS)			
	MLL4	MLL4 promotes transcriptional condensate formation, which recruits various TFs that regulate nuclear mechanics and chromatin compaction by balancing PcG condensates	Live-cell imaging, IF, STORM	Area: 0.0062–0.013 μm^2	Cluster lifetime: ~119 s	Yes 1–10 μM	[105]
	KLF4	KLF4 bridges DNA and initiates LLPS via tight and weak binding in an IDR-independent manner	Live-cell imaging, FRAP	Size: 0.5–3 μm *	FRAP half-time: $\tau_{1/2}$ ~10–15 s *	Yes 1.5–10 μM	[106]

Table 2. *Cont.*

Chromatin-Based Structure	Molecular Species Involved	Mechanistic Role(s) of Phase Separation	Imaging Methods Used	Quantitative Parameters In Vivo		In Vitro Validation /Concentration Range for LLPS	Refs.
				Spatial	Temporal		
Transcription-related hubs	HSF1	LLPS of HSF1 promotes chromatin binding and recruitment of transcription apparatus on HSOP gene loci to activate transcription upon heat stress	Live-cell imaging, FISH, FRAP, IF, STORM, SPT	Size: ~0.3 μm	FRAP half-time: $\tau_{1/2}$ ~10 s *	Yes 0.125–5 μM	[107]
	YY1	YY1 mediates LLPS to recruit coactivators and promote formation of enhancer clusters to activate FOXM1 gene expression	Live-cell imaging, FISH FRAP, IF	Area: 3–7 μm^2	FRAP half-time: $\tau_{1/2}$ ~3 s *	Yes 2–10 μM	[108]
Co-transcriptional/ splicing hubs	Cyclin T1	Cyclin T1 condensate promotes phosphorylation and recruitment of Pol II CTD, which transitions from transcription initiation condensate to transcription elongation/RNA splicing condensates consisting of SRSF2 or cyclin T1	Live-cell imaging, FISH, IF, LLSM, SPT	Size: 0.5–3 μm *	-	Yes 0.4–6 mg/mL	[109,110]
	SRSF2		Live-cell imaging, FISH, IF, LLSM	Size: ~0.2 μm *	FRAP half-time: $\tau_{1/2}$ ~1 s *	Yes 2.5–10 μM	[111]
ALT telomere-associated PML nuclear body (APB)	SUMO–SIM	APB condensates driven by SUMO–SIM LLPS promote telomere clustering in ALT DDR factors are recruited to APB condensates via DDR signaling for telomere elongation induced by DNA damage in ALT	Live-cell imaging, FISH, FRAP, IF	Size: 0.3–1 μm * aspect ratio: ~1 *	FRAP half-time: $\tau_{1/2}$ ~35–44 s	No	[45]

Abbreviations: AFM: atomic force microscopy; FCS: fluorescence correlation spectroscopy; FISH: fluorescence in situ hybridization; FLIP: fluorescence loss in photobleaching; FRAP: fluorescence recovery after photobleaching; FRET: Förster resonance energy transfer; IF: immunofluorescence; LLSM: lattice light-sheet microscopy; OT: optical tweezers; PALM: photoactivated localization microscopy; RICS: raster image correlation spectroscopy; SIM: structured illumination microscopy; SPT: single-particle tracking; STORM: stochastic optical reconstruction microscopy. * Estimated values from figures in the respective works.

4. Multi-Scale Chromatin Organization and Dynamics Mediated by Phase Separation

The eukaryotic genome is organized hierarchically in the nucleus across multiple length scales both in physical and sequence spaces [112,113]. At the finest scale (up to several kbs), DNA is first compacted into nucleosomes consisting of 147 bps of DNA wrapping around a histone octamer core. Nucleosomes are packed into 10 nm fibers and then organized into chromatin loop structures and topologically associating domains (TADs), which span several kbs to several Mbs [114,115]. Finally, chromatin domains assemble into A/B compartments (approximately corresponding to euchromatin and heterochromatin, respectively), which make up chromosomes that each occupy a distinct territory within the nuclear space, several μm in size and spanning hundreds to thousands of Mbs. Overall, phase separation has been implicated in the organization and dynamics of chromatin at each of these scales, which is consistent with its intrinsic propensity for phase separation as evidenced by the fact that nucleosome arrays are capable of condensing into liquid droplets in vitro under physiological conditions [41].

4.1. Large-Scale Chromatin Organization

In order to be packed into a certain nuclear domain or territory, chromatin needs to be organized into higher-order architectures, and the principles governing the process have been well-illustrated through theoretical modeling and computational simulations [116,117]. Starting from simple models that assume the chromatin fiber to be a self-avoiding polymer bead chain to account for the scaling properties of chromatin folding through binder-mediated interactions [47,118], the various models that have been developed to date are now capable of recapitulating the dynamic behaviors and properties of chromatin folding and its resulting architectural features, in good agreement with data from FISH and chromosome conformation capture experiments across different species [119–121]. Importantly, there is increasing in silico evidence supporting the role of phase separation in orchestrating genome compartmentalization by taking into account different types of biochemical interactions within a chromosome, where chromatin of the same epigenetic type tends to colocalize and adopt certain architectural conformations as a consequence of energetic stabilization [120–123]. These studies, coupled with the various experimental findings detailed below, are merging towards a unifying conception of phase separation as a key driving force of 3D genome organization, which has also been shown to be evolutionarily conserved across the three kingdoms of life [124].

Among the different packing states of chromatin, phase separation of the densely packed and largely transcriptionally inactive heterochromatin can be driven by heterochromatin protein 1 (HP1) and chromobox homolog 2 (CBX2), which recognize histone marks H3K9me3 and H3K27me3, respectively (Figure 3A,B) [40,84,85,90]. In particular, heterochromatin can be further categorized into constitutive and facultative heterochromatin, with the former being more densely packed and containing few genes but relatively large amounts of tandem repeats, while the latter contains genes that are often found in a transcriptionally repressed state in the absence of specific developmental cues [125,126]. The discovery of the role of HP1 in driving LLPS of constitutive heterochromatin (marked by H3K9me3) in Drosophila melanogaster and mammalian cells has expanded the conventional assembly mechanism for heterochromatin domains beyond chromatin compaction [84], although contrary evidence exists that suggests that heterochromatin foci can also form independently of HP1-driven LLPS [64]. Further supporting its role in the phase separation-driven formation of heterochromatin, HP1 binding has been found to increase the accessibility and dynamics of embedded histone residues within the nucleosome for more multivalent interaction sites [127], thereby promoting the bridging of multiple nucleosomes together through HP1 oligomerization and enhancing inter-nucleosome interactions. Heterochromatin condensate formation can also be further enhanced by linker histone H1 and post-translational modifications of HP1. H1, whose condensates colocalize with HP1α in vivo [40], has been shown to compartmentalize nucleosomes and reduce their dynamics within the condensate [41], while the phosphorylated N-terminal domain (NTD) of human HP1α possesses an enhanced propensity for driving LLPS through formation of higher-order oligomers that are more effective in bridging nucleosomes together [85]. In addition, the number of available chromodomains (CDs) that interact with H3K9me3-marked nucleosomes has also been found to serve as another driver for phase separation-mediated heterochromatin formation, in which synergetic interactions between HP1α/β and other heterochromatin-related proteins (e.g., TRIM28 and SUV39H1) in a complex lead to enhanced multivalent CD–H3K9me3 interactions that can drive heterochromatin condensation [86]. While these findings were mainly demonstrated in vitro with relatively short nucleosome arrays, the inter- and intra-molecular multivalent interactions between chromatin and its associated proteins, as well as the coalescence of heterochromatin condensates, could potentially drive the propagation of heterochromatin domains observed in live cells [128], beyond the conventional mechanism of protein–protein binding/oligomerization. Moreover, the intrinsic selectivity afforded by the combinations of macromolecular interactions through phase separation serves as a higher-level regulatory mechanism across different types of heterochromatin condensates. For instance, methyl-CpG-binding protein (MeCP2) condensates selectively incorporate HP1α and compete with H1 to form mutually exclusive and distinct heterochromatin foci [87,88]. DNA methylation (especially at CpG sites) is also a common feature of constitutive heterochromatin besides H3K9me3 marks and is known to negatively regulate transcription [129]. In this context (Figure 3A), the transcriptionally repressive effect could be attributed to phase separation of MeCP2, which binds strongly to highly methylated heterochromatic chromocenters and forms condensates via LLPS that are capable of excluding transcriptional machineries [87,88]. In addition, phase separation also kicks in when it comes to the overall maintenance of heterochromatin stability. Notably, condensates of the well-known DNA damage response (DDR) factor 53BP1 play an unexpected role in protecting heterochromatin from DNA damage in a HP1α-dependent manner [89].

Similarly, phase separation has also been implicated in the formation of facultative heterochromatin (marked by H3K27me3). The Polycomb repressive complex 1 (PRC1), which recognizes H3K27me3 marks, can mediate phase separation via its CBX2 and PHC subunits (Figure 3B), with the phosphorylation of CBX2's IDR domain and oligomerization of PHC's sterile alpha motif (SAM) being critical for driving the condensation process [90,91,130]. Intriguingly, CBX2 does not depend on H3K27me3 for phase separation, but rather nucleates on chromatin directly to assemble CBX2-PRC1 condensates to speed up the target search process of CBX2, thereby increasing its genomic occupancy to recruit more clients [131]. In fact, the chromatin compaction functionality of PRC1 is facilitated by CBX2, while other CBX proteins in PRC1 act as bridging factors that recognize and recruit H3K27me3-marked chromatin into CBX-PRC1 condensates [90,132]. In addition, PRC1 condensates nucleated at H3K27me3 have also been shown to drive ubiquitination of histone H2 for de novo recruitment of PRC2 [133], leading to the propagation of H3K27me3 marks that in turn recruit more PRC1 into the condensates and establish Polycomb domains in facultative heterochromatin via a positive feedback loop [90]. These distinct phase separation-based cofactor recruitment mechanisms and "scaffold–client" interactions function both independently and in synergy to establish dynamic and multifunctional heterochromatin domains. The collective effect of this complex interaction network might explain the observations from previous studies that heterochromatin droplets in vitro and in vivo often exhibit incomplete FRAP recovery and long recovery half-time [84,86,88,91], as well as incomplete dispersion upon 1,6-hexanediol treatment [84,89,90], properties that suggest that they are not purely liquid-like structures as predicted by the LLPS model.

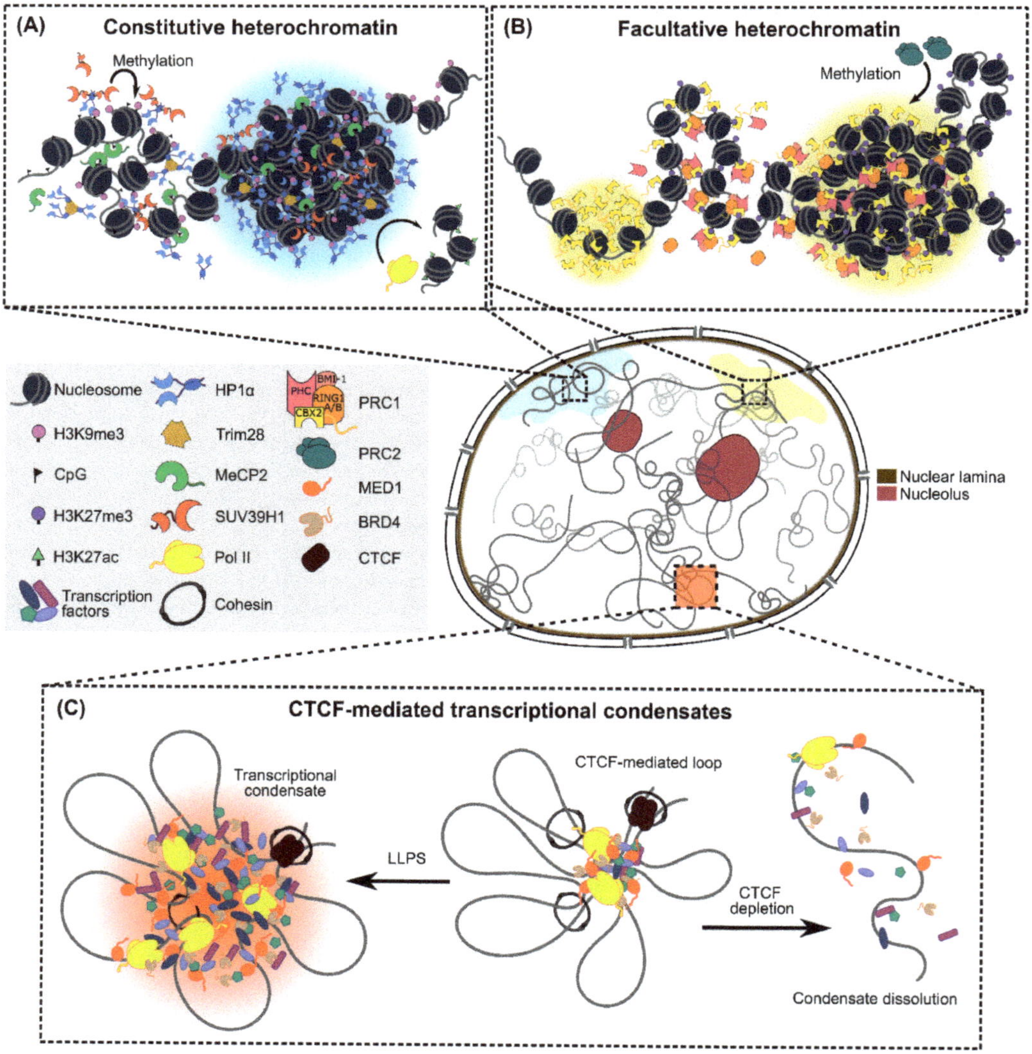

Figure 3. Intranuclear organization of chromatin via phase separation into large- and intermediate-scale condensates. (**A**) HP1α dimer binding to H3K9me3 on heterochromatin initiates condensate formation; further interactions with SUV39H1 and TRIM28 lead to higher-order oligomers that enhance LLPS to establish highly compact constitutive heterochromatin. Moreover, MeCP2 can also phase separate with HP1α, while the exclusion of H3K27ac and Pol II from the condensates further segregates the heterochromatin phase from the surrounding transcriptionally active regions. (**B**) Upon deposition of H3K27me3 marks on chromatin by PRC2, CBX2 binds to H3K27me3 and undergoes LLPS to establish facultative heterochromatin. The assembly of other PRC1 subunits further enhances the initial condensed phase into larger condensates. (**C**) CTCF-mediated chromatin looping provides an architectural framework for the local enrichment of various transcriptional machineries (e.g., Pol II, MED1 and BRD4) and drives the formation of transcriptional condensates via LLPS. Depletion of CTCF dissolves these condensates.

In contrast to heterochromatin, the role of phase separation in driving the formation of euchromatin as an intranuclear compartment has been studied less extensively. However, it is known that heterochromatin and active transcriptional condensates (which are mostly found in euchromatic regions) often form distinct phases from each other. Moreover, interactions between heterochromatin, but not between euchromatin, have been found to drive the compartmentalization of whole cell nucleus [134]. In line with the fact that acetylation generally enhances chromatin, allowing it to adopt an "open" euchromatic configuration for higher genomic accessibility [135], H3K27-acetylated chromatin only phase separates in the presence of multi-bromodomain proteins, such as BRD4, and is immiscible with H3K27me3 droplets [41]. In addition, given that many transcriptional regulators and RNA-binding proteins harbor high levels of IDRs and have a high propensity to phase separate in euchromatic regions [14], phase separation in euchromatin generally occurs at smaller length scales and is largely associated with transcription-related condensates, as discussed in the following sections.

4.2. Intermediate-Scale Chromatin Organization

Going further down the length scale, the organization of chromatin into smaller self-interacting TADs has been conventionally understood from a CTCF- and cohesin-mediated DNA loop extrusion mechanism [136,137]. However, TADs have been recently suggested to be far more dynamic than previously thought, as CTCF and cohesin form transient protein complexes with varying chromatin-binding dynamics to facilitate the formation and dissolution of chromatin loops throughout the cell cycle [138]. Notably, CTCF has been shown to be able to self-associate in an RNA-mediated manner via its RNA-binding region for chromatin loop formation [139], which also mediates CTCF clustering to speed up its nuclear target search by forming ~200 nm-sized "transiently trapping zones" [140]. Similar genomic reorganization is also observed in cells entering senescence, where such clusters are grouped into large senescence-induced CTCF clusters for chromatin loop reshuffling [141]. While the exact physico-chemical mechanism underlying CTCF clustering in these cases warrants further investigation, a recent study has shown that rather than undergoing phase separation itself, CTCF clusters can drive the local spatial confinement of chromatin and serve as a structural framework or nucleation site to facilitate the assembly of LLPS-mediated transcriptional condensates (Figure 3C) [92]. Furthermore, members of the SMC protein family, such as cohesion, can also induce the phase separation of DNA–cohesin clusters (~1 μm in size) in yeast cells through bridging of long DNA segments at least 3 kb in length [52]. This is the first experimental demonstration of PPPS in a biological system in vivo and suggests a potentially new mechanism for chromatin loop stabilization at transient CTCF-bound sites.

4.3. Small-Scale Chromatin Organization

At the local level, phase condensates can nucleate at regions with either low or high chromatin density and selectively compartmentalize their interacting partners in close proximity to regulate chromatin-based processes, particularly transcription. Super-enhancers (SEs), which consist of a large number of enhancer elements drawn from distinct genomic regions into close proximity, are one of the earliest observed examples of such transcriptional condensates (Figure 4A). Many of the enhancer-associated factors including transcription factors (TFs), coactivators and chromatin regulators that correlate with SEs, such as BRD4, OCT4, FUS and MED1 (a subunit of the Mediator complex), are IDR-rich and capable of driving phase separation at SEs to activate gene transcription [97,98,106,108]. Specificity of gene expression can be achieved through selective interactions between the various TF condensates. Hence, phase separation of transcription-related proteins not only impacts chromatin organization by drawing enhancer elements together within the condensates [108] but can also lead to synchronous transcriptional bursting of multiple genes controlled by a shared enhancer [142]. Importantly, in order to differentially regulate transcriptional outputs in a precise manner, each phase separation-mediated system has its own optimal stoichiometric window for the most productive gene expression, depending on the type, level and strength of IDR–IDR interactions; perturbing such a balance could lead to aberrant or repressed transcription of target genes [102,103,143]. As such, phase separation observed at SEs or target gene loci at endogenous protein levels likely occurs at a more local scale, with condensates in the order of ~100 s nm in size (see Table 2), and couples specific TF interactions to local chromatin organization. On the one hand, the mechanical stiffness of local chromatin networks has been shown to affect the growth of transcriptional condensates, and serve as selective chromatin filters that lead to genomic rearrangements [144]. On the other hand, specific TF–DNA interactions can also initiate and stabilize condensates by organizing chromatin interactions at SE loci, as exemplified by the FET (FUS/EWS/TAF15) family protein EWS, which can form transactivation hubs that target GGAA microsatellites for aberrant oncogene activation and expression associated with Ewing's sarcoma [104,145]. Furthermore, the key reprogramming factor KLF4, which recognizes specific promoter sequences, is able to mediate phase separation by bridging multiple DNA duplexes together, which in turn recruits other TFs and stabilizes long-range contacts of pluripotency-related genomic elements [106].

In addition to TFs, various components of the transcriptional machineries can also undergo phase separation to modulate the compartmentalization of chromatin and its interaction partners in the nuclear space. Firstly, RNA polymerase II (Pol II) and MED1 can both form small and transient (~100 nm in size and ~12 s in lifetime), as well as large and stable (>300 nm in size and >100 s in lifetime), clusters. The chromatin-associated stable clusters exhibit properties of phase condensates in which Pol II and MED1 colocalize at SEs that activate gene transcription [99], in line with earlier observations of the dynamic assembly of Pol II into heterogeneous populations of clusters that can correlate with transcription [83,146,147]. Moreover, the C-terminal domain (CTD) of Pol II can form condensates in both unphosphorylated and phosphorylated states that correspond to transcription initiation and elongation, respectively [42,111,148], and the phosphorylation status of Pol II CTD alters its selective partitioning into condensates for different transcriptional activities. Nascent Pol II CTD promotes the formation of TAF15 condensates by lowering the energetic barrier for its nucleation, which in turn recruits more Pol II into these transcription initiation hubs. In contrast, elongating Pol II CTD phosphorylated at Ser5 and Ser2 positions is excluded from TAF15 condensates but accumulates in concentrically adjacent regions [103]. These findings are in line with the earlier observation that phosphorylation dissolves Pol II CTD condensates, and the phosphorylated Pol II CTD is evicted from MED1 condensates [149]. Unphosphorylated Pol II CTD is also incorporated into MED1 condensates at SEs, while phosphorylation of Pol II CTD by CDK7/9 can drive its transition from the transcription initiation hubs to transcription elongation/splicing hubs [111]. Moreover, phosphorylated Pol II CTD is also recruited into cyclin T1 (a key component of nuclear speckles) condensates for enhanced phosphorylation of Pol II CTD and efficient transcription elongation [109,110] (Figure 4B). Taken together, these findings paint a general picture in which LLPS-mediated TF condensates draw a large number of enhancer elements together to stabilize the condensed phase, while mediating the formation of transcription initiation hubs by recruiting unphosphorylated Pol II CTD in the presence of short RNA transcripts produced from initial transcription. Upon phosphorylation, Pol II transitions to transcription elongation/splicing hubs located either concentrically to the initiation hubs or in nuclear speckles proximal to actively transcribed genes. Finally, the high number of long RNA transcripts produced during elongation helps dissolve these transcriptional condensates [42,103,111,150].

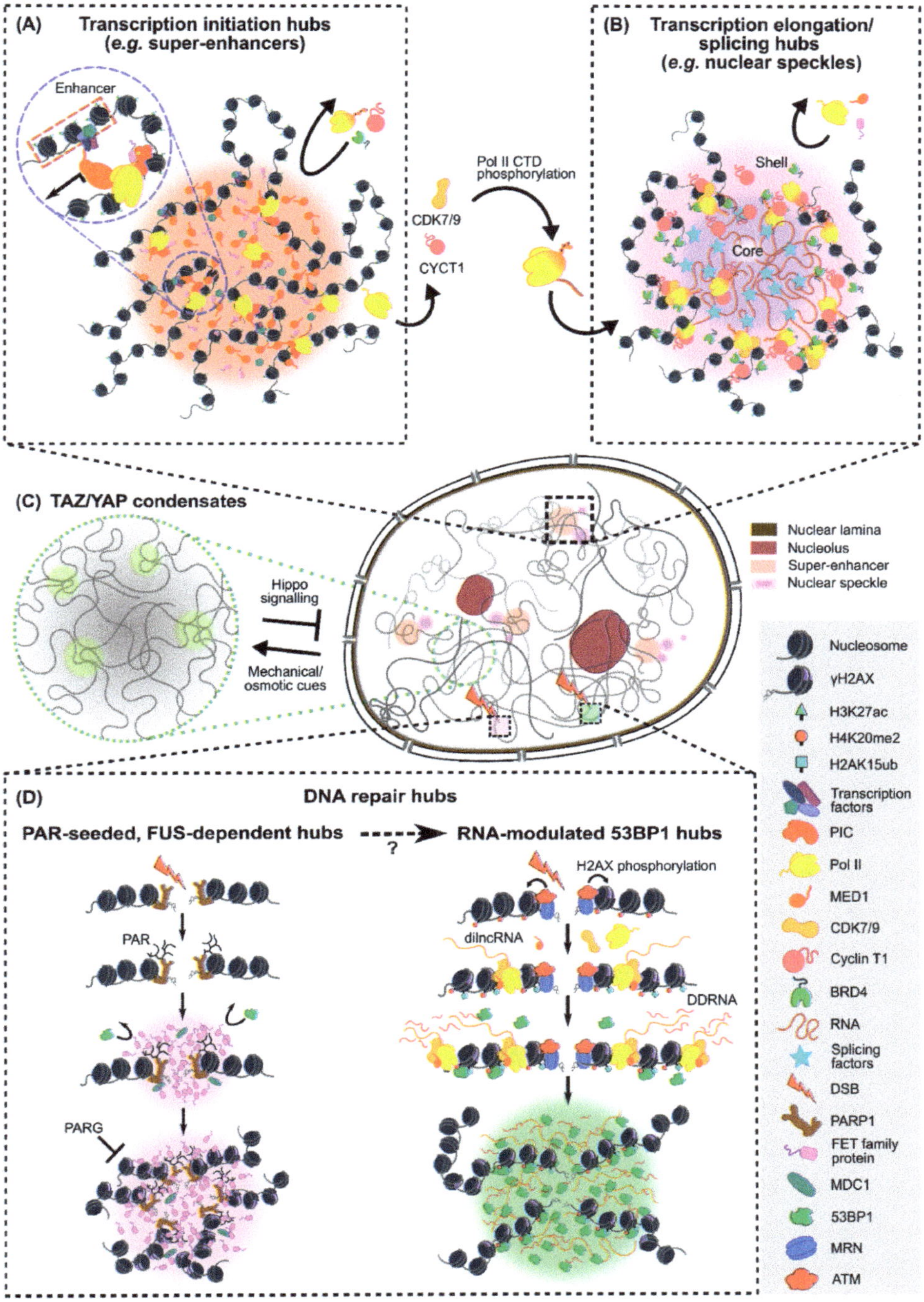

Figure 4. Phase separation-mediated small-scale chromatin-based condensates. (**A**) SE condensates serve as transcription initiation hubs that recruit TFs and coactivators (e.g., OCT4, c-MYC, KLF4 and MED1), which in turn recruit downstream transcriptional machineries. In particular, Pol II has a strong tendency to phase separate, but is excluded from these hubs upon phosphorylation of its CTD

by CDK7/9. (**B**) Transcription elongation/splicing hubs formed by phosphorylated Pol II CTD, BRD4, transcription elongation factors (e.g., cyclin T1) and splicing factors, which are also found in nuclear speckles consisting of a core formed by long RNA transcripts and a shell decorated by chromatin and associated transcriptional elongation machineries. (**C**) Mechanical stress triggers LLPS of TAZ to initiate the transcription of TAZ-specific genes, while osmotic stress can induce the redistribution of YAP into the nucleus and reorganize chromatin to form YAP condensates for downstream gene transcription. TAZ/YAP condensates behave similarly to the SEs shown in (**A**) and can incorporate the transcriptional machineries for effective gene activation and transcription. (**D**) FET family proteins nucleate to drive LLPS of PAR-induced DNA repair hubs (left) during early DDR. FUS is required for the recruitment of DDR factors (such as 53BP1) to DNA damage sites and reorganizes phosphorylated histone variant γH2AX nano-foci into higher-order clusters, which can be dissociated by PAR glycohydrolase (PARG). In RNA-modulated 53BP1 repair hubs (right), DSB recognition by MRN initiates DDR response by recruiting ATM protein to phosphorylate H2AX. In addition, dilncRNA synthesized by Pol II at DSB sites can be further processed into small DNA damage response RNAs (DDRNAs), which support the nucleation of DDR foci by promoting LLPS of DDR factors into 53BP1-phase separated repair hubs. The relationship between FUS-dependent repair hubs and 53BP1 repair hubs is, however, not fully understood.

An important functional role served by phase separation-mediated transcription of specific genes is rapid adaptation to extracellular stimuli for cell survival. Indeed, the transcriptional condensates observed in vivo are often short-lived, reflecting the highly dynamic nature of the cell's responses to various environmental signals. For instance, the transcriptional coactivator YAP forms condensates after redistributing into the nucleus upon hyperosmotic stress and reorganizes the genome into clusters of accessible chromatin regions. Such YAP condensates in turn enrich TFs, such as TEAD1, for downstream transcription of YAP target genes that regulate cell proliferation and survival (Figure 4C) [100]. Intranuclear condensates of the closely associated transcriptional coactivator TAZ, which differs from YAP in its ability to phase separate, can also compartmentalize transcriptional machineries, such as TEAD4, BRD4, MED1 and CDK9, to promote TAZ-specific gene expression implicated in growth, development and tumorigenesis, as well as harness the molecular selectivity afforded by LLPS to shield itself against upstream regulators [101]. Moreover, as the nucleo-cytoplasmic shuttling of YAP and TAZ is regulated mechanically, they can act as intranuclear mechano-effectors in conjunction with MLL4, which also promotes transcriptional condensate formation. Interestingly, in Kabuki syndrome, the loss of function of MLL4 disrupts the counter-balancing of Polycomb group (PcG) compartments needed for the proper maintenance of nuclear architecture, leading to increased mechanical stress, reduced nuclear YAP/TAZ levels and, hence, reduced condensate formation [105]. In the case of cellular heat stress, the intracellular heat-shock transcription factor 1 (HSF1) initiates a rapid response involving genome-wide transcriptional reprogramming (such as increased expression of genes encoding heat-shock proteins and chaperones) by forming phase condensates, which can be dissolved by the chaperone protein HSP70 when the cell recovers [107]. When the cell is under proteotoxic stress, HSF1 can also accumulate in nuclear stress bodies via phase separation, which can also be dissolved by HSP70 to increase transcriptional activities and ensure cell survival; those persistent bodies formed during prolonged stress, however, prime the cell for apoptosis [151]. Collectively, these diverse examples demonstrate that phase separation can organize genomic elements in a high-precision manner to serve as transcriptional hubs that activate specific genes in response to diverse biochemical/biophysical cues.

Lastly, apart from transcriptional condensates, DNA damage response (DDR), which is critical for maintaining genomic integrity and stability, can also be regulated by phase separation via DNA repair foci where large amounts of double-strand break (DSB) repair proteins interact at DNA damage sites [95,152,153]. A molecular marker of early DDR is the phosphorylation of histone variant γH2AX mediated by the ATM protein for downstream recruitment of early DDR factors (such as the sensor complex MRN and the DDR adaptor protein MDC1) to facilitate DNA repair [154]. In line with the earlier observation that DSBs in heterochromatic regions are actively relocated to outside the compartment for homologous repair [155], it has recently been found that RAD52 condensates coupled with various nuclear filaments can drive nucleoplasmic flow generation and DNA repair center formation. These RAD52 droplets can undergo fusion, move to the nuclear periphery and dissolve upon completion of repair, all of which are characteristics of LLPS [93]. Another example is the poly-(ADP-ribose) (PAR)-induced DNA repair hub, where PAR polymerase 1 (PARP1) binds to DSB sites to initiate the deposition of PAR, to which FET family proteins nucleate and drive LLPS of the repair hubs during early DDR (Figure 4D). In particular, FUS is involved in PAR-induced DNA repair hubs by undergoing phase separation to recruit key downstream DDR factors, such as 53BP1, KU80, NBS1 and SFPQ, and organize nano-foci of the phosphorylated histone variant γH2AX into higher-order clusters [12,94]. FET family protein condensates formed during early DDR at PAR-seeded repair hubs exclude 53BP1 but remain accessible for MDC1, which is responsible for phosphorylation signal propagation [12]. Given the fact that FUS is required for the relocation of 53BP1 to DNA damage sites and the accumulation of downstream DDR effectors [94], the post-modification state of DDR factors provides an additional layer of control for modulating DNA repair hubs, where the dissolution of PAR-seeded hubs by phosphorylation allows for the recruitment and accumulation of 53BP1 and other genome "caretakers" via ubiquitination [12]. LLPS of 53BP1 promoted by the synthesis of damage-induced long non-coding RNAs (dilncRNAs) can also drive DDR signaling upon DSB via the recruitment of Pol II pre-initiation complex (PIC), MED1 and CDK9 into the 53BP1 condensates [95]. Thus, LLPS enables the cell to achieve precise spatio-temporal control over a series of DDR events. Furthermore, LLPS of 53BP1 at DSBs can organize damaged chromatin and repair factors into larger repair hubs and shield the damaged sites from extensive nucleolytic processing. The formation of these repair hubs also promotes global p53 activation by incorporating p53 into the condensates, pointing to 53BP1's role in coordinating DNA lesions with global p53-dependent gene activation and cell fate decision in response to DNA damage [96].

5. Perspectives and Outlook

Since its initial demonstrations in biological systems more than a decade ago [7,23], phase separation has been found to play pervasive roles in organizing and regulating diverse chromatin-based molecular processes across a wide range of length scales (from the nucleosome level to higher-order chromatin domains), packing states (both heterochromatin and euchromatin) and intranuclear functions (such as transcription, splicing, DNA damage repair, chromatin loop stabilization and telomere maintenance). The unique physico-chemical properties of these phase condensates are harnessed by the cell to accomplish a wide range of chromatin-based regulatory functions in a spatially and temporally controlled manner. In addition to demonstrating the critical importance of intranuclear compartmentalization, these findings also add a new dimension to our existing understanding of the mechanistic modes and features that govern the hierarchical organization of the eukaryotic genome, such as polymer–polymer interactions, local chromatin motions and intranuclear architectural elements [112]. These insights are made possible through the interplay between conceptual advancements in the physics and chemistry of phase separation, comparative investigations across biological systems and the application of quantitative imaging techniques for the characterization of these phenomena with enhanced spatio-temporal resolutions and sensitivity.

Despite that, the quantitative parameters reported in the majority of the previous studies were primarily limited to the size, density, lifetime/recovery half-time, diffusion coefficient and in vitro concentration range/phase diagram associated with these phase condensates (Table 2). While these are certainly critical parameters that characterize a particular type of condensate, many other equally important and revealing physico-chemical properties, such as morphological features (e.g., aspect ratio), surface tension, viscosity, fusion kinetics and critical concentration for phase separation to occur in vivo were often not measured or specifically reported. This deficiency hence calls for more comprehensive and vigorous in vivo quantifications of chromatin-based phase condensates in future studies. In fact, measurements of some of these parameters have already been undertaken in a few of the recent studies; e.g., the aspect ratio [18,45,97], surface tension [95] and viscosity [18,95] of the condensate droplet, as well as the in vivo critical concentration for LLPS [103]. In addition, two recent studies have demonstrated the use of differential diffusion properties between the condensate and its surrounding cellular environment, as well as across the condensate boundary, as a quantitative criterion for validating phase separation in vivo [102,156]. Moreover, in addition to the fluorescence-based imaging methods conventionally used in most phase separation studies (Table 1), other complementary imaging modalities, such as atomic force microscopy (AFM), electron microscopy (EM) and optical tweezers (OTs), can potentially be used in conjunction with fluorescence imaging to probe specific aspects of chromatin-based condensates that otherwise cannot be easily accessed or measured accurately. To that end, EM has been used to quantify the size of FET protein aggregates induced by PAR chains or the degree of nucleosome array compaction by MeCP2 and its mutants with superior spatial resolution [12,88], OTs have been employed to quantitatively probe the dynamics of FUS droplets fusion in vitro [18] and AFM has enabled the direct visualization of DNA–cohesin holocomplex clusters with unprecedented morphological details [52]. These complementary approaches, combined with the potential usage of novel fluorescent probes (such as the recently developed fluorophores AggFluor capable of quantifying a wide range of local viscosity changes with uniform sensitivity [157]), can constitute a systematic, multi-parametric characterization to substantially enhance our confidence when validating phase separation as the mechanism at work in a specific biological system. In addition to imaging only the protein components, novel strategies for imaging and tracking RNAs (such as those based on fluorescent RNAs [158]) or DNA loci (such as those based on CRISPR/Cas labeling [159]) in living cells can also be simultaneously employed to reveal the in situ interactions between the different types of molecular players involved in various intranuclear condensates.

Moreover, despite the seeming "omnipresence" and "omnipotence" of phase separation, it is also important to exercise caution in not over-interpreting findings that could otherwise be attributed to alternative mechanisms. In particular, merely exhibiting phenomenological characteristics of phase condensates may not automatically mean that phase separation is indeed at work until definitive evidence is obtained. For example, it has been shown that herpes simplex virus replication compartments, while possessing many macroscopic properties of liquid-like condensates, are in fact mediated by a transient DNA-binding mechanism distinct from LLPS [156]. Another recent study has found that the formation of condensate-like TF droplets does not enhance transcription activation for a variety of tested synthetic TFs, hence demonstrating that phase separation is not the default multivalent interaction-based mechanism which the cell uses to regulate biological functions, such as transcription [160].

Another key deficiency in many of the previous studies of phase separation has been the inadequate effort in linking quantitative characterizations with the functional consequences of phase separation. For example, there has been evidence showing that enzymatic reaction rates can be significantly increased within condensates as compared to the surrounding milieu [161]. This example, albeit demonstrated using an in vitro model system, highlights that phase separation-mediated sequestration of biomolecules can enhance substrate-specific reactivity beyond that predicted by the law of mass action, which could in turn lead to far-reaching functional consequences. However, given the complexity of the biochemical pathways and interactions involved, correlating phase separation to its in vivo functional outcomes is challenging and often not performed in a sufficiently quantitative manner. A representative effort in this direction is the recent demonstration that the transcriptional activation of an endogenous oncogene requires a narrow optimal working window of IDR–IDR interactions; artificially inducing phase separation by tuning these interactions beyond the optimum will in fact lead to repressed transcription [102]. The importance of this finding is further underscored in light of the fact that many of the previous phase separation studies were performed at concentrations or expression levels far above the endogenous or physiologically relevant ranges (Table 2). Such potential caveats caution against the temptation to over-attribute the significance of phase separation and call for rigorous quantitative interrogations with close-to-native cellular environments and abundance levels when ascribing phase separation as the underlying mechanism, as has already been eloquently argued by others [162].

Finally, the complexity of chromatin-based condensates involving a variety of distinct biomolecular species, as opposed to purely protein-based condensates, also raises concerns as to whether LLPS is still a sufficiently accurate mechanistic model for describing these systems, or if certain modifications are needed to account for the size and structural features of chromatin, as well as the distinct types of interactions involved. For example, the finding that Pol II clusters adopt a variety of shapes in zebrafish is consistent with a model in which regulatory chromatin provides surfaces for liquid condensation at concentrations too low for LLPS to occur and points to an alternative surface condensation mechanism distinct from canonical LLPS [163]. In other cases where chromatin bridging is necessary to initiate condensate formation, PPPS is perhaps more suitable as an alternative mechanism [46], although definitive experimental evidence for PPPS in actual biological systems remains very limited to date. Moreover, the fact that some of these condensates formed via alternative mechanisms can also exhibit liquid-like macroscopic properties [52] poses a challenge to our current notion of the defining characteristics associated with LLPS. As such, there is a strong need for a comprehensive set of quantitative, universally applicable, yet mechanism-specific criteria that can be applied to ascertain the exact physico-chemical process at work in a specific intranuclear system.

Needless to say, the list of chromatin-based condensates surveyed here is certainly not exhaustive, and new discoveries are constantly emerging. Growing evidence also demonstrates that dysregulation of phase separation-mediated mechanisms could lead to various ailments, such as neurodegenerative diseases and cancers [164–166]. As such, another future challenge for the field is to go beyond cultured cell systems and probe phase-separation mediated chromatin organization and dynamics in more physiologically relevant contexts, such as developing embryos, live organoids or disease models [167], in order to solidly place phase separation as a versatile regulatory paradigm underlying diverse intranuclear processes in vivo.

Author Contributions: Conceptualization and supervision, Z.W.Z.; graphics, W.S.N.; writing, W.S.N., Z.W.Z. and H.S. All authors have read and agreed to the published version of the manuscript.

Funding: This work was supported by a National Medical Research Council Open Fund—Young Individual Research Grant (MOH-000227-00), a Ministry of Education Academic Research Fund Tier 1 Grant (A-0008484-00-00) and a National University of Singapore Presidential Young Professorship Start-up Fund (R-143-000-B02-133) to Z.W.Z.

Acknowledgments: We thank members of the Zhao Group for their suggestions and critical reading of the manuscript. We apologize to those colleagues whose work we were unable to review here.

Conflicts of Interest: The authors declare no conflict of interest.

References

1. Cajal, S.R. Un Sencillo Método de Coloración Seletiva del Reticulo Protoplasmático y Sus Efectos en Los Diversos Organos Nerviosos de Vertebrados e Invertebrados. *Trab. Lab. Invest. Biol.* **1903**, *2*, 129–221.
2. Wagner, R. Einige Bemerkungen und Fragen über das Keimbläschen (Vesicular Germinativa). *Müller's Arch. Anat. Physiol. Wiss. Med.* **1835**, *268*, 373–377.
3. Valentin, G. *Repertorium Für Anatomie Und Physiologie*; Verlag von Veit und Comp: Berlin, Germany, 1836.
4. Dundr, M.; Hebert, M.D.; Karpova, T.S.; Stanek, D.; Xu, H.; Shpargel, K.B.; Meier, U.T.; Neugebauer, K.M.; Matera, A.G.; Misteli, T. In Vivo Kinetics of Cajal Body Components. *J. Cell Biol.* **2004**, *164*, 831–842. [CrossRef] [PubMed]

5. Platani, M.; Goldberg, I.; Swedlow, J.R.; Lamond, A.I. In Vivo Analysis of Cajal Body Movement, Separation, and Joining in Live Human Cells. *J. Cell Biol.* **2000**, *151*, 1561–1574. [CrossRef] [PubMed]
6. Phair, R.D.; Misteli, T. High Mobility of Proteins in the Mammalian Cell Nucleus. *Nature* **2000**, *404*, 604–609. [CrossRef]
7. Brangwynne, C.P.; Eckmann, C.R.; Courson, D.S.; Rybarska, A.; Hoege, C.; Gharakhani, J.; Jülicher, F.; Hyman, A.A. Germline P Granules Are Liquid Droplets That Localize by Controlled Dissolution/Condensation. *Science* **2009**, *324*, 1729–1732. [CrossRef]
8. Hennig, S.; Kong, G.; Mannen, T.; Sadowska, A.; Kobelke, S.; Blythe, A.; Knott, G.J.; Iyer, S.S.; Ho, D.; Newcombe, E.A.; et al. Prion-like Domains in RNA Binding Proteins Are Essential for Building Subnuclear Paraspeckles. *J. Cell Biol.* **2015**, *210*, 529–539. [CrossRef]
9. Yamazaki, T.; Souquere, S.; Chujo, T.; Kobelke, S.; Chong, Y.S.; Fox, A.H.; Bond, C.S.; Nakagawa, S.; Pierron, G.; Hirose, T. Functional Domains of NEAT1 Architectural LncRNA Induce Paraspeckle Assembly through Phase Separation. *Mol. Cell* **2018**, *70*, 1038–1053. [CrossRef]
10. Marzahn, M.R.; Marada, S.; Lee, J.; Nourse, A.; Kenrick, S.; Zhao, H.; Ben-Nissan, G.; Kolaitis, R.; Peters, J.L.; Pounds, S.; et al. Higher-order Oligomerization Promotes Localization of SPOP to Liquid Nuclear Speckles. *EMBO J.* **2016**, *35*, 1254–1275. [CrossRef]
11. Banani, S.F.; Rice, A.M.; Peeples, W.B.; Lin, Y.; Jain, S.; Parker, R.; Rosen, M.K. Compositional Control of Phase-Separated Cellular Bodies. *Cell* **2016**, *166*, 651–663. [CrossRef]
12. Altmeyer, M.; Neelsen, K.J.; Teloni, F.; Pozdnyakova, I.; Pellegrino, S.; Grøfte, M.; Rask, M.B.D.; Streicher, W.; Jungmichel, S.; Nielsen, M.L.; et al. Liquid Demixing of Intrinsically Disordered Proteins Is Seeded by Poly(ADP-Ribose). *Nat. Commun.* **2015**, *6*, 8088. [CrossRef] [PubMed]
13. Hyman, A.A.; Weber, C.A.; Jülicher, F. Liquid-Liquid Phase Separation in Biology. *Annu. Rev. Cell Dev. Biol.* **2014**, *30*, 39–58. [CrossRef] [PubMed]
14. Banani, S.F.; Lee, H.O.; Hyman, A.A.; Rosen, M.K. Biomolecular Condensates: Organizers of Cellular Biochemistry. *Nat. Rev. Mol. Cell Biol.* **2017**, *18*, 285–298. [CrossRef]
15. Sabari, B.R.; Dall'Agnese, A.; Young, R.A. Biomolecular Condensates in the Nucleus. *Trends Biochem. Sci.* **2020**, *45*, 961–977. [CrossRef] [PubMed]
16. King, J.T.; Shakya, A. Phase Separation of DNA: From Past to Present. *Biophys. J.* **2021**, *120*, 1139–1149. [CrossRef]
17. Shin, Y.; Brangwynne, C.P. Liquid Phase Condensation in Cell Physiology and Disease. *Science* **2017**, *357*, eaaf4382. [CrossRef] [PubMed]
18. Patel, A.; Lee, H.O.; Jawerth, L.; Maharana, S.; Jahnel, M.; Hein, M.Y.; Stoynov, S.; Mahamid, J.; Saha, S.; Franzmann, T.M.; et al. A Liquid-to-Solid Phase Transition of the ALS Protein FUS Accelerated by Disease Mutation. *Cell* **2015**, *162*, 1066–1077. [CrossRef] [PubMed]
19. Berry, J.; Weber, S.C.; Vaidya, N.; Haataja, M.; Brangwynne, C.P.; Weitz, D.A. RNA Transcription Modulates Phase Transition-Driven Nuclear Body Assembly. *Proc. Natl. Acad. Sci. USA* **2015**, *112*, E5237–E5245. [CrossRef] [PubMed]
20. Maharana, S.; Wang, J.; Papadopoulos, D.K.; Richter, D.; Pozniakovsky, A.; Poser, I.; Bickle, M.; Rizk, S.; Guillén-Boixet, J.; Franzmann, T.M.; et al. RNA Buffers the Phase Separation Behavior of Prion-like RNA Binding Proteins. *Science* **2018**, *360*, 918–921. [CrossRef]
21. Zhang, H.; Elbaum-Garfinkle, S.; Langdon, E.M.; Taylor, N.; Occhipinti, P.; Bridges, A.A.; Brangwynne, C.P.; Gladfelter, A.S. RNA Controls PolyQ Protein Phase Transitions. *Mol. Cell* **2015**, *60*, 220–230. [CrossRef]
22. Conicella, A.E.; Zerze, G.H.; Mittal, J.; Fawzi, N.L. ALS Mutations Disrupt Phase Separation Mediated by α-Helical Structure in the TDP-43 Low-Complexity C-Terminal Domain. *Structure* **2016**, *24*, 1537–1549. [CrossRef] [PubMed]
23. Brangwynne, C.P.; Mitchison, T.J.; Hyman, A.A. Active Liquid-like Behavior of Nucleoli Determines Their Size and Shape in Xenopus Laevis Oocytes. *Proc. Natl. Acad. Sci. USA* **2011**, *108*, 4334–4339. [CrossRef]
24. Elbaum-Garfinkle, S.; Kim, Y.; Szczepaniak, K.; Chen, C.C.H.; Eckmann, C.R.; Myong, S.; Brangwynne, C.P. The Disordered P Granule Protein LAF-1 Drives Phase Separation into Droplets with Tunable Viscosity and Dynamics. *Proc. Natl. Acad. Sci. USA* **2015**, *112*, 7189–7194. [CrossRef] [PubMed]
25. Mitrea, D.M.; Cika, J.A.; Guy, C.S.; Ban, D.; Banerjee, P.R.; Stanley, C.B.; Nourse, A.; Deniz, A.A.; Kriwacki, R.W. Nucleophosmin Integrates within the Nucleolus via Multi-Modal Interactions with Proteins Displaying R-Rich Linear Motifs and RRNA. *eLife* **2016**, *5*, e13571. [CrossRef] [PubMed]
26. Lin, Y.; Protter, D.S.W.; Rosen, M.K.; Parker, R. Formation and Maturation of Phase-Separated Liquid Droplets by RNA-Binding Proteins. *Mol. Cell* **2015**, *60*, 208–219. [CrossRef]
27. Yao, R.W.; Xu, G.; Wang, Y.; Shan, L.; Luan, P.F.; Wang, Y.; Wu, M.; Yang, L.Z.; Xing, Y.H.; Yang, L.; et al. Nascent Pre-RRNA Sorting via Phase Separation Drives the Assembly of Dense Fibrillar Components in the Human Nucleolus. *Mol. Cell* **2019**, *76*, 767–783. [CrossRef]
28. Gueroussov, S.; Weatheritt, R.J.; O'Hanlon, D.; Lin, Z.Y.; Narula, A.; Gingras, A.C.; Blencowe, B.J. Regulatory Expansion in Mammals of Multivalent HnRNP Assemblies That Globally Control Alternative Splicing. *Cell* **2017**, *170*, 324–339. [CrossRef]
29. Wang, A.; Conicella, A.E.; Schmidt, H.B.; Martin, E.W.; Rhoads, S.N.; Reeb, A.N.; Nourse, A.; Ramirez Montero, D.; Ryan, V.H.; Rohatgi, R.; et al. A Single N-terminal Phosphomimic Disrupts TDP-43 Polymerization, Phase Separation, and RNA Splicing. *EMBO J.* **2018**, *37*, e97452. [CrossRef]

30. Ishimoto, C.; Tanaka, T. Critical Behavior of a Binary Mixture of Protein and Salt Water. *Phys. Rev. Lett.* **1977**, *39*, 474–477. [CrossRef]
31. Tanaka, T.; Benedek, G.B. Observation of Protein Diffusivity in Intact Human and Bovine Lenses with Application to Cataract. *Investig. Ophthalmol.* **1975**, *14*, 449–456.
32. Clark, J.I.; Carper, D. Phase Separation in Lens Cytoplasm Is Genetically Linked to Cataract Formation in the Philly Mouse. *Proc. Natl. Acad. Sci. USA* **1987**, *84*, 122–125. [CrossRef] [PubMed]
33. Strom, A.R.; Brangwynne, C.P. The Liquid Nucleome—Phase Transitions in the Nucleus at a Glance. *J. Cell Sci.* **2019**, *132*, jcs235093. [CrossRef] [PubMed]
34. Gomes, E.; Shorter, J. The Molecular Language of Membraneless Organelles. *J. Biol. Chem.* **2019**, *294*, 7115–7127. [CrossRef] [PubMed]
35. Li, P.; Banjade, S.; Cheng, H.C.; Kim, S.; Chen, B.; Guo, L.; Llaguno, M.; Hollingsworth, J.V.; King, D.S.; Banani, S.F.; et al. Phase Transitions in the Assembly of Multivalent Signalling Proteins. *Nature* **2012**, *483*, 336–340. [CrossRef]
36. Dignon, G.L.; Best, R.B.; Mittal, J. Biomolecular Phase Separation: From Molecular Driving Forces to Macroscopic Properties. *Annu. Rev. Phys. Chem.* **2020**, *71*, 53–75. [CrossRef]
37. Mitrea, D.M.; Cika, J.A.; Stanley, C.B.; Nourse, A.; Onuchic, P.L.; Banerjee, P.R.; Phillips, A.H.; Park, C.G.; Deniz, A.A.; Kriwacki, R.W. Self-Interaction of NPM1 Modulates Multiple Mechanisms of Liquid-Liquid Phase Separation. *Nat. Commun.* **2018**, *9*, 842. [CrossRef]
38. Mittag, T.; Parker, R. Multiple Modes of Protein–Protein Interactions Promote RNP Granule Assembly. *J. Mol. Biol.* **2018**, *430*, 4636–4649. [CrossRef]
39. Carter, G.C.; Hsiung, C.H.; Simpson, L.; Yang, H.; Zhang, X. N-Terminal Domain of TDP43 Enhances Liquid-Liquid Phase Separation of Globular Proteins. *J. Mol. Biol.* **2021**, *433*, 166948. [CrossRef]
40. Shakya, A.; Park, S.; Rana, N.; King, J.T. Liquid-Liquid Phase Separation of Histone Proteins in Cells: Role in Chromatin Organization. *Biophys. J.* **2020**, *118*, 753–764. [CrossRef]
41. Gibson, B.A.; Doolittle, L.K.; Schneider, M.W.G.; Jensen, L.E.; Gamarra, N.; Henry, L.; Gerlich, D.W.; Redding, S.; Rosen, M.K. Organization of Chromatin by Intrinsic and Regulated Phase Separation. *Cell* **2019**, *179*, 470–484. [CrossRef]
42. Henninger, J.E.; Oksuz, O.; Shrinivas, K.; Sagi, I.; LeRoy, G.; Zheng, M.M.; Andrews, J.O.; Zamudio, A.V.; Lazaris, C.; Hannett, N.M.; et al. RNA-Mediated Feedback Control of Transcriptional Condensates. *Cell* **2021**, *184*, 207–225. [CrossRef] [PubMed]
43. Mendenhall, E.M.; Koche, R.P.; Truong, T.; Zhou, V.W.; Issac, B.; Chi, A.S.; Ku, M.; Bernstein, B.E. GC-Rich Sequence Elements Recruit PRC2 in Mammalian ES Cells. *PLoS Genet.* **2010**, *6*, e1001244. [CrossRef] [PubMed]
44. Min, J.; Wright, W.E.; Shay, J.W. Clustered Telomeres in Phase-Separated Nuclear Condensates Engage Mitotic DNA Synthesis through BLM and RAD52. *Genes Dev.* **2019**, *33*, 814–827. [CrossRef]
45. Zhang, H.; Zhao, R.; Tones, J.; Liu, M.; Dilley, R.L.; Chenoweth, D.M.; Greenberg, R.A.; Lampson, M.A. Nuclear Body Phase Separation Drives Telomere Clustering in ALT Cancer Cells. *Mol. Biol. Cell* **2020**, *31*, 2048–2056. [CrossRef] [PubMed]
46. Erdel, F.; Rippe, K. Formation of Chromatin Subcompartments by Phase Separation. *Biophys. J.* **2018**, *114*, 2262–2270. [CrossRef]
47. Barbieri, M.; Chotalia, M.; Fraser, J.; Lavitas, L.M.; Dostie, J.; Pombo, A.; Nicodemi, M. Complexity of Chromatin Folding Is Captured by the Strings and Binders Switch Model. *Proc. Natl. Acad. Sci. USA* **2012**, *109*, 16173–16178. [CrossRef]
48. Conte, M.; Fiorillo, L.; Bianco, S.; Chiariello, A.M.; Esposito, A.; Nicodemi, M. Polymer Physics Indicates Chromatin Folding Variability across Single-Cells Results from State Degeneracy in Phase Separation. *Nat. Commun.* **2020**, *11*, 3289. [CrossRef]
49. Brackley, C.A.; Taylor, S.; Papantonis, A.; Cook, P.R.; Marenduzzo, D. Nonspecific Bridging-Induced Attraction Drives Clustering of DNA-Binding Proteins and Genome Organization. *Proc. Natl. Acad. Sci. USA* **2013**, *110*, E3605–E3611. [CrossRef]
50. Johnson, J.; Brackley, C.A.; Cook, P.R.; Marenduzzo, D. A Simple Model for DNA Bridging Proteins and Bacterial or Human Genomes: Bridging-Induced Attraction and Genome Compaction. *J. Phys. Condens. Matter* **2015**, *27*, 064119. [CrossRef]
51. Brackley, C.A.; Liebchen, B.; Michieletto, D.; Mouvet, F.; Cook, P.R.; Marenduzzo, D. Ephemeral Protein Binding to DNA Shapes Stable Nuclear Bodies and Chromatin Domains. *Biophys. J.* **2017**, *112*, 1085–1093. [CrossRef]
52. Ryu, J.K.; Bouchoux, C.; Liu, H.W.; Kim, E.; Minamino, M.; de Groot, R.; Katan, A.J.; Bonato, A.; Marenduzzo, D.; Michieletto, D.; et al. Bridging-Induced Phase Separation Induced by Cohesin SMC Protein Complexes. *Sci. Adv.* **2021**, *7*, eabe5905. [CrossRef] [PubMed]
53. Brangwynne, C.P.; Tompa, P.; Pappu, R.V. Polymer Physics of Intracellular Phase Transitions. *Nat. Phys.* **2015**, *11*, 899–904. [CrossRef]
54. Berry, J.; Brangwynne, C.P.; Haataja, M. Physical Principles of Intracellular Organization via Active and Passive Phase Transitions. *Reports Prog. Phys.* **2018**, *81*, 046601. [CrossRef] [PubMed]
55. Nakashima, K.K.; Vibhute, M.A.; Spruijt, E. Biomolecular Chemistry in Liquid Phase Separated Compartments. *Front. Mol. Biosci.* **2019**, *6*, 21. [CrossRef] [PubMed]
56. Laghmach, R.; Di Pierro, M.; Potoyan, D. A Liquid State Perspective on Dynamics of Chromatin Compartments. *Front. Mol. Biosci.* **2022**, *8*, 781981. [CrossRef] [PubMed]
57. Zhao, Z.W.; White, M.D.; Bissiere, S.; Levi, V.; Plachta, N. Quantitative Imaging of Mammalian Transcriptional Dynamics: From Single Cells to Whole Embryos. *BMC Biol.* **2016**, *14*, 115. [CrossRef]
58. Keppler, A.; Gendreizig, S.; Gronemeyer, T.; Pick, H.; Vogel, H.; Johnsson, K. A General Method for the Covalent Labeling of Fusion Proteins with Small Molecules in Vivo. *Nat. Biotechnol.* **2003**, *21*, 86–89. [CrossRef]

59. Gautier, A.; Juillerat, A.; Heinis, C.; Corrêa, I.R.; Kindermann, M.; Beaufils, F.; Johnsson, K. An Engineered Protein Tag for Multiprotein Labeling in Living Cells. *Chem. Biol.* **2008**, *15*, 128–136. [CrossRef]
60. Los, G.V.; Encell, L.P.; McDougall, M.G.; Hartzell, D.D.; Karassina, N.; Zimprich, C.; Wood, M.G.; Learish, R.; Ohana, R.F.; Urh, M.; et al. HaloTag: A Novel Protein Labeling Technology for Cell Imaging and Protein Analysis. *ACS Chem. Biol.* **2008**, *3*, 373–382. [CrossRef]
61. Elson, E.L. Fluorescence Correlation Spectroscopy: Past, Present, Future. *Biophys. J.* **2011**, *101*, 2855–2870. [CrossRef]
62. Macháň, R.; Wohland, T. Recent Applications of Fluorescence Correlation Spectroscopy in Live Systems. *FEBS Lett.* **2014**, *588*, 3571–3584. [CrossRef] [PubMed]
63. Zhao, Z.W.; White, M.D.; Alvarez, Y.D.; Zenker, J.; Bissiere, S.; Plachta, N. Quantifying Transcription Factor-DNA Binding in Single Cells in Vivo with Photoactivatable Fluorescence Correlation Spectroscopy. *Nat. Protoc.* **2017**, *12*, 1458–1471. [CrossRef] [PubMed]
64. Erdel, F.; Rademacher, A.; Vlijm, R.; Tünnermann, J.; Frank, L.; Weinmann, R.; Schweigert, E.; Yserentant, K.; Hummert, J.; Bauer, C.; et al. Mouse Heterochromatin Adopts Digital Compaction States without Showing Hallmarks of HP1-Driven Liquid-Liquid Phase Separation. *Mol. Cell* **2020**, *78*, 236–249. [CrossRef] [PubMed]
65. Digman, M.A.; Dalal, R.; Horwitz, A.F.; Gratton, E. Mapping the Number of Molecules and Brightness in the Laser Scanning Microscope. *Biophys. J.* **2008**, *94*, 2320–2332. [CrossRef]
66. Cutrale, F.; Rodriguez, D.; Hortigüela, V.; Chiu, C.L.; Otterstrom, J.; Mieruszynski, S.; Seriola, A.; Larrañaga, E.; Raya, A.; Lakadamyali, M.; et al. Using Enhanced Number and Brightness to Measure Protein Oligomerization Dynamics in Live Cells. *Nat. Protoc.* **2019**, *14*, 616–638. [CrossRef]
67. Rossow, M.J.; Sasaki, J.M.; Digman, M.A.; Gratton, E. Raster Image Correlation Spectroscopy in Live Cells. *Nat. Protoc.* **2010**, *5*, 1761–1774. [CrossRef]
68. Krieger, J.W.; Singh, A.P.; Bag, N.; Garbe, C.S.; Saunders, T.E.; Langowski, J.; Wohland, T. Imaging Fluorescence (Cross-) Correlation Spectroscopy in Live Cells and Organisms. *Nat. Protoc.* **2015**, *10*, 1948–1974. [CrossRef]
69. Kusumi, A.; Tsunoyama, T.A.; Hirosawa, K.M.; Kasai, R.S.; Fujiwara, T.K. Tracking Single Molecules at Work in Living Cells. *Nat. Chem. Biol.* **2014**, *10*, 524–532. [CrossRef]
70. Tokunaga, M.; Imamoto, N.; Sakata-Sogawa, K. Highly Inclined Thin Illumination Enables Clear Single-Molecule Imaging in Cells. *Nat. Methods* **2008**, *5*, 159–161. [CrossRef]
71. Ritter, J.G.; Veith, R.; Veenendaal, A.; Siebrasse, J.P.; Kubitscheck, U. Light Sheet Microscopy for Single Molecule Tracking in Living Tissue. *PLoS ONE* **2010**, *5*, 11639. [CrossRef]
72. Gebhardt, J.C.M.; Suter, D.M.; Roy, R.; Zhao, Z.W.; Chapman, A.R.; Basu, S.; Maniatis, T.; Xie, X.S. Single-Molecule Imaging of Transcription Factor Binding to DNA in Live Mammalian Cells. *Nat. Methods* **2013**, *10*, 421–426. [CrossRef] [PubMed]
73. Chen, B.C.; Legant, W.R.; Wang, K.; Shao, L.; Milkie, D.E.; Davidson, M.W.; Janetopoulos, C.; Wu, X.S.; Hammer, J.A.; Liu, Z.; et al. Lattice Light-Sheet Microscopy: Imaging Molecules to Embryos at High Spatiotemporal Resolution. *Science* **2014**, *346*, 1257998. [CrossRef] [PubMed]
74. Axelrod, D.; Koppel, D.E.; Schlessinger, J.; Elson, E.; Webb, W.W. Mobility Measurement by Analysis of Fluorescence Photobleaching Recovery Kinetics. *Biophys. J.* **1976**, *16*, 1055–1069. [CrossRef]
75. Lippincott-Schwartz, J.; Altan-Bonnet, N.; Patterson, G.H. Photobleaching and Photoactivation: Following Protein Dynamics in Living Cells. *Nat. Cell Biol.* **2003**, *4*, S7–S14.
76. Betzig, E.; Patterson, G.H.; Sougrat, R.; Lindwasser, O.W.; Olenych, S.; Bonifacino, J.S.; Davidson, M.W.; Lippincott-Schwartz, J.; Hess, H.F. Imaging Intracellular Fluorescent Proteins at Nanometer Resolution. *Science* **2006**, *313*, 1642–1645. [CrossRef]
77. Rust, M.J.; Bates, M.; Zhuang, X. Sub-Diffraction-Limit Imaging by Stochastic Optical Reconstruction Microscopy (STORM). *Nat. Methods* **2006**, *3*, 793–795. [CrossRef]
78. Hess, S.T.; Girirajan, T.P.K.; Mason, M.D. Ultra-High Resolution Imaging by Fluorescence Photoactivation Localization Microscopy. *Biophys. J.* **2006**, *91*, 4258–4272. [CrossRef]
79. Gustafsson, M.G.L. Nonlinear Structured-Illumination Microscopy: Wide-Field Fluorescence Imaging with Theoretically Unlimited Resolution. *Proc. Natl. Acad. Sci. USA* **2005**, *102*, 13081–13086. [CrossRef]
80. Klar, T.A.; Jakobs, S.; Dyba, M.; Egner, A.; Hell, S.W. Fluorescence Microscopy with Diffraction Resolution Barrier Broken by Stimulated Emission. *Proc. Natl. Acad. Sci. USA* **2000**, *97*, 8206–8210. [CrossRef]
81. Cella Zanacchi, F.; Lavagnino, Z.; Perrone Donnorso, M.; Del Bue, A.; Furia, L.; Faretta, M.; Diaspro, A. Live-Cell 3D Super-Resolution Imaging in Thick Biological Samples. *Nat. Methods* **2011**, *8*, 1047–1050. [CrossRef]
82. Gao, L.; Shao, L.; Higgins, C.D.; Poulton, J.S.; Peifer, M.; Davidson, M.W.; Wu, X.; Goldstein, B.; Betzig, E. Noninvasive Imaging beyond the Diffraction Limit of 3D Dynamics in Thickly Fluorescent Specimens. *Cell* **2012**, *151*, 1370–1385. [CrossRef] [PubMed]
83. Zhao, Z.W.; Roy, R.; Gebhardt, J.C.M.; Suter, D.M.; Chapman, A.R.; Xie, X.S. Spatial Organization of RNA Polymerase II Insidea Mammalian Cell Nucleus Revealed by Reflectedlight-Sheet Superresolution Microscopy. *Proc. Natl. Acad. Sci. USA* **2014**, *111*, 681–686. [CrossRef] [PubMed]
84. Strom, A.R.; Emelyanov, A.V.; Mir, M.; Fyodorov, D.V.; Darzacq, X.; Karpen, G.H. Phase Separation Drives Heterochromatin Domain Formation. *Nature* **2017**, *547*, 241–245. [CrossRef] [PubMed]

85. Larson, A.G.; Elnatan, D.; Keenen, M.M.; Trnka, M.J.; Johnston, J.B.; Burlingame, A.L.; Agard, D.A.; Redding, S.; Narlikar, G.J. Liquid Droplet Formation by HP1α Suggests a Role for Phase Separation in Heterochromatin. *Nature* **2017**, *547*, 236–240. [CrossRef]
86. Wang, L.; Gao, Y.; Zheng, X.; Liu, C.; Dong, S.; Li, R.; Zhang, G.; Wei, Y.; Qu, H.; Li, Y.; et al. Histone Modifications Regulate Chromatin Compartmentalization by Contributing to a Phase Separation Mechanism. *Mol. Cell* **2019**, *76*, 646–659. [CrossRef]
87. Li, C.H.; Coffey, E.L.; Dall'Agnese, A.; Hannett, N.M.; Tang, X.; Henninger, J.E.; Platt, J.M.; Oksuz, O.; Zamudio, A.V.; Afeyan, L.K.; et al. MeCP2 Links Heterochromatin Condensates and Neurodevelopmental Disease. *Nature* **2020**, *586*, 440–444. [CrossRef]
88. Wang, L.; Hu, M.; Zuo, M.Q.; Zhao, J.; Wu, D.; Huang, L.; Wen, Y.; Li, Y.; Chen, P.; Bao, X.; et al. Rett Syndrome-Causing Mutations Compromise MeCP2-Mediated Liquid–Liquid Phase Separation of Chromatin. *Cell Res.* **2020**, *30*, 393–407. [CrossRef]
89. Zhang, L.; Geng, X.; Wang, F.; Tang, J.; Ichida, Y.; Sharma, A.; Jin, S.; Chen, M.; Tang, M.; Pozo, F.M.; et al. 53BP1 Regulates Heterochromatin through Liquid Phase Separation. *Nat. Commun.* **2022**, *13*, 360. [CrossRef]
90. Tatavosian, R.; Kent, S.; Brown, K.; Yao, T.; Duc, H.N.; Huynh, T.N.; Zhen, C.Y.; Ma, B.; Wang, H.; Ren, X. Nuclear Condensates of the Polycomb Protein Chromobox 2 (CBX2) Assemble through Phase Separation. *J. Biol. Chem.* **2019**, *294*, 1451–1463. [CrossRef]
91. Plys, A.J.; Davis, C.P.; Kim, J.; Rizki, G.; Keenen, M.M.; Marr, S.K.; Kingston, R.E. Phase Separation of Polycomb-Repressive Complex 1 Is Governed by a Charged Disordered Region of CBX2. *Genes Dev.* **2019**, *33*, 799–813. [CrossRef]
92. Lee, R.; Kang, M.K.; Kim, Y.J.; Yang, B.; Shim, H.; Kim, S.; Kim, K.; Yang, C.M.; Min, B.G.; Jung, W.J.; et al. CTCF-Mediated Chromatin Looping Provides a Topological Framework for the Formation of Phase-Separated Transcriptional Condensates. *Nucleic Acids Res.* **2022**, *50*, 207–226. [CrossRef] [PubMed]
93. Oshidari, R.; Huang, R.; Medghalchi, M.; Tse, E.Y.W.; Ashgriz, N.; Lee, H.O.; Wyatt, H.; Mekhail, K. DNA Repair by Rad52 Liquid Droplets. *Nat. Commun.* **2020**, *11*, 695. [CrossRef] [PubMed]
94. Levone, B.R.; Lenzken, S.C.; Antonaci, M.; Maiser, A.; Rapp, A.; Conte, F.; Reber, S.; Mechtersheimer, J.; Ronchi, A.E.; Mühlemann, O.; et al. FUS-Dependent Liquid-Liquid Phase Separation Is Important for DNA Repair Initiation. *J. Cell Biol.* **2021**, *220*, e202008030. [CrossRef]
95. Pessina, F.; Giavazzi, F.; Yin, Y.; Gioia, U.; Vitelli, V.; Galbiati, A.; Barozzi, S.; Garre, M.; Oldani, A.; Flaus, A.; et al. Functional Transcription Promoters at DNA Double-Strand Breaks Mediate RNA-Driven Phase Separation of Damage-Response Factors. *Nat. Cell Biol.* **2019**, *21*, 1286–1299. [CrossRef] [PubMed]
96. Kilic, S.; Lezaja, A.; Gatti, M.; Bianco, E.; Michelena, J.; Imhof, R.; Altmeyer, M. Phase Separation of 53 BP 1 Determines Liquid-like Behavior of DNA Repair Compartments. *EMBO J.* **2019**, *38*, e101379. [CrossRef] [PubMed]
97. Sabari, B.R.; Dall'Agnese, A.; Boija, A.; Klein, I.A.; Coffey, E.L.; Shrinivas, K.; Abraham, B.J.; Hannett, N.M.; Zamudio, A.V.; Manteiga, J.C.; et al. Coactivator Condensation at Super-Enhancers Links Phase Separation and Gene Control. *Science* **2018**, *361*, eaar3958. [CrossRef]
98. Boija, A.; Klein, I.A.; Sabari, B.R.; Dall'Agnese, A.; Coffey, E.L.; Zamudio, A.V.; Li, C.H.; Shrinivas, K.; Manteiga, J.C.; Hannett, N.M.; et al. Transcription Factors Activate Genes through the Phase-Separation Capacity of Their Activation Domains. *Cell* **2018**, *175*, 1842–1855. [CrossRef]
99. Cho, W.K.; Spille, J.H.; Hecht, M.; Lee, C.; Li, C.; Grube, V.; Cisse, I.I. Mediator and RNA Polymerase II Clusters Associate in Transcription-Dependent Condensates. *Science* **2018**, *361*, 412–415. [CrossRef]
100. Cai, D.; Feliciano, D.; Dong, P.; Flores, E.; Gruebele, M.; Porat-Shliom, N.; Sukenik, S.; Liu, Z.; Lippincott-Schwartz, J. Phase Separation of YAP Reorganizes Genome Topology for Long-Term YAP Target Gene Expression. *Nat. Cell Biol.* **2019**, *21*, 1578–1589. [CrossRef]
101. Lu, Y.; Wu, T.; Gutman, O.; Lu, H.; Zhou, Q.; Henis, Y.I.; Luo, K. Phase Separation of TAZ Compartmentalizes the Transcription Machinery to Promote Gene Expression. *Nat. Cell Biol.* **2020**, *22*, 453–464. [CrossRef]
102. Chong, S.; Graham, T.G.W.; Dugast-Darzacq, C.; Dailey, G.M.; Darzacq, X.; Tjian, R. Tuning Levels of Low-Complexity Domain Interactions to Modulate Endogenous Oncogenic Transcription. *Mol. Cell* **2022**, *82*, 2084–2097. [CrossRef] [PubMed]
103. Wei, M.T.; Chang, Y.C.; Shimobayashi, S.F.; Shin, Y.; Strom, A.R.; Brangwynne, C.P. Nucleated Transcriptional Condensates Amplify Gene Expression. *Nat. Cell Biol.* **2020**, *22*, 1187–1196. [CrossRef]
104. Chong, S.; Dugast-Darzacq, C.; Liu, Z.; Dong, P.; Dailey, G.M.; Cattoglio, C.; Heckert, A.; Banala, S.; Lavis, L.; Darzacq, X.; et al. Imaging Dynamic and Selective Low-Complexity Domain Interactions That Control Gene Transcription. *Science* **2018**, *361*, eaar2555. [CrossRef] [PubMed]
105. Fasciani, A.; D'Annunzio, S.; Poli, V.; Fagnocchi, L.; Beyes, S.; Michelatti, D.; Corazza, F.; Antonelli, L.; Gregoretti, F.; Oliva, G.; et al. MLL4-Associated Condensates Counterbalance Polycomb-Mediated Nuclear Mechanical Stress in Kabuki Syndrome. *Nat. Genet.* **2020**, *52*, 1397–1411. [CrossRef]
106. Sharma, R.; Choi, K.J.; Quan, M.D.; Sharma, S.; Sankaran, B.; Park, H.; LaGrone, A.; Kim, J.J.; MacKenzie, K.R.; Ferreon, A.C.M.; et al. Liquid Condensation of Reprogramming Factor KLF4 with DNA Provides a Mechanism for Chromatin Organization. *Nat. Commun.* **2021**, *12*, 5579. [CrossRef] [PubMed]
107. Zhang, H.; Shao, S.; Zeng, Y.; Wang, X.; Qin, Y.; Ren, Q.; Xiang, S.; Wang, Y.; Xiao, J.; Sun, Y. Reversible Phase Separation of HSF1 Is Required for an Acute Transcriptional Response during Heat Shock. *Nat. Cell Biol.* **2022**, *24*, 340–352. [CrossRef]
108. Wang, W.; Qiao, S.; Li, G.; Cheng, J.; Yang, C.; Zhong, C.; Stovall, D.B.; Shi, J.; Teng, C.; Li, D.; et al. A Histidine Cluster Determines YY1-Compartmentalized Coactivators and Chromatin Elements in Phase-Separated Enhancer Clusters. *Nucleic Acids Res.* **2022**, *50*, 4917–4937. [CrossRef]

109. Fu, H.; Liu, R.; Jia, Z.; Li, R.; Zhu, F.; Zhu, W.; Shao, Y.; Jin, Y.; Xue, Y.; Huang, J.; et al. Poly(ADP-Ribosylation) of P-TEFb by PARP1 Disrupts Phase Separation to Inhibit Global Transcription after DNA Damage. *Nat. Cell Biol.* **2022**, *24*, 513–525. [CrossRef]
110. Lu, H.; Liu, R.; Zhou, Q. Balanced between Order and Disorder: A New Phase in Transcription Elongation Control and Beyond. *Transcription* **2019**, *10*, 157–163. [CrossRef]
111. Guo, Y.E.; Manteiga, J.C.; Henninger, J.E.; Sabari, B.R.; Dall'Agnese, A.; Hannett, N.M.; Spille, J.H.; Afeyan, L.K.; Zamudio, A.V.; Shrinivas, K.; et al. Pol II Phosphorylation Regulates a Switch between Transcriptional and Splicing Condensates. *Nature* **2019**, *572*, 543–548. [CrossRef]
112. Misteli, T. The Self-Organizing Genome: Principles of Genome Architecture and Function. *Cell* **2020**, *183*, 28–45. [CrossRef] [PubMed]
113. Hansen, A.S.; Cattoglio, C.; Darzacq, X.; Tjian, R. Recent Evidence That TADs and Chromatin Loops Are Dynamic Structures. *Nucleus* **2018**, *9*, 20–32. [CrossRef] [PubMed]
114. Filion, G.J.; van Bemmel, J.G.; Braunschweig, U.; Talhout, W.; Kind, J.; Ward, L.D.; Brugman, W.; de Castro, I.J.; Kerkhoven, R.M.; Bussemaker, H.J.; et al. Systematic Protein Location Mapping Reveals Five Principal Chromatin Types in Drosophila Cells. *Cell* **2010**, *143*, 212–224. [CrossRef] [PubMed]
115. Rao, S.S.P.; Huntley, M.H.; Durand, N.C.; Stamenova, E.K.; Bochkov, I.D.; Robinson, J.T.; Sanborn, A.L.; Machol, I.; Omer, A.D.; Lander, E.S.; et al. A 3D Map of the Human Genome at Kilobase Resolution Reveals Principles of Chromatin Looping. *Cell* **2014**, *159*, 1665–1680. [CrossRef] [PubMed]
116. Esposito, A.; Abraham, A.; Conte, M.; Vercellone, F.; Prisco, A.; Bianco, S.; Chiariello, A.M. The Physics of DNA Folding: Polymer Models and Phase-Separation. *Polymers* **2022**, *14*, 1918. [CrossRef] [PubMed]
117. Esposito, A.; Chiariello, A.M.; Conte, M.; Fiorillo, L.; Musella, F.; Sciarretta, R.; Bianco, S. Higher-Order Chromosome Structures Investigated by Polymer Physics in Cellular Morphogenesis and Differentiation. *J. Mol. Biol.* **2020**, *432*, 701–711. [CrossRef] [PubMed]
118. Nicodemi, M.; Panning, B.; Prisco, A. A Thermodynamic Switch for Chromosome Colocalization. *Genetics* **2008**, *179*, 717–721. [CrossRef]
119. Jost, D.; Carrivain, P.; Cavalli, G.; Vaillant, C. Modeling Epigenome Folding: Formation and Dynamics of Topologically Associated Chromatin Domains. *Nucleic Acids Res.* **2014**, *42*, 9553. [CrossRef]
120. Di Pierro, M.; Cheng, R.R.; Aiden, E.L.; Wolynes, P.G.; Onuchic, J.N. De Novo Prediction of Human Chromosome Structures: Epigenetic Marking Patterns Encode Genome Architecture. *Proc. Natl. Acad. Sci. USA* **2017**, *114*, 12126–12131. [CrossRef]
121. Di Pierro, M.; Zhang, B.; Aiden, E.L.; Wolynes, P.G.; Onuchic, J.N. Transferable Model for Chromosome Architecture. *Proc. Natl. Acad. Sci. USA* **2016**, *113*, 12168–12173. [CrossRef]
122. Nuebler, J.; Fudenberg, G.; Imakaev, M.; Abdennur, N.; Mirny, L.A. Chromatin Organization by an Interplay of Loop Extrusion and Compartmental Segregation. *Proc. Natl. Acad. Sci. USA* **2018**, *115*, E6697–E6706. [CrossRef] [PubMed]
123. Cheng, R.R.; Contessoto, V.G.; Aiden, E.L.; Wolynes, P.G.; Di Pierro, M.; Onuchic, J.N. Exploring Chromosomal Structural Heterogeneity across Multiple Cell Lines. *eLife* **2020**, *9*, e60312. [CrossRef]
124. Feric, M.; Misteli, T. Phase Separation in Genome Organization across Evolution. *Trends Cell Biol.* **2021**, *31*, 671–685. [CrossRef] [PubMed]
125. Saksouk, N.; Simboeck, E.; Déjardin, J. Constitutive Heterochromatin Formation and Transcription in Mammals. *Epigenet. Chromatin* **2015**, *8*, 3. [CrossRef] [PubMed]
126. Trojer, P.; Reinberg, D. Facultative Heterochromatin: Is There a Distinctive Molecular Signature? *Mol. Cell* **2007**, *28*, 1–13. [CrossRef]
127. Sanulli, S.; Trnka, M.J.; Dharmarajan, V.; Tibble, R.W.; Pascal, B.D.; Burlingame, A.L.; Griffin, P.R.; Gross, J.D.; Narlikar, G.J. HP1 Reshapes Nucleosome Core to Promote Phase Separation of Heterochromatin. *Nature* **2019**, *575*, 390–394. [CrossRef]
128. Janssen, A.; Colmenares, S.U.; Karpen, G.H. Heterochromatin: Guardian of the Genome. *Annu. Rev. Cell Dev. Biol.* **2018**, *34*, 265–288. [CrossRef]
129. Moore, L.D.; Le, T.; Fan, G. DNA Methylation and Its Basic Function. *Neuropsychopharmacology* **2012**, *38*, 23–38. [CrossRef]
130. Seif, E.; Kang, J.J.; Sasseville, C.; Senkovich, O.; Kaltashov, A.; Boulier, E.L.; Kapur, I.; Kim, C.A.; Francis, N.J. Phase Separation by the Polyhomeotic Sterile Alpha Motif Compartmentalizes Polycomb Group Proteins and Enhances Their Activity. *Nat. Commun.* **2020**, *11*, 5609. [CrossRef]
131. Kent, S.; Brown, K.; Yang, C.H.; Alsaihati, N.; Tian, C.; Wang, H.; Ren, X. Phase-Separated Transcriptional Condensates Accelerate Target-Search Process Revealed by Live-Cell Single-Molecule Imaging. *Cell Rep.* **2020**, *33*, 108248. [CrossRef]
132. Lau, M.S.; Schwartz, M.G.; Kundu, S.; Savol, A.J.; Wang, P.I.; Marr, S.K.; Grau, D.J.; Schorderet, P.; Sadreyev, R.I.; Tabin, C.J.; et al. Mutation of a Nucleosome Compaction Region Disrupts Polycomb-Mediated Axial Patterning. *Science* **2017**, *355*, 1081–1084. [CrossRef] [PubMed]
133. Guo, Y.; Zhao, S.; Wang, G.G. Polycomb Gene Silencing Mechanisms: PRC2 Chromatin Targeting, H3K27me3 "Readout", and Phase Separation-Based Compaction. *Trends Genet.* **2021**, *37*, 547–565. [CrossRef] [PubMed]
134. Falk, M.; Feodorova, Y.; Naumova, N.; Imakaev, M.; Lajoie, B.R.; Leonhardt, H.; Joffe, B.; Dekker, J.; Fudenberg, G.; Solovei, I.; et al. Heterochromatin Drives Compartmentalization of Inverted and Conventional Nuclei. *Nature* **2019**, *570*, 395–399. [CrossRef]
135. Mobley, A.S. *Neural Stem Cells and Adult Neurogenesis*; Academic Press: Cambridge, MA, USA, 2019; pp. 67–94.

136. Serizay, J.; Ahringer, J. Genome Organization at Different Scales: Nature, Formation and Function. *Curr. Opin. Cell Biol.* **2018**, *52*, 145–153. [CrossRef] [PubMed]
137. Fudenberg, G.; Imakaev, M.; Lu, C.; Goloborodko, A.; Abdennur, N.; Mirny, L.A. Formation of Chromosomal Domains by Loop Extrusion. *Cell Rep.* **2016**, *15*, 2038–2049. [CrossRef]
138. Hansen, A.S.; Pustova, I.; Cattoglio, C.; Tjian, R.; Darzacq, X. CTCF and Cohesin Regulate Chromatin Loop Stability with Distinct Dynamics. *eLife* **2017**, *6*, e25776. [CrossRef]
139. Hansen, A.S.; Hsieh, T.H.S.; Cattoglio, C.; Pustova, I.; Saldaña-Meyer, R.; Reinberg, D.; Darzacq, X.; Tjian, R. Distinct Classes of Chromatin Loops Revealed by Deletion of an RNA-Binding Region in CTCF. *Mol. Cell* **2019**, *76*, 395–411. [CrossRef]
140. Hansen, A.S.; Amitai, A.; Cattoglio, C.; Tjian, R.; Darzacq, X. Guided Nuclear Exploration Increases CTCF Target Search Efficiency. *Nat. Chem. Biol.* **2020**, *16*, 257–266. [CrossRef]
141. Zirkel, A.; Nikolic, M.; Sofiadis, K.; Mallm, J.P.; Brackley, C.A.; Gothe, H.; Drechsel, O.; Becker, C.; Altmüller, J.; Josipovic, N.; et al. HMGB2 Loss upon Senescence Entry Disrupts Genomic Organization and Induces CTCF Clustering across Cell Types. *Mol. Cell* **2018**, *70*, 730–744. [CrossRef]
142. Hnisz, D.; Shrinivas, K.; Young, R.A.; Chakraborty, A.K.; Sharp, P.A. A Phase Separation Model for Transcriptional Control. *Cell* **2017**, *169*, 13–23. [CrossRef]
143. Schneider, N.; Wieland, F.G.; Kong, D.; Fischer, A.A.M.; Hörner, M.; Timmer, J.; Ye, H.; Weber, W. Liquid-Liquid Phase Separation of Light-Inducible Transcription Factors Increases Transcription Activation in Mammalian Cells and Mice. *Sci. Adv.* **2021**, *7*, eabd3568. [CrossRef] [PubMed]
144. Shin, Y.; Chang, Y.C.; Lee, D.S.W.; Berry, J.; Sanders, D.W.; Ronceray, P.; Wingreen, N.S.; Haataja, M.; Brangwynne, C.P. Liquid Nuclear Condensates Mechanically Sense and Restructure the Genome. *Cell* **2018**, *175*, 1481–1491. [CrossRef] [PubMed]
145. Zuo, L.; Zhang, G.; Massett, M.; Cheng, J.; Guo, Z.; Wang, L.; Gao, Y.; Li, R.; Huang, X.; Li, P.; et al. Loci-Specific Phase Separation of FET Fusion Oncoproteins Promotes Gene Transcription. *Nat. Commun.* **2021**, *12*, 1491. [CrossRef]
146. Cho, W.K.; Jayanth, N.; English, B.P.; Inoue, T.; Andrews, J.O.; Conway, W.; Grimm, J.B.; Spille, J.H.; Lavis, L.D.; Lionnet, T.; et al. RNA Polymerase II Cluster Dynamics Predict MRNA Output in Living Cells. *eLife* **2016**, *5*, e13617. [CrossRef] [PubMed]
147. Cisse, I.I.; Izeddin, I.; Causse, S.Z.; Boudarene, L.; Senecal, A.; Muresan, L.; Dugast-Darzacq, C.; Hajj, B.; Dahan, M.; Darzacq, X. Real-Time Dynamics of RNA Polymerase II Clustering in Live Human Cells. *Science* **2013**, *341*, 664–667. [CrossRef] [PubMed]
148. Lu, H.; Yu, D.; Hansen, A.S.; Ganguly, S.; Liu, R.; Heckert, A.; Darzacq, X.; Zhou, Q. Phase-Separation Mechanism for C-Terminal Hyperphosphorylation of RNA Polymerase II. *Nature* **2018**, *558*, 318–323. [CrossRef] [PubMed]
149. Boehning, M.; Dugast-Darzacq, C.; Rankovic, M.; Hansen, A.S.; Yu, T.; Marie-Nelly, H.; McSwiggen, D.T.; Kokic, G.; Dailey, G.M.; Cramer, P.; et al. RNA Polymerase II Clustering through Carboxy-Terminal Domain Phase Separation. *Nat. Struct. Mol. Biol.* **2018**, *25*, 833–840. [CrossRef] [PubMed]
150. Chen, Y.; Zhang, Y.; Wang, Y.; Zhang, L.; Brinkman, E.K.; Adam, S.A.; Goldman, R.; Van Steensel, B.; Ma, J.; Belmont, A.S. Mapping 3D Genome Organization Relative to Nuclear Compartments Using TSA-Seq as a Cytological Ruler. *J. Cell Biol.* **2018**, *217*, 4025–4048. [CrossRef]
151. Gaglia, G.; Rashid, R.; Yapp, C.; Joshi, G.N.; Li, C.G.; Lindquist, S.L.; Sarosiek, K.A.; Whitesell, L.; Sorger, P.K.; Santagata, S. HSF1 Phase Transition Mediates Stress Adaptation and Cell Fate Decisions. *Nat. Cell Biol.* **2020**, *22*, 151–158. [CrossRef]
152. Bekker-Jensen, S.; Lukas, C.; Kitagawa, R.; Melander, F.; Kastan, M.B.; Bartek, J.; Lukas, J. Spatial Organization of the Mammalian Genome Surveillance Machinery in Response to DNA Strand Breaks. *J. Cell Biol.* **2006**, *173*, 195–206. [CrossRef]
153. Schrank, B.; Gautier, J. Assembling Nuclear Domains: Lessons from DNA Repair. *J. Cell Biol.* **2019**, *218*, 2444–2455. [CrossRef] [PubMed]
154. Stucki, M.; Jackson, S.P. ΓH2AX and MDC1: Anchoring the DNA-Damage-Response Machinery to Broken Chromosomes. *DNA Repair* **2006**, *5*, 534–543. [CrossRef] [PubMed]
155. Chiolo, I.; Minoda, A.; Colmenares, S.U.; Polyzos, A.; Costes, S.V.; Karpen, G.H. Double-Strand Breaks in Heterochromatin Move Outside of a Dynamic HP1a Domain to Complete Recombinational Repair. *Cell* **2011**, *144*, 732–744. [CrossRef]
156. McSwiggen, D.T.; Hansen, A.S.; Teves, S.S.; Marie-Nelly, H.; Hao, Y.; Heckert, A.B.; Umemoto, K.K.; Dugast-Darzacq, C.; Tjian, R.; Darzacq, X. Evidence for DNA-Mediated Nuclear Compartmentalization Distinct from Phase Separation. *eLife* **2019**, *8*, e47098. [CrossRef]
157. Wolstenholme, C.H.; Hu, H.; Ye, S.; Funk, B.E.; Jain, D.; Hsiung, C.H.; Ning, G.; Liu, Y.; Li, X.; Zhang, X. AggFluor: Fluorogenic Toolbox Enables Direct Visualization of the Multi-Step Protein Aggregation Process in Live Cells. *J. Am. Chem. Soc.* **2020**, *142*, 17515–17523. [CrossRef] [PubMed]
158. Chen, X.; Zhang, D.; Su, N.; Bao, B.; Xie, X.; Zuo, F.; Yang, L.; Wang, H.; Jiang, L.; Lin, Q.; et al. Visualizing RNA Dynamics in Live Cells with Bright and Stable Fluorescent RNAs. *Nat. Biotechnol.* **2019**, *37*, 1287–1293. [CrossRef] [PubMed]
159. Chen, B.; Gilbert, L.A.; Cimini, B.A.; Schnitzbauer, J.; Zhang, W.; Li, G.W.; Park, J.; Blackburn, E.H.; Weissman, J.S.; Qi, L.S.; et al. Dynamic Imaging of Genomic Loci in Living Human Cells by an Optimized CRISPR/Cas System. *Cell* **2013**, *155*, 1479–1491. [CrossRef]
160. Trojanowski, J.; Frank, L.; Rademacher, A.; Grigaitis, P.; Rippe, K. Transcription Activation Is Enhanced by Multivalent Interactions Independent of Liquid-Liquid Phase Separation. *Mol. Cell* **2020**, *82*, 1878–1893. [CrossRef]
161. Peeples, W.; Rosen, M.K. Mechanistic Dissection of Increased Enzymatic Rate in a Phase-Separated Compartment. *Nat. Chem. Biol.* **2021**, *17*, 693–702. [CrossRef]

162. McSwiggen, D.T.; Mir, M.; Darzacq, X.; Tjian, R. Evaluating Phase Separation in Live Cells: Diagnosis, Caveats, and Functional Consequences. *Genes Dev.* **2019**, *33*, 1619–1634. [CrossRef]
163. Pancholi, A.; Klingberg, T.; Zhang, W.; Prizak, R.; Mamontova, I.; Noa, A.; Sobucki, M.; Kobitski, A.Y.; Nienhaus, G.U.; Zaburdaev, V.; et al. RNA Polymerase II Clusters Form in Line with Surface Condensation on Regulatory Chromatin. *Mol. Syst. Biol.* **2021**, *17*, e10272. [PubMed]
164. Boija, A.; Klein, I.A.; Young, R.A. Biomolecular Condensates and Cancer. *Cancer Cell* **2021**, *39*, 174–192. [PubMed]
165. Cai, D.; Liu, Z.; Lippincott-Schwartz, J. Biomolecular Condensates and Their Links to Cancer Progression. *Trends Biochem. Sci.* **2021**, *46*, 535–549. [PubMed]
166. Ahn, J.H.; Davis, E.S.; Daugird, T.A.; Zhao, S.; Quiroga, I.Y.; Uryu, H.; Li, J.; Storey, A.J.; Tsai, Y.H.; Keeley, D.P.; et al. Phase Separation Drives Aberrant Chromatin Looping and Cancer Development. *Nature* **2021**, *595*, 591–595. [PubMed]
167. White, M.D.; Zhao, Z.W.; Plachta, N. In Vivo Imaging of Single Mammalian Cells in Development and Disease. *Trends Mol. Med.* **2018**, *24*, 278–293. [PubMed]

International Journal of *Molecular Sciences*

MDPI

Article

The Patterning and Proportion of Charged Residues in the Arginine-Rich Mixed-Charge Domain Determine the Membrane-Less Organelle Targeted by the Protein

Tamami Miyagi [1], Rio Yamazaki [1], Koji Ueda [2], Satoshi Narumi [3], Yuhei Hayamizu [4], Hiroshi Uji-i [5,6], Masahiko Kuroda [1,*] and Kohsuke Kanekura [1,*]

1 Department of Molecular Pathology, Tokyo Medical University, 6-1-1 Shinjuku, Shinjuku-ku, Tokyo 160-8402, Japan; s121060@tokyo-med.ac.jp (T.M.); s122066@tokyo-med.ac.jp (R.Y.)
2 Cancer Proteomics Group, Cancer Precision Medicine Center, Japanese Foundation for Cancer Research, 3-8-31 Ariake, Koto-ku, Tokyo 135-8550, Japan; koji.ueda@jfcr.or.jp
3 Department of Molecular Endocrinology, National Research Institute for Child Health and Development, 2-10-1 Okura, Setagaya-ku, Tokyo 157-8535, Japan; narumi-s@ncchd.go.jp
4 Department of Materials Science and Engineering, School of Materials and Chemical Technology, Tokyo Institute of Technology, 2-12-1, Ookayama, Meguro-ku, Tokyo 152-8550, Japan; hayamizu.y.aa@m.titech.ac.jp
5 Research Institute for Electronic Science (RIES), Hokkaido University, Kita 10 Nishi 20, Kita Ward, Sapporo 001-0020, Japan; hiroshi.ujii@es.hokudai.ac.jp
6 Division of Information Science and Technology, Graduate School of Information Science and Technology, Hokkaido University, Kita 14 Nishi 9, Kita Ward, Sapporo 001-0020, Japan
* Correspondence: kuroda@tokyo-med.ac.jp (M.K.); kanekura@tokyo-med.ac.jp (K.K.)

Abstract: Membrane-less organelles (MLOs) are formed by biomolecular liquid–liquid phase separation (LLPS). Proteins with charged low-complexity domains (LCDs) are prone to phase separation and localize to MLOs, but the mechanism underlying the distributions of such proteins to specific MLOs remains poorly understood. Recently, proteins with Arg-enriched mixed-charge domains (R-MCDs), primarily composed of R and Asp (D), were found to accumulate in nuclear speckles via LLPS. However, the process by which R-MCDs selectively incorporate into nuclear speckles is unknown. Here, we demonstrate that the patterning of charged amino acids and net charge determines the targeting of specific MLOs, including nuclear speckles and the nucleolus, by proteins. The redistribution of R and D residues from an alternately sequenced pattern to uneven blocky sequences caused a shift in R-MCD distribution from nuclear speckles to the nucleolus. In addition, the incorporation of basic residues in the R-MCDs promoted their localization to the MLOs and their apparent accumulation in the nucleolus. The R-MCD peptide with alternating amino acids did not undergo LLPS, whereas the blocky R-MCD peptide underwent LLPS with affinity to RNA, acidic poly-Glu, and the acidic nucleolar protein nucleophosmin, suggesting that the clustering of R residues helps avoid their neutralization by D residues and eventually induces R-MCD migration to the nucleolus. Therefore, the distribution of proteins to nuclear speckles requires the proximal positioning of D and R for the mutual neutralization of their charges.

Keywords: liquid–liquid phase separation; membrane-less organelle; nuclear speckle; nucleolus

Citation: Miyagi, T.; Yamazaki, R.; Ueda, K.; Narumi, S.; Hayamizu, Y.; Uji-i, H.; Kuroda, M.; Kanekura, K. The Patterning and Proportion of Charged Residues in the Arginine-Rich Mixed-Charge Domain Determine the Membrane-Less Organelle Targeted by the Protein. *Int. J. Mol. Sci.* **2022**, *23*, 7658. https://doi.org/10.3390/ijms23147658

Academic Editor: Vladimir N. Uversky

Received: 13 June 2022
Accepted: 8 July 2022
Published: 11 July 2022

1. Introduction

Eukaryotic cells carry out a wide variety of biochemical reactions simultaneously. Some of these reactions lead to toxic byproducts and are mutually incompatible. Hence, these biochemical reactions occur in parallel in isolated compartments in typical lipid-membrane-bound organelles. In contrast, cells form dynamic ribonucleoprotein compartments to respond rapidly to various external and internal stimuli [1,2]. These ribonucleoprotein bodies, formed through the liquid–liquid phase separation (LLPS) of proteins and RNA, are known as membrane-less organelles (MLOs) [3]. MLOs such as nuclear speckles, stress granules, and the nucleolus are formed according to the requirements of the cell and are eliminated when unused. These structures can change their size, fluidity, and protein composition according to requirements, to mediate a dynamic cellular response. The most important factor in distinguishing between membrane-bound organelles and MLOs is the ability of material to move freely across the boundaries of the organelle. This is attributed to the presence or absence of a lipid membrane. In addition, due to this feature, disease-associated proteins such as amyotrophic lateral sclerosis (ALS)-causing *C9ORF72* dipeptides are often incorporated into MLOs and influence their dynamics [4]. MLOs have constituent proteins with low-complexity domains (LCDs) and/or intrinsically disordered regions (IDRs). The repetitive motifs of LCDs promote multivalent interaction, and the flexibility of IDRs makes them strong driving factors of phase separation [5,6]. Multivalent interactions occurring through LCD-IDRs promote the polymerization of macromolecules, leading to phase separation and MLO formation. However, the process of distribution of MLO-targeting proteins to specific MLOs remains elusive.

The localization of molecules to the nucleolus is the most widely studied process in MLO targeting. The consensus regarding the nucleolus localization signals (NoLS) is a sequence with a cluster of basic amino acids such as arginine (R) and lysine (K) [7,8]. Findings from the comparison of various NoLS and proteomic studies on the nucleolus suggest that NoLS do not contain any specific sequences that act as a recognition motif for binding to specific receptor molecules [9,10]; instead, they undergo high-affinity interactions with nucleolar molecules such as proteins and nucleic acids, thus facilitating nucleolar localization. However, the precise mechanism remains unknown. The targeting of nuclear speckles by proteins has also been investigated. A nuclear speckle is an MLO with an essential role in mRNA splicing, the splicing factor storage, and modification [11]. SON protein and serine/arginine repetitive matrix protein 2 (SRRM2), both natively unfolded proteins with large LCDs, have been identified as the scaffolding molecules for nuclear speckles [12]. Recently, the R-enriched mixed-charge domain (R-MCD) was shown to determine whether proteins will localize to nuclear speckles [13]. Interestingly, the combination of aspartate (D) and R in proteins triggers their incorporation into nuclear speckles, whereas the presence of glutamate (E) and K exerts a limited effect on protein localization. However, the role of the R-MCD structure–function relationship in protein distribution to nuclear speckles is yet to be fully characterized.

In this study, we first tested a series of artificial R-MCD variants with different periodicities or different proportions to determine how the charge position and net charge affect the distribution of proteins to the MLOs. When the positive and negative charges were separated, the localization of R-MCDs shifted from nuclear speckles to the nucleolus, even though the net charge remained neutral. Furthermore, we showed that the incorporation of R-MCD into MLOs was enhanced when their net charge was positive. We performed interactome analysis by proximity-dependent biotin labeling and showed that proteins from the nucleolar granular components (GCs) were enriched in the interactome of R-MCD with blocky positive/negative charges. Lastly, the R-MCD peptide with opposite charges segregated was found to be susceptible to phase separation and associated with RNA, poly-E peptide, and nucleophosmin (NPM1), a nucleolar protein. These results indicate that the net charge of the LCD and the position of the charged amino acids in the LCD determine the degree of phase separation and the localization of proteins to specific MLOs.

2. Results

2.1. Charge Patterning in the R-MCDs Influences Their Distribution to MLOs

To investigate how the position of oppositely charged residues in the R-MCDs affects their subcellular localization, we synthesized $(DR)_{50}$ variants with different charge distributions as models for R-MCD (Figure 1A). All $(DR)_{50}$ variants carried the same number of R ($n = 50$) and D ($n = 50$) residues, and the charge patterning was the only difference, as indicated by the charge patterning parameter κ (Figures 1A and S1) [14].

As previously reported, when overexpressed in HeLa cells, green fluorescent protein (GFP)-$(DR)_{50}$ colocalized with serine/arginine-rich splicing factor 1 (SRSF1), a nuclear speckle marker (Figure 1B) [15]. When we changed the charge patterning without changing the net charge, $(D_8R_8)_6$ tended to accumulate in the nucleolus, and $(D_{16}R_{16})_3$ localized almost exclusively to the nucleolus, as shown by a good Pearson correlation coefficient (PCC) with a nucleolar marker, NPM1-DsRed [16] (Figure 1B–F). Reportedly, the localization of proteins to the nucleolus requires the presence of basic amino acids such as R and K at high levels [7–9]. However, in the case of $(DR)_{50}$ variants, even though both $(DR)_{50}$ and $(D_{16}R_{16})_3$ had the same number of D and R residues and equal net charges, $(D_{16}R_{16})_3$ migrated to the nucleolus, whereas $(DR)_{50}$ accumulated in nuclear speckles. This suggests that uneven intramolecular charge distribution may be a key determinant of protein localization to MLOs. To elucidate the roles of the acidic and basic blocks of $(D_{16}R_{16})_3$ in the nucleolar localization, we tested the localization of contiguous acidic and basic amino acids. When we overexpressed R_{10}, R_{20}, K_{10}, or K_{20}, contiguous basic amino acid chains of different lengths showed nucleolar localization (Figure 2A,B). However, D_{10}, D_{20}, E_{10}, or E_{20} expressed in HeLa cells showed no specific localization to MLOs (Figure 2C,D), suggesting that the localization of $(D_{16}R_{16})_3$ to the nucleolus is owing to the presence of contiguous basic amino acids. These results indicate that the primary sequence of oppositely charged residues determines the MLO protein targets.

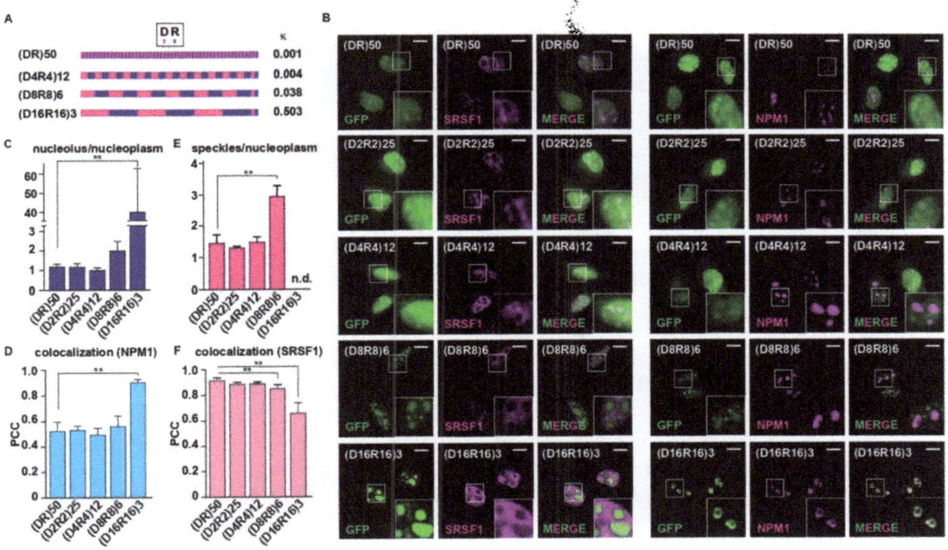

Figure 1. The periodicity of the $(DR)_{50}$ repeat determines the MLOs targeted by the R-MCD. (**A**) Schematic structures of $(DR)_{50}$ variants with different repeat periodicities. D_2R_2 was fused to the C-terminus of $(D_4R_4)_{12}$, $(D_8R_8)_6$, and $(D_{16}R_{16})_3$ to equalize the number of D and R residues. Kappa (κ: charge patterning parameter) was calculated by CIDER [6]. (**B**) Subcellular localization of GFP-fused R-MCD variants in HeLa cells. Nuclear speckles were visualized by co-expressed serine/arginine-rich splicing factor 1 (SRSF1)-mCherry, and the nucleolus was visualized by co-expressed nucleophosmin (NPM1)-DsRed. Scale bar = 10 μm. The inset images are high-magnification images of squared areas. (**C**) Ratio of GFP signals in the nucleolus/nucleoplasm, with n = 8~10 cells/condition. (**D**) The degree of colocalization between GFP-$(DR)_{50}$ variants and NPM1 quantified by the Pearson correlation coefficient (PCC). (**E**) The ratio of GFP signals in nuclear speckles/nucleoplasm. The ratio for $(D_{16}R_{16})_3$ was not determined because only a few cells showed speckle incorporation. N = 8~10 cells/condition. (**F**) The degree of colocalization between GFP-$(DR)_{50}$ variants and SRSF1, quantified by the PCC. The error bars show ± SD. The asterisks indicate significant differences obtained via one-way analysis of variance (ANOVA) with Dunnett's test; ** $p < 0.01$.

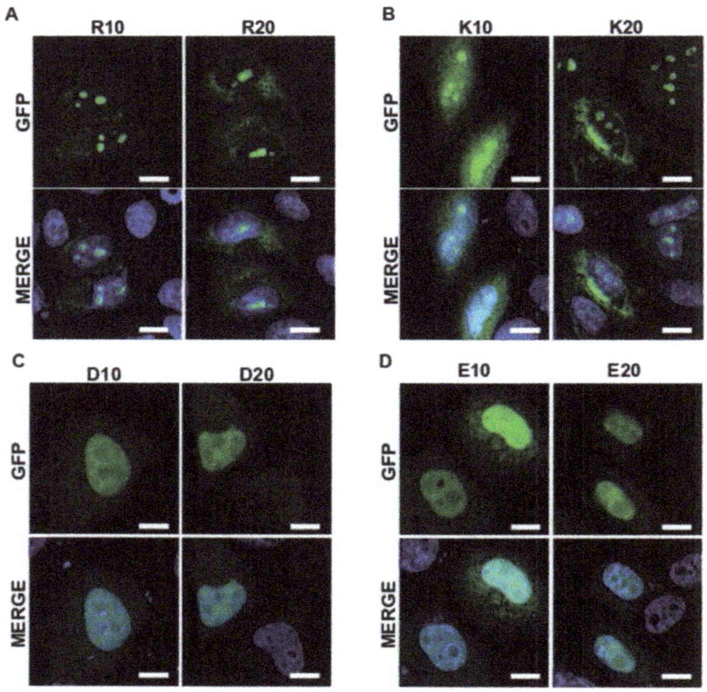

Figure 2. The presence of consecutive basic amino acids is responsible for the nucleolar distribution of proteins. (**A–D**) The subcellular localization of ten or twenty consecutive basic amino acids (R for (**A**) and K for (**B**)) or acidic amino acids (D for (**C**) and E for (**D**)) fused to the C-terminus of GFP expressed in HeLa cells. Scale bar = 10 μm.

2.2. Charge Patterning Affects the Interactome of R-MCDs

As indicated by the incorporation of $(DR)_{50}$ into the nuclear speckle and the localization of $(D_{16}R_{16})_3$ to the nucleolus, charge patterning affects the protein–protein interaction of R-MCDs. Since R and D are highly charged residues, conventional immunoprecipitation/proteomics analysis, which requires cell lysis, may lead to artificial interactions during the lysis/immunoprecipitation process. To investigate the protein–protein interactions while maintaining spatial information, we performed TurboID-mediated proximity biotin labeling of the proteomes in close proximity to each R-MCD (Figure 3A) [17,18]. The visualization of biotinylated proteins using AlexaFluor488-conjugated streptavidin revealed that TurboID-$(DR)_{50}$ exclusively biotinylated proteins in nuclear speckles, and TurboID-$(D_{16}R_{16})_3$ biotinylated the proteins in the nucleolus, as expected (Figure 3B). Biotinylated proteomes were analyzed using quantitative liquid chromatography–mass spectrometry (LC–MS) [19]. Signals from nucleolar markers and speckle markers showed that the segregation of oppositely charged residues in the R-MCD increased the cohesion of the nucleolar GC, where rRNA and proteins assemble in the ribosomal subunit, and reduced the cohesion to speckle markers (Figure 3C). Some nuclear and cytoplasmic MLOs have been shown to require RNA to maintain their integrity [20]. RNA has a negative charge and can interact with basic amino acids via electrostatic interactions, and RNA depletion by intranuclear expression of RNAse induces morphological changes in MLOs [20]. Furthermore, the inhibition of RNA transcription by actinomycin D (ActD) is known to cause the enlargement of nuclear speckles and dwarfing of the nucleolus [21–23]. A decrease in the levels of negatively charged RNAs alters the subcellular localization of proteins with a high basic amino acid content [24]. We investigated how the ActD-mediated inhibition of RNA synthesis affects the localization of R-MCD. The distribution of $(DR)_{50}$, which localizes to nuclear speckles, showed no significant change in colocalization with SRSF1 after ActD treatment (Figure S2A–C). However, the distribution of $(D_{16}R_{16})_3$, which localizes to the nucleolus, was significantly affected upon ActD treatment and showed a granular pattern in the nucleoplasm (Figure S2D–F), suggesting that electrostatic interaction of $(D_{16}R_{16})_3$ with RNA may be one of the drivers of its distribution to the nucleolus.

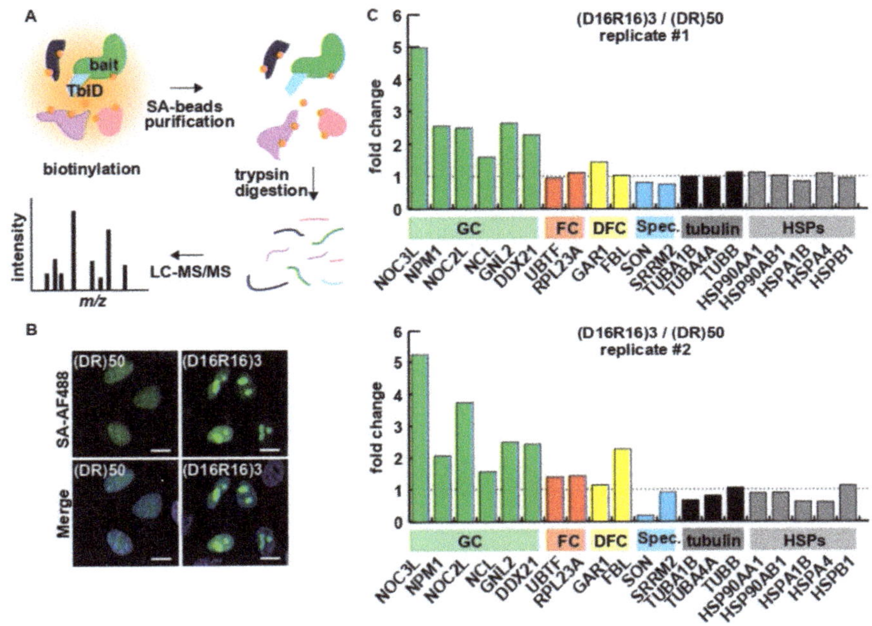

Figure 3. Proximity labeling analysis revealed that charge patterning affects the interactome of R-MCDs. (**A**) Schematic illustration of TurboID (TbID)-mediated proximity biotin labeling. $(DR)_{50}$ or $(D_{16}R_{16})_3$ was used as the bait in the experiment. SA: streptavidin. (**B**) Protein biotinylation by TurboID-$(DR)_{50}$ or TurboID-$(D_{16}R_{16})_3$, followed by visualization with AlexaFluor488-labeled streptavidin (SA-AF488). Scale bar = 10 µm. (**C**) Fold change in the signals of markers for the granular component (GC), fibrillar center (FC), or dense fibrillar component (DFC) of the nucleolus, nuclear speckles (Spec.), tubulin species, and heat-shock proteins (HSPs). Proximity labeling analysis followed by liquid chromatography–tandem mass spectrometry (LC–MS/MS) was performed twice for independently prepared sample sets.

2.3. A Net Positive Charge in R-MCDs Enhances Their Incorporation into MLOs

Next, we examined how the net charge of an R-MCD affects its subcellular localization. To this end, we synthesized R-MCD variants containing the same number of R residues ($n = 50$) with different net charges (Figure 4A). The net charge per residue (NCPR) of these R-MCD variants ranged from −0.333 to 0.493. An increase in the NCPR through R up to +0.2 was shown to enhance the incorporation of $(DR)_{50}$ mutants into nuclear speckles [13]. GFP-$(D_2R_1)_{50}$ (NPCR = −0.333) showed a diffused distribution pattern and low cohesion with nuclear speckles (Figure 4B). When the proportion of R-MCD changed and the number of acidic amino acids was reduced to impart a net positive charge, the condensation of R-MCDs increased not only in nuclear speckles but also in the nucleolus (Figure 4B–G). This effect was more evident when the net charge increased, which suggests that an increase in the net positive charge of R-MCDs enhances their condensation in MLOs (Figure 4C,E). Note that when the net charge of R-MCDs was negative, their nuclear localization ceased and they diffused into the nucleus and cytoplasm (Figure 4G).

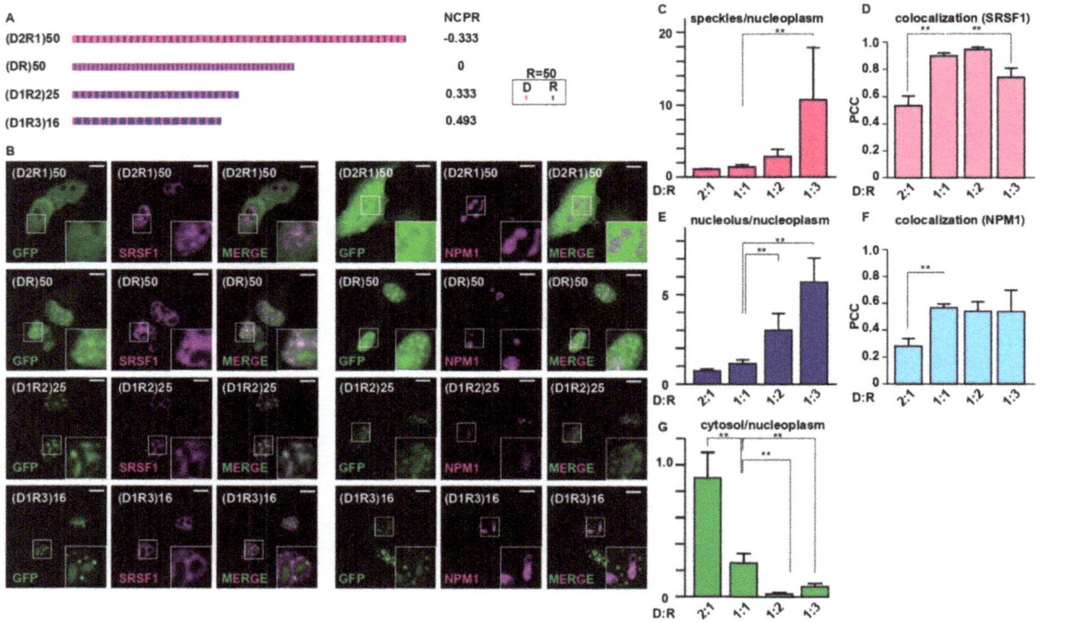

Figure 4. Imparting a net positive charge to the R-MCD enhances its ability to target MLOs. (**A**) Schematic structures of $(DR)_{50}$ proportional variants with different net charges per residue (NCPR) [6]. (**B**) Subcellular localization of GFP-fused $(DR)_{50}$ proportional variants in HeLa cells. Nuclear speckles were visualized by co-expressed SRSF1-mCherry, and the nucleolus was visualized by co-expressed NPM1-DsRed. Scale bar = 10 μm. The inset images are high-magnification images of squared areas. (**C**) The ratio of GFP signals in nuclear speckles/nucleoplasm, with n = 10~12 cells/condition. (**D**) The degree of colocalization between GFP-$(DR)_{50}$ proportional variants and SRSF1 quantified by PCC. (**E**) The ratio of GFP signals in the nucleolus/nucleoplasm, with n = 10~12 cells/condition. (**F**) The degree of colocalization between GFP-$(DR)_{50}$ proportional variants and NPM1 quantified by PCC. (**G**) The ratio of GFP signals in the cytosol/nucleoplasm, with n = 10~12 cells/condition. The error bars show ± SD. The asterisks indicate significant differences derived using one-way ANOVA with Dunnett's test; ** $p < 0.01$.

2.4. Charge Patterning in Natural R-MCDs Determines Protein Distribution to Specific MLOs

Next, we tested the effect of charge patterning on the R-MCDs of natural nuclear speckle-associated proteins. Wild-type (wt) full lengths of small nuclear ribonucleoprotein U1 subunit 70 (SNRNP70-FL) and wt-SNRNP70-MCD were localized to nuclear speckles (Figure 5A–C). When oppositely charged amino acids were segregated without changing the net charge, the site of localization shifted from the nuclear speckles to the nucleolus. Similarly, the site of localization of the full-length protein and MCD of negative elongation factor complex member E (NELFE) also changed from the nuclear speckles to the nucleolus when the distribution of charged amino acids was changed (Figure 5D–F). These results show that the charge patterning effects observed in the artificial R-MCD are similar to those in natural proteins and are a means of controlling the localization of proteins to MLOs in cells.

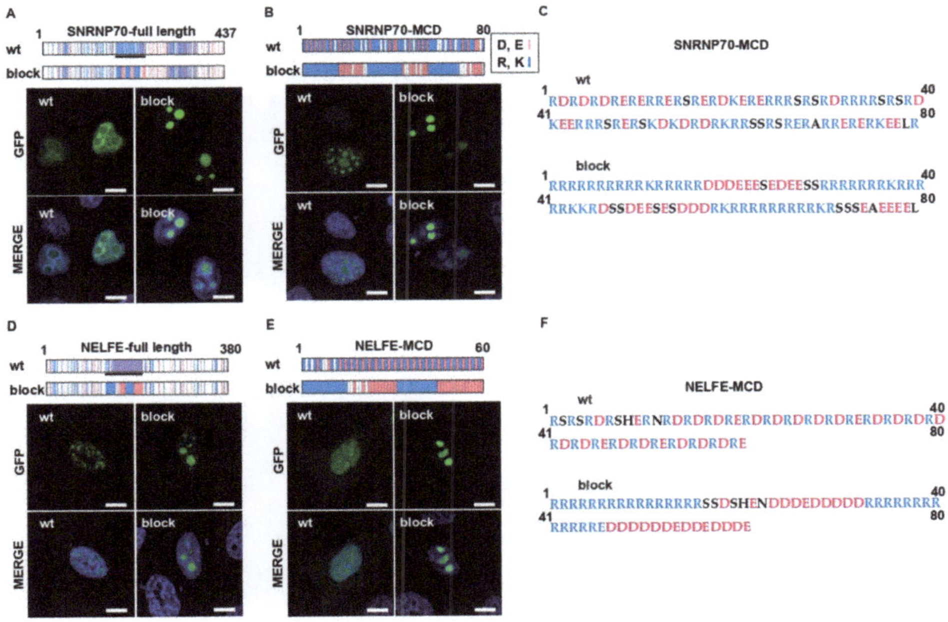

Figure 5. Charge patterning in a natural R-MCD determines the distribution of proteins to specific MLOs. (**A**) The schemes (top) show the structures of wild-type (wt) full-length SNRNP70 or full-length SNRNP70 with the R-MCD of blocky charges (block). The pink lines indicate acidic amino acids (D and E), and the blue lines indicate basic amino acids (R and K). The black line represents the R-MCD analyzed in this study. The lower panels show the localization of GFP-wt-SNRNP70 or GFP-block-SNRNP70. (**B**) The structure and subcellular localization of the wt-R-MCD of SNRNP70 or the block-R-MCD of SNRNP70. (**C**) The amino acid sequence of the wt or block of SNRNP70-MCD. (**D**) The structure and subcellular localization of wt full-length NELFE (wt) or full-length NELFE with the block-R-MCD (block). (**E**) The structure and subcellular localization of the wt-R-MCD of NELFE or the block-R-MCD of NELFE. (**F**) The amino acid sequence of the wt or block of NELFE-MCD. Scale bar = 10 μm.

2.5. Charge Patterning Determines the Phase-Separating Properties of R-MCD Peptides

The molecular mechanism by which charge segregation in the MCD causes a shift in localization from nuclear speckles to the nucleolus remains unknown. We synthesized $(DR)_{12}$ mutants with different charge patterns and tested their biochemical characteristics with respect to phase separation (Figure 6A). Among the mutants, $D_{12}R_{12}$ underwent simple coacervation and formed droplets (Figure 6B). $(DR)_{12}$ and $(D_4R_4)_3$ did not undergo phase separation, even when their concentrations were increased (Figure 6C). When a fluorescently labeled polyE peptide or RNA (rA15) was added to the droplets formed by $D_{12}R_{12}$, these acidic molecules were incorporated into the droplets (Figure 6D). Furthermore, when we added the $(DR)_{12}$ mutants to recombinant NPM1 protein, a nucleolar protein with acidic motifs, $D_{12}R_{12}$ underwent phase separation, but other peptides did not (Figure 6E). Thus, the positioning of opposite charges in an alternating pattern in the R-MCDs causes mutual charge neutralization and leads to the loss of affinity of R-MCDs for neighboring molecules. The marginal unevenness in local charge is important for the localization of proteins to nuclear speckles (Figure 6F). However, the segregation of oppositely charged residues reduces the likelihood of neutralization of positive and negative charges, and as a result, positively charged amino acids may bind to RNAs in a loop-like structure and localize to the nucleolus. To prove that the electrostatic force is retained more strongly in $(D_{16}R_{16})_3$ compared with $(DR)_{50}$, we challenged the HeLa cells expressing GFP-$(DR)_{50}$ or

GFP-$(D_{16}R_{16})_3$ with a digitonin solution containing different concentrations of NaCl. After cell permeabilization, $(D_{16}R_{16})_3$ localized to the nucleolus even in the presence of 200 mM NaCl, but $(DR)_{50}$ lost its localization to nuclear speckles in the presence of 100 mM NaCl, suggesting that $(D_{16}R_{16})_3$ binds to the surrounding molecules via stronger electrostatic forces than $(DR)_{50}$ (Figure S3A,B).

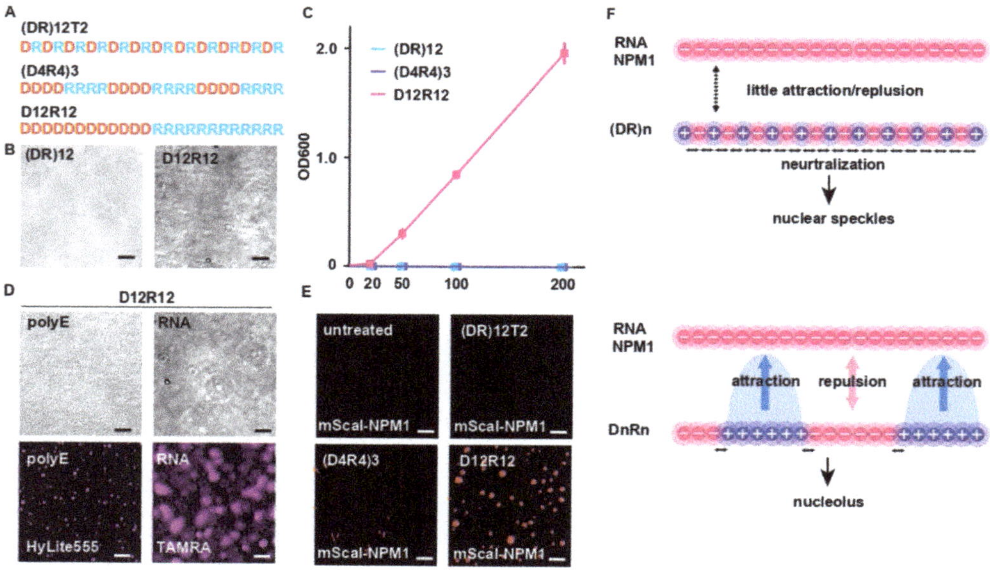

Figure 6. Charge patterning determines the phase-separation properties of R-MCD peptides. (**A**) The structure of $(DR)_{12}$ variants with different periodicities. (**B**) $D_{12}R_{12}$, but not $(DR)_{12}$, underwent simple coacervation. Scale bar: 10 μm. (**C**) The turbidity of $(DR)_{12}$ variants with different periodicities at different concentrations determined using the OD_{600}, with $n = 3$. The error bars show ± SD. (**D**) The $D_{12}R_{12}$ droplets mixed with 10 nM TAMRA-labeled rA15 RNA or 10 ng/μL of HyLite555-labeled polyE peptide. (**E**) Recombinant mScarlet-I-NPM1 protein mixed with each $(DR)_{12}$ variant peptide (10 μM). (**F**) Scheme of intramolecular neutralization by oppositely charged amino acids in the vicinity of (DR)n and DnRn.

3. Discussion

MLOs form and disappear dynamically in the nucleus and cytoplasm and play essential roles in cellular homeostasis. The appropriate distribution of each MLO-specific component to the designated MLOs is essential for the maintenance of their functions, but the mechanism underlying their distribution is unknown. In this study, we showed that the charge pattern in R-MCD determines the phase separation of proteins to MLOs and that segregation of residues with opposite charges influences the interactome of R-MCD, promoting phase separation and shifting their distribution from nuclear speckles to the nucleolus.

The $(DR)_{50}$ variants with different periodicities used in this study were fused with enhanced GFP and had an isoelectric point of 5.6. Although EGFP and EGFP-$(DR)_{50}$, with different periodicities, have similar isoelectric points (pI = 5.6), the EGFP-$(DR)_{50}$ variants are localized to nuclear speckles or the nucleolus. Thus, even though the net charge of the R-MCD is zero, the abundance of basic amino acids acts as a nuclear-localizing signal and induces nuclear protein distribution.

Nuclear speckles are the nuclear MLOs with an important role in mRNA splicing, storage, maturation, and splicing-factor modification [25]. The mechanism of nuclear speckle formation remains unknown. Recently, SRRM2 and SON were identified as speckle scaffold proteins [12]. While SRRM2 and SON were shown to be essential for nuclear speckle formation, the process by which SRRM2 and SON form nuclear speckles has not been determined. Many splicing factors that localize to nuclear speckles harbor D/R or S/R repeats [26–28]. The D/R and S/R repeats are reportedly essential for the localization of factors to nuclear speckles [13]. Poly(PR) and poly(GR) dipeptides produced from the mutant *C9ORF72* gene, which is associated with ALS [29,30], are structurally similar to $(DR)_{50}$ in that they contain alternating R residues; however, their net charge is positive and their targeted MLO is the nucleolus [31], suggesting that the neutralization of the charge on R by neighboring acidic amino acids is important for the localization of factors to nuclear speckles. This is consistent with the fact that when phosphorylated, serine residues in the S/R repeat sequence act as acidic amino acids, to neutralize R-MCDs and induce their localization to nuclear speckles [13].

Peptides with an alternating sequence of acidic and basic amino acids are known as zwitterionic peptides. Zwitterionic peptides are easily hydrated and interact poorly with proteins and charged molecules [32,33]. Because many strongly charged molecules such as RNA are present in the nucleus, natural R-MCDs use their D/R-alternating zwitterionic structure to reduce their susceptibility to the strong attractive force of RNA and localize to nuclear speckles. Conversely, once the oppositely charged residues are segregated and the local charge unevenness is established, the zwitterionic peptide loses its properties and accumulates in the nucleolus by changing its molecular binding mode. Although both the nucleolus and nuclear speckle contain RNA, the nucleolus contains much more abundant RNA than the nuclear speckles, as shown by strong signals when intracellular RNA is stained [34], and the abundant RNA-derived electrostatic forces might attract the blocky charged sequences, which $(DR)_{50}$ can escape. On the other hand, the mechanism of how $(DR)_{50}$ specifically condenses to nuclear speckles remains unclear. The zwitterionic structure may be advantageous for specific interaction with SON, SRRM2, or other components of nuclear speckles. The intra-organellar environmental differences such as hydrophobicity and density of aromatic compounds might also have an influence. Supporting the idea that the $(D_{16}R_{16})_3$ has stronger electrostatic interactions than $(DR)_{50}$, $(D_{16}R_{16})_3$ remained in the nucleolus when cells were permeabilized with digitonin and treated with salts that shield electrostatic forces, whereas $(DR)_{50}$ lost its localization to the nuclear speckle (Figure S3). Nevertheless, the possibility that blocked charged residues localize to the nucleolus by interactions with specific nucleolar molecules cannot be ruled out. It also remains unclear why a net positive charge in $(DR)_{50}$ variants enhances their incorporation into nuclear speckles as well as the nucleolus. The NCPR of $(D_1R_3)_{50}$ is 0.493 (Figure S1), which is similar to that of the ALS-causing *C9ORF72*-encoded R-rich dipeptides, poly(GR) and poly(PR) (NPCR = 0.5 for both), which exclusively localize to the nucleolus [31]. Therefore, we believe that the positioning of the anionic D residue next to the cationic R residue is necessary for protein localization to the speckle; however, further studies are warranted.

Only few studies have reported the role of the charge pattern of R-MCD in the protein localization to MLOs. Greig et al. reported that modest unevenness of charge distribution in the R-MCD enhances cohesion for both the nucleolus and nuclear speckles, and also a shift in the net charge of R-MCD to positive increases the size of the nuclear speckle and triggers retention of mRNA in the nucleus [13]. Therefore, the charge distribution of the R-MCD plays an important role not only in the subcellular localization but also in the protein function.

Research on the mechanism of localization to each MLO has only commenced recently. MLOs are involved in various intracellular functions, and their disruption has been implicated in various diseases such as neurodegenerative diseases and malignant tumors. However, only a few drugs target MLOs. Klein et al. recently showed that small molecules such as cisplatin are incorporated into the droplets of super-enhancers; however, their controlled delivery to specific MLOs is yet to be achieved [35]. The targeting of specific MLOs by charge patterning, as shown here, may help develop novel drug delivery systems (DDS) in the future.

In conclusion, we found that charge patterning is important for the targeted distribution of proteins in specific MLOs and that it regulates the protein interactome as well as the manner of phase separation in MCDs. We believe that these data are of significance from both biological and medical perspectives and will help develop DDS that can target specific MLOs.

4. Materials and Methods

4.1. Recombinant Plasmids Construction

All MCD constructs and cDNAs for SNRNP70 (full length and MCD), NELFE (full length and MCD), and TurboID were synthesized and subcloned into pcDNA3.1-N-eGFP vector or pcDNA3.1 vector (for TurboID) by GenScript (Piscataway, NJ, USA). NPM1-DsRed was a gift from Mary Dasso (Addgene plasmid # 34,553; http://n2t.net/addgene:34553 (accessed on 6 July 2022); RRID:Addgene_34553). For bacterial expression, human NPM1 tagged with mScarlet-I cDNA was synthesized and subcloned into pET28a vector by GenScript. Human SRSF1 cDNA was PCR-amplified from a HeLa-cell cDNA pool with a sense primer (GGATCCATGTCGGGAGGTGGTGTG) and an antisense primer (GAATTCTTATGTACGAGAGCGAGATCTG) using PrimeSTAR Max DNA polymerase (Takara Bio, Tokyo, Japan) and subcloned into pmCherry-C1 vector (Clontech, Mountain View, CA, USA).

4.2. Transfection of HeLa Cells and Image Acquisition by Confocal Microscopy

HeLa cells (RIKEN BRC, Tsukuba, Japan) plated on a chambered glass slide (Matsunami, Osaka, Japan) were transfected with each GFP-MCD construct in the presence or absence of an MLO marker using Lipofectamine 2000 (Thermo Fisher Scientific, Waltham, MA, USA), in accordance with the manufacturer's instructions. Twenty-four hours after transfection, the cells were fixed with formalin-PBS, and coverslips were mounted using ProLong Gold-DAPI antifade reagent (Thermo Fisher Scientific). For ActD treatment, the transfected HeLa cells were treated with 100 nM of ActD (Sigma-Aldrich, St. Louis, MO, USA) for the indicated period. For the digitonin permeabilization assay, the transfected HeLa cells were treated with a buffer containing 10 mM HEPES (pH 7.4), 0.1% digitonin (Nacalai, Kyoto, Japan), 300 mM sucrose, and 100 mM or 200 mM NaCl for 10 min before fixation. The cells were imaged with FV10i confocal microscopy (Olympus, Tokyo, Japan). Image analyses were performed using the ImageJ Fiji software (http://rsbweb.nih.gov/ij, accessed on 6 July 2022).

4.3. Peptide Synthesis and Purification of Recombinant mScarletI-NPM1

All $(DR)_{12}$ variant peptides were synthesized by GenScript. Trifluoroacetic acid was substituted with acetic acid. For the expression of the recombinant mScarlet-I-NPM1 protein, *Escherichia coli* BL21 cells were transformed with pET28a-mScarlet-I-NPM1, and expression of NPM1 was induced by incubation of the cells in 1mM IPTG for 16 h at 25 °C. The recombinant protein was purified using a HisTalon gravity column (Clontech) and dialyzed with Tris-buffered saline.

4.4. Phase Separation of Peptides

The $(DR)_{12}$ or $D_{12}R_{12}$ peptide (1 mM) was dissolved in water and diluted in a phase separation buffer (composed of 10 mM HEPES (PH7.4) and 100 mM NaCl) at 100 μM. The droplets were observed by FV10i confocal microscopy, and the OD_{600} value was measured using a NanoDrop One (Thermo Fisher). The droplets were mixed with 10 nM TAMRA-labeled rA15 RNA (Fasmac, Kanagawa, Japan) or 10 ng/μL of HyLite555-labeled polyE peptide [34]. For phase separation with mScarlet-I-NPM1, 1 μM mScarlet-I-NPM1 was mixed with each $(DR)_{12}$ variant peptide (10 μM) in the phase separation buffer.

4.5. Proximity Labeling and Quantitative LC–MS

TurboID-mediated proximity labeling was performed as previously described [17]. Briefly, HeLa cells plated on a 15 cm dish were transfected with TurboID-$(DR)_{50}$ or TurboID-$(D_{16}R_{16})_3$. Twenty-four hours after transfection, the cells were incubated with 50 μM biotin for 2 h. After the cells were washed four times with ice-cold PBS, they were lysed using RIPA buffer containing cOmplete protease inhibitor cocktail (Roche). Excess biotin was removed by ultrafiltration using an Amicon Ultra filter. For purifying biotinylated proteins, cell lysates were incubated with Dynabeads MyOne streptavidin beads (Thermo Fisher Scientific) in PBS for 2 h. The beads were sequentially washed with RIPA buffer, 1 M KCl, 0.1 M Na_2CO_3, 2 M urea in 10 mM Tris-HCl (pH 7.5), and RIPA buffer. The proteins were eluted by boiling with 2× Laemmli sample buffer (Bio-Rad, Hercules, CA, USA) and subjected to quantitative LC–MS [19]. Briefly, the extracted samples were reduced with 10 mM TCEP at 100 °C for 10 min, alkylated with 50 mM iodoacetamide at room temperature for 45 min, and separated using SDS-PAGE. Protein bands, visualized via Coomassie brilliant blue staining, were excised, destained, and cut finely before in-gel digestion with Trypsin/Lys-C Mix (Promega) at 37°C for 12 h. The digested peptides were analyzed with an Orbitrap Fusion Lumos mass spectrometer (Thermo Fisher Scientific) in combination with an UltiMate 3000 RSLC nano-flow HPLC system (Thermo Fisher Scientific) in the HCD MS/MS mode. Peptides were identified and quantified using Proteome Discoverer 2.4 software (Thermo Fisher Scientific), where the MS/MS spectra were searched against the *Homo sapiens* protein database in SwissProt (https://www.uniprot.org/, accessed on 6 July 2022), with a false discovery rate of 1% as an identification cutoff.

4.6. Colocalization Analysis

The colocalization analysis of GFP-$(DR)_{50}$ variants with NPM1 or SRSF1 was performed with Fiji ImageJ software using the EzColocalization plugin [36], and the PCC for each combination was calculated.

4.7. Bioinformatic Analysis of Biochemical Properties of Peptides and Proteins

The biochemical properties of peptides and proteins used in this study were analyzed using the CIDER web server (http://pappulab.wustl.edu/CIDER/analysis/, accessed on 6 July 2022) [6]. The following components were analyzed and are shown in Figure S1: FCR, fraction of charged residues; NCPR, net charge per residue; Kappa (κ), charge patterning parameter; and hydropathy, the Kyte–Doolittle hydropathy score for the sequence on a scale of 0–9.

4.8. Statistic Analysis

Data are represented as means ± standard deviations. Statistical analysis of the data was performed using SPSS software 28 (IBM).

Supplementary Materials: The following supporting information can be downloaded at: https://www.mdpi.com/article/10.3390/ijms23147658/s1.

Author Contributions: Conceptualization, T.M. and K.K.; data curation, T.M., K.U. and K.K.; formal analysis, T.M. and K.K.; funding acquisition, S.N., Y.H., H.U.-i., M.K. and K.K.; investigation, T.M., R.Y., K.U. and K.K.; methodology, T.M., K.U., S.N., Y.H., H.U.-i. and K.K.; project administration, K.K.; resources, K.K.; supervision, M.K. and K.K.; validation, K.K.; visualization, K.K.; writing—original draft, T.M. and K.K.; writing—review and editing, T.M., S.N., Y.H., H.U.-i., M.K. and K.K. All authors have read and agreed to the published version of the manuscript.

Funding: This work was financially supported by the JSPS KAKENHI (20H03593 to K.K., 21H02706 to M.K., 21H04634 to H.U., 19H03627 to S.N., 20H02564 to Y.H.).

Institutional Review Board Statement: Not applicable.

Informed Consent Statement: Not applicable.

Data Availability Statement: The raw data supporting the conclusions of this article will be provided by the authors without undue reservation.

Acknowledgments: We thank Mary Dasso for providing us with NPM1-DsRed (Addgene: # 34553).

Conflicts of Interest: The authors declare no conflict of interest.

References

1. Su, Q.; Mehta, S.; Zhang, J. Liquid-liquid phase separation: Orchestrating cell signaling through time and space. *Mol. Cell* **2021**, *81*, 4137–4146. [CrossRef] [PubMed]
2. Wright, P.E.; Dyson, H.J. Intrinsically disordered proteins in cellular signalling and regulation. *Nat. Rev. Mol. Cell Biol.* **2015**, *16*, 18–29. [CrossRef] [PubMed]
3. Alberti, S.; Gladfelter, A.; Mittag, T. Considerations and Challenges in Studying Liquid-Liquid Phase Separation and Biomolecular Condensates. *Cell* **2019**, *176*, 419–434. [CrossRef] [PubMed]
4. Lee, K.-H.; Zhang, P.; Kim, H.J.; Mitrea, D.M.; Sarkar, M.; Freibaum, B.D.; Cika, J.; Coughlin, M.; Messing, J.; Molliex, A.; et al. C9orf72 Dipeptide Repeats Impair the Assembly, Dynamics, and Function of Membrane-Less Organelles. *Cell* **2016**, *167*, 774–788.e17. [CrossRef]
5. Babinchak, W.M.; Surewicz, W.K. Liquid-Liquid Phase Separation and Its Mechanistic Role in Pathological Protein Aggregation. *J. Mol. Biol.* **2020**, *432*, 1910–1925. [CrossRef]
6. Holehouse, A.S.; Das, R.K.; Ahad, J.N.; Richardson, M.O.; Pappu, R.V. CIDER: Resources to Analyze Sequence-Ensemble Relationships of Intrinsically Disordered Proteins. *Biophys. J.* **2017**, *112*, 16–21. [CrossRef]
7. Martin, R.M.; Ter-Avetisyan, G.; Herce, H.D.; Ludwig, A.K.; Lattig-Tunnemann, G.; Cardoso, M.C. Principles of protein targeting to the nucleolus. *Nucleus* **2015**, *6*, 314–325. [CrossRef]
8. Musinova, Y.R.; Kananykhina, E.Y.; Potashnikova, D.M.; Lisitsyna, O.M.; Sheval, E.V. A charge-dependent mechanism is responsible for the dynamic accumulation of proteins inside nucleoli. *Biochim. Biophys. Acta* **2015**, *1853*, 101–110. [CrossRef]
9. Emmott, E.; Hiscox, J.A. Nucleolar targeting: The hub of the matter. *EMBO Rep.* **2009**, *10*, 231–238. [CrossRef]
10. Stenstrom, L.; Mahdessian, D.; Gnann, C.; Cesnik, A.J.; Ouyang, W.; Leonetti, M.D.; Uhlen, M.; Cuylen-Haering, S.; Thul, P.J.; Lundberg, E. Mapping the nucleolar proteome reveals a spatiotemporal organization related to intrinsic protein disorder. *Mol. Syst. Biol.* **2020**, *16*, e9469. [CrossRef]
11. Galganski, L.; Urbanek, M.O.; Krzyzosiak, W.J. Nuclear speckles: Molecular organization, biological function and role in disease. *Nucleic Acids Res.* **2017**, *45*, 10350–10368. [CrossRef]
12. Ilik, I.A.; Malszycki, M.; Lubke, A.K.; Schade, C.; Meierhofer, D.; Aktas, T. SON and SRRM2 are essential for nuclear speckle formation. *Elife* **2020**, *9*, e60579. [CrossRef]
13. Greig, J.A.; Nguyen, T.A.; Lee, M.; Holehouse, A.S.; Posey, A.E.; Pappu, R.V.; Jedd, G. Arginine-Enriched Mixed-Charge Domains Provide Cohesion for Nuclear Speckle Condensation. *Mol. Cell* **2020**, *77*, 1237–1250.e4. [CrossRef]
14. Das, R.K.; Pappu, R.V. Conformations of intrinsically disordered proteins are influenced by linear sequence distributions of oppositely charged residues. *Proc. Natl. Acad. Sci. USA* **2013**, *110*, 13392–13397. [CrossRef]
15. Tripathi, V.; Song, D.Y.; Zong, X.; Shevtsov, S.P.; Hearn, S.; Fu, X.D.; Dundr, M.; Prasanth, K.V. SRSF1 regulates the assembly of pre-mRNA processing factors in nuclear speckles. *Mol. Biol. Cell* **2012**, *23*, 3694–3706. [CrossRef]
16. Yun, C.; Wang, Y.; Mukhopadhyay, D.; Backlund, P.; Kolli, N.; Yergey, A.; Wilkinson, K.D.; Dasso, M. Nucleolar protein B23/nucleophosmin regulates the vertebrate SUMO pathway through SENP3 and SENP5 proteases. *J. Cell Biol.* **2008**, *183*, 589–595. [CrossRef]
17. Branon, T.C.; Bosch, J.A.; Sanchez, A.D.; Udeshi, N.D.; Svinkina, T.; Carr, S.A.; Feldman, J.L.; Perrimon, N.; Ting, A.Y. Efficient proximity labeling in living cells and organisms with TurboID. *Nat. Biotechnol.* **2018**, *36*, 880–887. [CrossRef]
18. Liu, F.; Morderer, D.; Wren, M.C.; Vettleson-Trutza, S.A.; Wang, Y.; Rabichow, B.E.; Salemi, M.R.; Phinney, B.S.; Oskarsson, B.; Dickson, D.W.; et al. Proximity proteomics of C9orf72 dipeptide repeat proteins identifies molecular chaperones as modifiers of poly-GA aggregation. *Acta Neuropathol. Commun.* **2022**, *10*, 22. [CrossRef]
19. Ueda, K.; Saichi, N.; Takami, S.; Kang, D.; Toyama, A.; Daigo, Y.; Ishikawa, N.; Kohno, N.; Tamura, K.; Shuin, T.; et al. A Comprehensive Peptidome Profiling Technology for the Identification of Early Detection Biomarkers for Lung Adenocarcinoma. *PLoS ONE* **2011**, *6*, e18567. [CrossRef]
20. Decker, C.J.; Burke, J.M.; Mulvaney, P.K.; Parker, R. RNA is required for the integrity of multiple nuclear and cytoplasmic membrane-less RNP granules. *EMBO J.* **2022**, *41*, e110137. [CrossRef]
21. Boulon, S.; Westman, B.J.; Hutten, S.; Boisvert, F.M.; Lamond, A.I. The nucleolus under stress. *Mol. Cell* **2010**, *40*, 216–227. [CrossRef] [PubMed]
22. Shav-Tal, Y.; Blechman, J.; Darzacq, X.; Montagna, C.; Dye, B.T.; Patton, J.G.; Singer, R.H.; Zipori, D. Dynamic sorting of nuclear components into distinct nucleolar caps during transcriptional inhibition. *Mol. Biol. Cell* **2005**, *16*, 2395–2413. [CrossRef] [PubMed]
23. Raina, K.; Rao, B.J. Mammalian nuclear speckles exhibit stable association with chromatin: A biochemical study. *Nucleus* **2022**, *13*, 58–73. [CrossRef] [PubMed]

24. Bursac, S.; Brdovcak, M.C.; Pfannkuchen, M.; Orsolic, I.; Golomb, L.; Zhu, Y.; Katz, C.; Daftuar, L.; Grabusic, K.; Vukelic, I.; et al. Mutual protection of ribosomal proteins L5 and L11 from degradation is essential for p53 activation upon ribosomal biogenesis stress. *Proc. Natl. Acad. Sci. USA* **2012**, *109*, 20467–20472. [CrossRef] [PubMed]
25. Spector, D.L.; Lamond, A.I. Nuclear speckles. *Cold Spring Harb. Perspect. Biol.* **2011**, *3*, a000646. [CrossRef] [PubMed]
26. Zheng, X.; Peng, Q.; Wang, L.; Zhang, X.; Huang, L.; Wang, J.; Qin, Z. Serine/arginine-rich splicing factors: The bridge linking alternative splicing and cancer. *Int. J. Biol. Sci.* **2020**, *16*, 2442–2453. [CrossRef] [PubMed]
27. Cazalla, D.; Zhu, J.; Manche, L.; Huber, E.; Krainer, A.R.; Caceres, J.F. Nuclear export and retention signals in the RS domain of SR proteins. *Mol. Cell. Biol.* **2002**, *22*, 6871–6882. [CrossRef]
28. Staknis, D.; Reed, R. Members of a family of proteins (the RD family) detected by a U1 70K monoclonal antibody are present in spliceosomal complexes. *Nucleic Acids Res.* **1995**, *23*, 4081–4086. [CrossRef]
29. Renton, A.E.; Majounie, E.; Waite, A.; Simón-Sánchez, J.; Rollinson, S.; Gibbs, J.R.; Schymick, J.C.; Laaksovirta, H.; van Swieten, J.C.; Myllykangas, L.; et al. A hexanucleotide repeat expansion in C9ORF72 is the cause of chromosome 9p21-linked ALS-FTD. *Neuron* **2011**, *72*, 257–268. [CrossRef]
30. DeJesus-Hernandez, M.; Mackenzie, I.R.; Boeve, B.F.; Boxer, A.L.; Baker, M.; Rutherford, N.J.; Nicholson, A.M.; Finch, N.A.; Flynn, H.; Adamson, J.; et al. Expanded GGGGCC Hexanucleotide Repeat in Noncoding Region of C9ORF72 Causes Chromosome 9p-Linked FTD and ALS. *Neuron* **2011**, *72*, 245–256. [CrossRef]
31. Kanekura, K.; Harada, Y.; Fujimoto, M.; Yagi, T.; Hayamizu, Y.; Nagaoka, K.; Kuroda, M. Characterization of membrane penetration and cytotoxicity of C9orf72-encoding arginine-rich dipeptides. *Sci. Rep.* **2018**, *8*, 12740. [CrossRef]
32. Ederth, T.; Lerm, M.; Orihuela, B.; Rittschof, D. Resistance of Zwitterionic Peptide Monolayers to Biofouling. *Langmuir* **2019**, *35*, 1818–1827. [CrossRef]
33. Beyer, C.D.; Thavalingam, S.; Guseva, T.; Schardt, L.; Zimmermann, R.; Werner, C.; Dietze, P.; Bandow, J.E.; Metzler-Nolte, N.; Rosenhahn, A. Zwitterionic Peptides Reduce Accumulation of Marine and Freshwater Biofilm Formers. *ACS Appl. Mater. Interfaces* **2021**, *13*, 49682–49691. [CrossRef]
34. Chen, C.; Yamanaka, Y.; Ueda, K.; Li, P.; Miyagi, T.; Harada, Y.; Tezuka, S.; Narumi, S.; Sugimoto, M.; Kuroda, M.; et al. Phase separation and toxicity of C9orf72 poly(PR) depends on alternate distribution of arginine. *J. Cell Biol.* **2021**, *220*, e202103160. [CrossRef]
35. Klein, I.A.; Boija, A.; Afeyan, L.K.; Hawken, S.W.; Fan, M.; Dall'Agnese, A.; Oksuz, O.; Henninger, J.E.; Shrinivas, K.; Sabari, B.R.; et al. Partitioning of cancer therapeutics in nuclear condensates. *Science* **2020**, *368*, 1386–1392. [CrossRef]
36. Stauffer, W.; Sheng, H.; Lim, H.N. EzColocalization: An ImageJ plugin for visualizing and measuring colocalization in cells and organisms. *Sci. Rep.* **2018**, *8*, 15764. [CrossRef]

International Journal of
Molecular Sciences

Article

The Multivalent Polyampholyte Domain of Nst1, a P-Body-Associated *Saccharomyces cerevisiae* Protein, Provides a Platform for Interacting with P-Body Components

Yoon-Jeong Choi [1], Yujin Lee [1], Yuxi Lin [2], Yunseok Heo [2], Young-Ho Lee [2,3,4] and Kiwon Song [1,*]

1 Department of Biochemistry, College of Life Science and Biotechnology, Yonsei University, Seoul 03722, Korea; yoon3004@yonsei.ac.kr (Y.-J.C.); 2019162016@yonsei.ac.kr (Y.L.)
2 Research Center for Bioconvergence Analysis, Korea Basic Science Institute (KBSI), Chungbuk 28119, Korea; linyuxi@kbsi.re.kr (Y.L.); uoonsek1@kbsi.re.kr (Y.H.); mr0505@kbsi.re.kr (Y.-H.L.)
3 Department of Bio-Analytical Science, University of Science and Technology (UST), Daejeon 34113, Korea
4 Graduate School of Analytical Science and Technology (GRAST), Chungnam National University (CNU), Daejeon 34134, Korea
* Correspondence: bc5012@yonsei.ac.kr; Tel.: +82-2-2123-2705; Fax: +82-2-362-9897

Abstract: The condensation of nuclear promyelocytic leukemia bodies, cytoplasmic P-granules, P-bodies (PBs), and stress granules is reversible and dynamic via liquid–liquid phase separation. Although each condensate comprises hundreds of proteins with promiscuous interactions, a few key scaffold proteins are required. Essential scaffold domain sequence elements, such as poly-Q, low-complexity regions, oligomerizing domains, and RNA-binding domains, have been evaluated to understand their roles in biomolecular condensation processes. However, the underlying mechanisms remain unclear. We analyzed Nst1, a PB-associated protein that can intrinsically induce PB component condensations when overexpressed. Various Nst1 domain deletion mutants with unique sequence distributions, including intrinsically disordered regions (IDRs) and aggregation-prone regions, were constructed based on structural predictions. The overexpression of Nst1 deletion mutants lacking the aggregation-prone domain (APD) significantly inhibited self-condensation, implicating APD as an oligomerizing domain promoting self-condensation. Remarkably, cells overexpressing the Nst1 deletion mutant of the polyampholyte domain (PD) in the IDR region ($Nst1_{\Delta PD}$) rarely accumulate endogenous enhanced green fluorescent protein (EGFP)-tagged Dcp2. However, $Nst1_{\Delta PD}$ formed self-condensates, suggesting that Nst1 requires PD to interact with Dcp2, regardless of its self-condensation. In $Nst1_{\Delta PD}$-overexpressing cells treated with cycloheximide (CHX), Dcp2, Xrn1, Dhh1, and Edc3 had significantly diminished condensation compared to those in CHX-treated Nst1-overexpressing cells. These observations suggest that the PD of the IDR in Nst1 functions as a hub domain interacting with other PB components.

Keywords: P-body; liquid–liquid phase separation; Nst1; polyampholyte domain; aggregation-prone domain; *Saccharomyces cerevisiae*

Citation: Choi, Y.-J.; Lee, Y.; Lin, Y.; Heo, Y.; Lee, Y.-H.; Song, K. The Multivalent Polyampholyte Domain of Nst1, a P-Body-Associated *Saccharomyces cerevisiae* Protein, Provides a Platform for Interacting with P-Body Components. *Int. J. Mol. Sci.* **2022**, *23*, 7380. https://doi.org/10.3390/ijms23137380

Academic Editor: Vladimir N. Uversky

Received: 23 May 2022
Accepted: 29 June 2022
Published: 2 July 2022

1. Introduction

The phenomenon of biomolecular phase separation has expanded our understanding of biomolecular condensation in cells [1]. Biomolecular condensates include many non-membranous cellular structures, such as Cajal bodies, nuclear speckles, histone-locus bodies, promyelocytic leukemia (PML) nuclear bodies (NB) in the nucleus [2–5], P-bodies (PBs), stress granules (SGs), and germ granules in the cytoplasm [1,2,4,6–9]. These membrane-less cellular structures are not random biomolecule mixtures. Some components are shared in different condensates, but each membraneless organelle contains a specific group of proteins and RNA/DNA, differentiating it from others. Not all of the condensate components are critical for inducing condensation, but a few components, so-called scaffolds, play crucial roles [10].

Ribonucleoprotein (RNP) granules are among the most representative biomolecular condensates and are an efficient model for studying biomolecular condensation in cells. RNA, the general component of these condensates, is essential for these condensation processes [11]. RNase treatment disperses isolated messenger RNPs (mRNPs) in vitro [11]. Additionally, decreased free ribosomal mRNA influx alleviates mRNP granule condensation in cells [12]. These phenomena strongly support the notion that RNA is a critical factor for molecular condensation.

mRNP granules also contain hundreds of proteins [10,13–17]. Scaffold proteins that function as nodes for protein–protein interaction networks are typically sufficient to form condensates [10]. Scaffold proteins have the intrinsic potential to induce condensation, while client elements are concentrated within the structure often by direct interactions with scaffolds but are not required for condensate formation [18]. The scaffold proteins show a few distinctive characteristics that distinguish them from client proteins. First, numerous scaffold proteins exhibit self-oligomerizing properties. The RING finger-B box-coiled coil (RBCC) motif [19–21] contains an N-terminal RING, B1-box, B2-box, and a C-terminal coiled-coil (CC) domain and is considered essential for PML oligomerization to form PML NBs [22–24]. In the case of Ras-GTPase-activating protein (SH3 domain)-binding protein (G3BP), the dimerization domain nuclear transport factor 2 (NTF2) is insufficient but necessary for SG formation [25].

Scaffold proteins also have multivalent regions that participate in weak interactions with numerous binding partners [26,27]. Intrinsically disordered regions (IDRs) with few three-dimensional (3D) structures and little specificity [28–33] are reportedly necessary for liquid–liquid phase separation (LLPS) dynamics and multivalency. IDRs of the heterogeneous RNP family, including hbRNPA1 in SGs [31], hbRNPA2B1 [33], RNA helicase Ddx4 in nuage [28], and Laf-1 in P-granules [34,35], are sufficient to mediate phase separation. Low-complexity domains (LCDs) [33,36] such as the poly-Q/N prion-like domain (PrD) and the arginine-glycine-rich (RGG) motif [4,34,37] are also known as critical modifiers for generating LLPS. The polyampholyte or polyelectrolyte region of the IDR may function as a sticker to promote LLPS [34]. Although previous research has established the link between IDR and multivalency, it has not elucidated the syntax of molecular condensation.

PBs of the budding yeast *Saccharomyces cerevisiae* provide an excellent system for studying the elements and mechanisms to form cellular condensates. The predominant components of yeast PBs are mRNA decapping protein Dcp1 and Dcp2, which constitute the decapping enzyme, enhancer of mRNA decapping protein 3 (Edc3), Pat1, Dhh1, and the Lsm1-7 complex, all of which are mRNA-binding proteins that stimulate mRNA decapping [37–39]. Predominantly, multivalent Edc3 interactions appear to drive PB formation. Edc3 serves as a scaffold for PB assembly, primarily under glucose deprivation when PB formation is robust. In a previous study, we identified Nst1 as a novel PB component. Nst1 accumulates in PBs more densely in stationary phase cells and under glucose deprivation. Ectopically overexpressed Nst1 is self-condensed and induces the condensation of other PB components, such as Dcp2, indicating that Nst1 has the intrinsic potential to self-condensate and accumulate other PB components [40].

Here, we dissected Nst1 by overexpressing various Nst1 domain deletion mutants to understand the functions of distinctive Nst1 sequence elements in its self-condensation and recruitment of other PB components and improve our knowledge of molecular condensation in cells.

2. Results

2.1. The Nst1 C-Terminal Domain (CTD) Contains Polyampholyte and Aggregation-Prone Regions

We previously reported that Nst1, similarly to Edc3, induced Dcp2 accumulation via self-condensation and physical interactions with other PB components [40]. These observations strongly suggest that Nst1 contains an oligomerizing domain similar to Edc3, with the intrinsic potential to drive self-condensation. Nst1 is a 141 kDa protein consisting of 1240 amino acids with a unique sequence distribution (Supplemental Figure S1A). To determine the properties of Nst1 in self-generating condensates and the induced condensation of other PB components, we analyzed the Nst1 sequence using multifaceted sequence prediction tools: protein structure prediction using GalaxyWEB (http://galaxy.seoklab.org/, accessed on 12 October 2018) (Supplemental Figure S1A–C), IDR prediction with IUPRED2A, PONDR, and DISOPRED3, and aggregation-prone region prediction with AGGRESCAN, Tango, and PASTA 2.0 (Figure 1A,B).

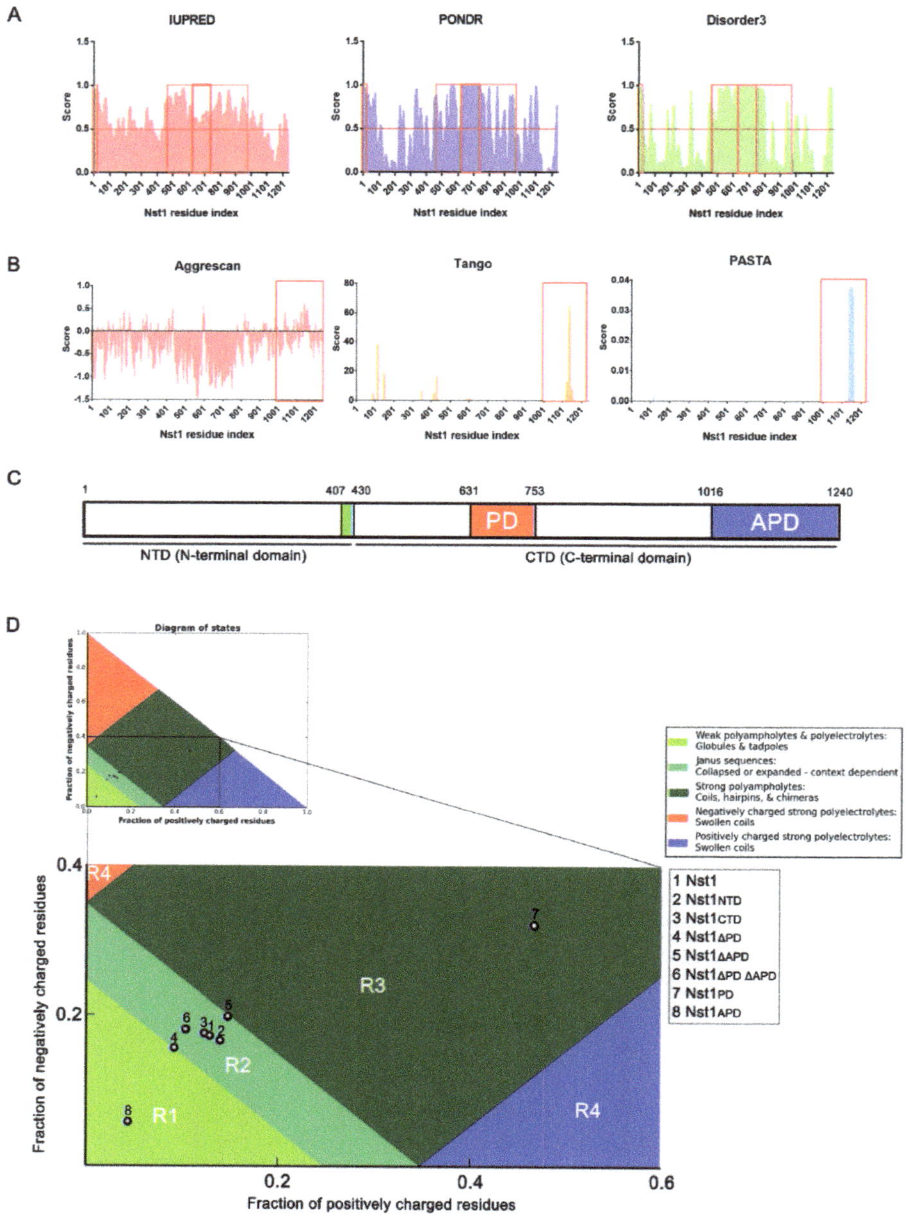

Figure 1. Nst1 contains a polyampholyte domain (PD) and a predicted aggregation-prone region. (**A**) Predicting the Nst1 intrinsically disordered regions (IDRs). IUPRED2A (red) [41], PONDR

(blue) [42], and DISOPRED3 (green) [43] algorithms were used for the prediction. Disorder scores were calculated and presented. Scores exceeding the 0.5 threshold indicate the amino acid residues in the Nst1 disordered regions. The disordered regions with scores >0.5 in all three algorithms used are identified as IDRs and highlighted in red-lined boxes. A length threshold for the disordered regions is also set to >30 residues [44]. The PD in predicted disordered regions is marked with thick red-lined boxes. PD corresponding residues are labeled in Supplemental Figure S1. (**B**) Predicting the Nst1 aggregation-prone regions. The regions with high aggregation propensities were calculated using AGGRESCAN (purple) [45], Tango (orange) [46], and PASTA 2.0 (blue) [47] algorithms. Amino acid residues with scores greater than the threshold value in all three algorithms were aggregation-prone and marked with a red box in the Nst1 sequence. Corresponding residues are labeled in Supplemental Figure S1. (**C**) A diagram of Nst1 with domain architectures predicted by (**A**,**B**), and GalaxyWEB. Each color in the schematic corresponds to a particular domain in the sequence. (**D**) Das–Pappu diagrams of the full-length Nst1 and its domain deletion mutants. The full-length Nst1 and its various domain deletion mutants are numbered in the box: 1. Nst1 (residues 1–1240), 2. $Nst1_{NTD}$ (N-terminal domain (NTD) Nst1 residues 1–429), 3. $Nst1_{CTD}$ (C-terminal domain (CTD) Nst1 residues 430–1240), 4. $Nst1_{\Delta PD}$ (residues 1–630 and 753–1240), 5. $Nst1_{\Delta APD}$ (residues 1–1015), 6. $Nst1_{\Delta PD \Delta APD}$ (residues 1–630 and 753–1015), 7. $Nst1_{PD}$ (residues 631–752), and 8. $Nst1_{APD}$ (residues 1016–1240). The x- and y-axes represent the fraction of positively and negatively charged residues, respectively. The four zones (R1–R4) of the diagram are colored in bright green (R1), emerald (R2), forest (R3), and red/blue (R4), respectively. The physicochemical properties of each colored zone are explained in the inset. The numbers for the full-length Nst1 and its domain deletion mutants are assigned to a corresponding region from R1–R4 with a circle.

The 980-amino acid sequence (from amino acid 131 to 1110), excluding 130 amino acids of each Nst1 N- and C-terminus, was analyzed because of the 1000 amino acid limit of GalaxyWEB (Supplemental Figure S1B). N-terminus (residues 1–429) (data not shown) and C-terminus (residues 430–1240) (Supplemental Figure S1C) structures were also predicted independently. Nst1 was expected to be low-ordered and to not form a globular 3D structure (Supplemental Figure S1B). Based on the prediction of the secondary structure by GalaxyWEB, Nst1 could be divided mainly into two domains: the N-terminal domain (NTD) (1–406) and CTD (430–1240), with a short 23-amino acid unstructured region (UR) between them (Figure 1C).

The 225 amino acids (residues 1016–1240) in the Nst1 C-terminus contain particularly high scores in aggregation propensity prediction (Figure 1B). Considering the aggregation propensity and protein secondary structure predictions, we designated this region as an aggregation-prone domain (APD) (Figure 1C and Figure S1C).

We found that amino acids 1–32 in the NTD and 491–980 in the CTD scored highly in all three IDR predictions (Figure 1A). The polyampholyte sequence, including charged amino acid clusters D, E, R, and K with sparse hydrophobic amino acid L, was embedded in the predicted IDR (Figure 1A and Figure S1A). Considering that the polyampholytic sequence was predicted as coiled-coil (CC) helices in the secondary structure prediction by GalaxyWEB, we designated this predicted region as the polyampholyte domain (PD) (Supplemental Figure S1B).

Polyampholyte sequences are commonly present in many IDRs [48]. Charged amino acids, such as D, E, K, and R, function as inter- and intra-molecular stickers to generate LLPS [34,49]. The molecular conformation of proteins can be deduced based on the fraction of the charged amino acids in the Das–Pappu diagram [50]. Fundamentally, the fraction of charged residues (FCR) and the net charge per residue (NCPR) determine the four regions, R1, R2, R3, and R4, in the Das–Pappu diagram. The proteins showing an FCR value smaller than 0.35 were assigned to R1 or R2. The protein sequences presented in R1 and R2 were expected to have a globular conformation of weak polyampholytes and an alternative globular conformation of a context-dependent polyampholyte, respectively.

Proteins with an FCR value greater than 0.35 were classified as R3 or R4. These protein sequences were strong polyampholytes or polyelectrolytes that were expected to be coiled. We projected the sequence of Nst1 and each Nst1 deletion mutant of the predicted domain onto the Das–Pappu diagram (Figure 1D). Full-length Nst1 (1) was projected in R2, where the fraction of negatively or positively charged residues was between 0.25 and 0.35. The zone of the context-dependent polyampholyte indicates that Nst1 may not have a stable globular protein structure, and its composition may be altered in a context-dependent manner. Additionally, the Nst1 N-terminal (Nst1$_{NTD}$) (2) and Nst1 C-terminal (Nst1$_{CTD}$) (3) projections were close to the full-length Nst1 in the same R2 region of the diagram. This indicated that the ratio of the charged Nst1$_{NTD}$ and Nst1$_{CTD}$ residues was similar to that of full-length Nst1. However, as expected due to its IDR predictions, Nst1$_{\Delta PD}$ (4), the mutant lacking the polyampholyte region, was projected on the border of R2 and R1, in which the fraction of negatively or positively charged residues was below 0.25. This Nst1$_{\Delta PD}$ prediction indicated that deleting the Nst1 polyampholyte region could severely alter the full-length Nst1 FCR. In contrast, Nst1$_{\Delta APD}$ (5) was projected onto the R3 zone, demonstrating that deleting the aggregation-prone domain (APD) increased the Nst1 FCR. The Nst1$_{\Delta PD\Delta APD}$ projection (6) showed the offset effect of PD and APD deletions. Collectively, these predictions suggest that the unique sequence distribution of Nst1, especially the PD and APD, may enhance Nst1 self-condensation and the condensation of other PB components.

2.2. The Nst1 CTD Is Sufficient for Nst1 Self-Condensation

To identify the specific regions of Nst1 responsible for the self-condensation and condensation of other PB components, we designed various Nst1 domain deletion mutants with different domain combinations based on the predictions (Figure 1). Each green fluorescent protein (GFP)-tagged Nst1 domain deletion mutant was overexpressed under the galactose inducible (*GAL*) promoter of pMW20 in the wild-type cells, and its expression was confirmed by Western blot analysis (Supplemental Figure S2A). As reported in a previous study [40], overexpressed GFP-tagged Nst1 formed bright puncta (Figure 2A). The Nst1 mutant, Nst1$_{NTD}$, was completely dispersed throughout the cytoplasm when overexpressed (Figure 2A). In contrast, Nst1$_{CTD}$ formed clear puncta upon overexpression (Figure 2A). These observations demonstrate that the Nst1 CTD was sufficient to form self-condensates upon overexpression (Figure 2A).

The puncta formed by Nst1 overexpression were closely correlated with the physical LLPS properties obtained via 1,6-hexanediol treatment [27]. 1,6-hexanediol is reported to eradicate the nuclear pore permeability barrier by interfering with hydrophobic interactions in the pores and is generally used to interfere with the integrity of reversible condensates with liquid-like properties [51,52]. In budding yeast cells, treatment with 5–10% 1,6-hexanediol for 30 min can impede PB integrity but cannot disperse irreversible amyloids [27]. When cells were treated with 1,6-hexanediol, we observed that condensates of overexpressed GFP-tagged Nst1$_{CTD}$ dispersed as those of full-length Nst1, exhibiting the liquid-like property of both condensates (Figure 2A).

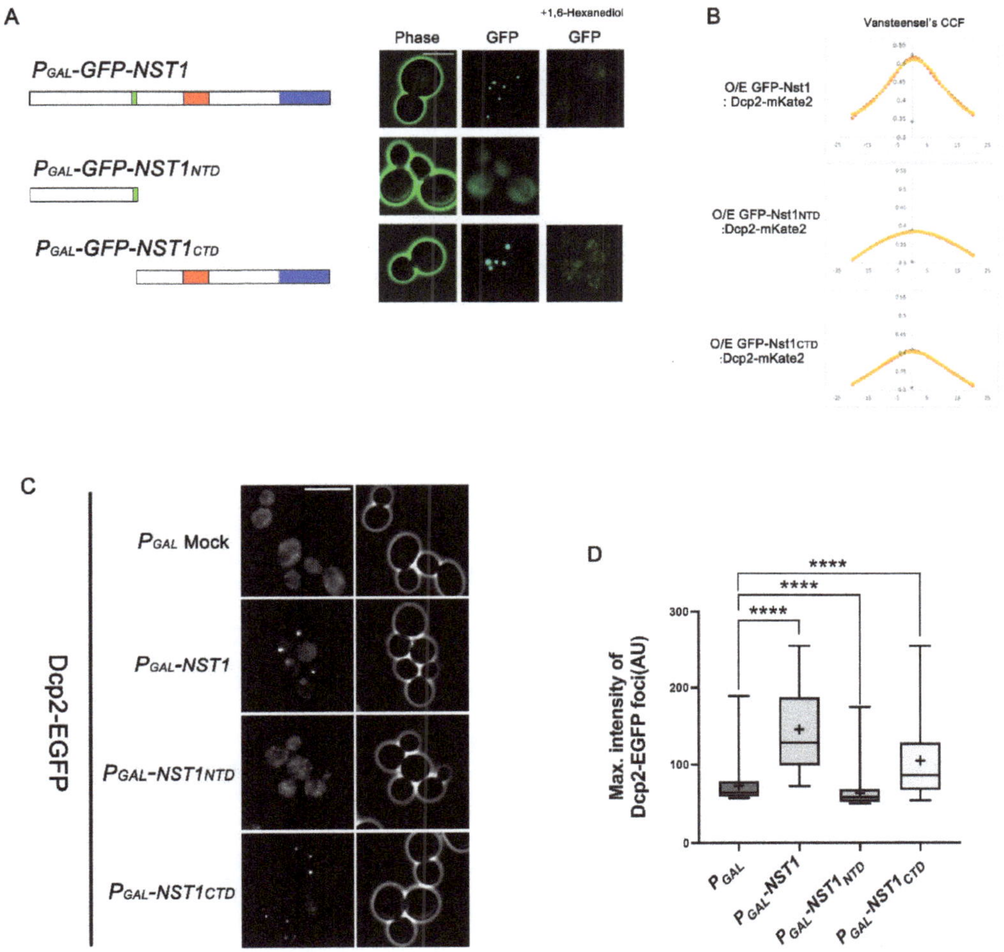

Figure 2. The Nst1 CTD is sufficient for Nst1 self-condensation. (**A**) Fluorescence microscopy of the cells overexpressing full-length enhanced green fluorescent protein (EGFP)-tagged Nst1 and the NTD ($Nst1_{NTD}$, 1429) and CTD ($Nst1_{CTD}$, 430–1240) of Nst1. Overexpression of each EGFP-tagged Nst1 deletion mutant was induced in wildtype cells, then observed before and after 1,6-hexanediol treatment. Scale bar: 5 μm. Schematic diagrams of the designed Nst1 domain deletion mutants are shown on the left. (**B**) The van Steensel's crosscorrelation coefficients (CCFs) between each overexpressed Nst1 deletion mutant used in (**A**). The endogenous mRNA decapping protein 2 (Dcp2)-mKate2 signals were analyzed and presented. Overexpression of each EGFP-tagged Nst1 domain deletion mutant was induced in wild-type cells whose chromosomal *DCP2* was tagged with mKate2. Each Nst1 domain deletion mutant (n = total observed cell number): P_{GAL}-*GFP-NST1* (n = 257), P_{GAL}-*GFP-NST1*$_{NTD}$ (n = 158), and P_{GAL}-*GFP-NST1*$_{CTD}$ (n = 161). All images were analyzed by FIJI (https://imagej.net/Fiji, accessed on 9 August 2020). (**C**,**D**) Each Nst1 deletion mutant was overexpressed in the wild-type cells with EGFP-tagged *DCP2* (YSK3485). (**C**) Fluorescence microscopy of cells expressing endogenous EGFP-Dcp2 that overexpress the NTD (1–429), CTD (430–1240), and full-length of Nst1. Scale bar: 10 μm. (**D**) Quantification of the endogenous EGFP-Dcp2 puncta analysis of (**C**). The pixels of the top 0.1% EGFP-Dcp2 signal intensities were segmented

for puncta analysis. The maximal intensity of each segmented punctum was plotted. '+' in the boxplot indicates the mean value of maximal intensities of foci. Each Nst1 domain deletion mutant (n = total observed cell number): P_{GAL} vector—only control (n = 218), P_{GAL}-*NST1* (n = 307), P_{GAL}-$NST1_{NTD}$ (n = 333), and P_{GAL}-$NST1_{CTD}$ (n = 300). All measurements and analyses were performed by FIJI (https://imagej.net/Fiji, accessed on 31 March 2022). Statistical significance was determined by a Mann–Whitney test **** $p < 0.0001$).

Previously, we demonstrated that the accumulated GFP-tagged Nst1 condensates co-localized with endogenous Dcp2-mKate2 [40]. To quantitatively investigate the correlation between Nst1 self-condensation and its association with PBs, we analyzed the co-localization of each overexpressed Nst1 domain deletion mutant and endogenous Dcp2 (a PB marker) using van Steensel's cross-correlation function (CCF) (Figure 2B). Here, van Steensel's CCF determines the degree of co-localization between two different signals (red and green) by crossing the Pearson coefficients of each image signal [53]. Endogenous Dcp2-mKate2 was captured for analysis in wild-type cells whose chromosomal *DCP2* was tagged with mKate2 after each GFP-tagged Nst1 mutant was overexpressed. Van Steensel's CCF of overexpressed GFP-tagged $Nst1_{NTD}$ did not show a bell-shaped curve with Dcp2-mKate2, indicating that the red and green signals did not overlap (Figure 2B). However, van Steensel's CCF of overexpressed GFP-tagged Nst1 and $Nst1_{CTD}$ for Dcp2-mKate2 showed a bell-shaped curve, although GFP-tagged $Nst1_{CTD}$ and Dcp2-mKate2 showed a weaker correlation than the wild-type Nst1 for Dcp2-mKate2 (Figure 2B). GFP-tagged $Nst1_{CTD}$ and Dcp2-mKate2 correlated more closely than GFP-tagged $Nst1_{NTD}$ and Dcp2-mKate2 (Figure 2B). These data suggest that Nst1 self-condensation is correlated with the accumulation of the Dcp2 PB marker.

To demonstrate the Nst1 domain responsible for EGFP-Dcp2 condensation, we monitored endogenous EGFP-Dcp2 in cells overexpressing Nst1, $Nst1_{NTD}$, and $Nst1_{CTD}$. As expected, $Nst1_{NTD}$ overexpression did not increase EGFP-Dcp2 condensation, whereas overexpression of full-length Nst1 induced EGFP-Dcp2 condensation (Figure 2C). EGFP-Dcp2 condensation induced by $Nst1_{CTD}$ overexpression was enhanced compared to the vector control and overexpressed $Nst1_{NTD}$ (Figure 2C). However, EGFP-Dcp2 condensation in cells overexpressing $Nst1_{CTD}$ was reduced compared with that in cells overexpressing full-length Nst1 (Figure 2C). To quantify the degree of EGFP-Dcp2 puncta generated, we segmented pixels of the top 0.05% intensity for puncta analysis, and the maximum intensities of the segmented puncta scaled from 0–255 were analyzed using a boxplot. The measuring method is described in detail in the Materials and Methods section and our previous study [40]. Consistent with Figure 2C, full-length Nst1 and $Nst1_{CTD}$ overexpression increased the maximum intensities of EGFP-Dcp2 condensates, while $Nst1_{NTD}$ overexpression did not (Figure 2D). Instead, the EGFP-Dcp2 condensates were decreased in $Nst1_{NTD}$-overexpressing cells compared to the vector control cells (Figure 2C,D). The endogenous EGFP-Dcp2 expression level in each mutant overexpressing cell was monitored by Western blotting to confirm that the overexpression of each mutant did not affect EGFP-Dcp2 expression levels (Supplemental Figure S2B).

These data suggest that an intrinsic sequence factor responsible for Nst1 self-condensation is present in the Nst1 CTD. In addition, the condensation of PB components in the Nst1-overexpressed cells was produced based on Nst1 self-condensation through LLPS.

2.3. The APD in the Nst1 CTD Is Insufficient but Crucial for Inducing Nst1 Self-Condensation

Next, we investigated whether the predicted APD in the C-terminus induced Nst1 condensation. We constructed GFP-tagged $Nst1_{\Delta APD}$ and $Nst1_{CTD\Delta APD}$ mutants and compared their overexpression phenotypes with those of GFP-tagged wild-type Nst1 and $Nst1_{CTD}$ in BY4741 wild-type cells. Overexpression of GFP-tagged Nst1 and $Nst1_{CTD}$ generated self-condensation (Figure 3A). However, when $Nst1_{\Delta APD}$ was overexpressed, major GFP signals were dispersed throughout the cytoplasm. We observed the same dispersed phenotype in cells overexpressing GFP-tagged $Nst1_{CTD\Delta APD}$, where the APD was deleted in $Nst1_{CTD}$, although $Nst1_{CTD}$ displayed discrete puncta (Figure 3A). These observations suggest that the APD plays a critical role in Nst1 condensation. To further test the sufficiency of the APD inducing condensation, only the APD was overexpressed. This did not result in the assembly of any condensates (Figure 3A). These observations demonstrate that the Nst1 APD is crucial but insufficient for Nst1 condensation to form condensates. Consistently, the CCF of overexpressed GFP-tagged $Nst1_{\Delta APD}$ and $Nst1_{CTD\Delta APD}$ versus endogenous Dcp2-mKate2 did not show a bell-shaped curve (Figure 3B).

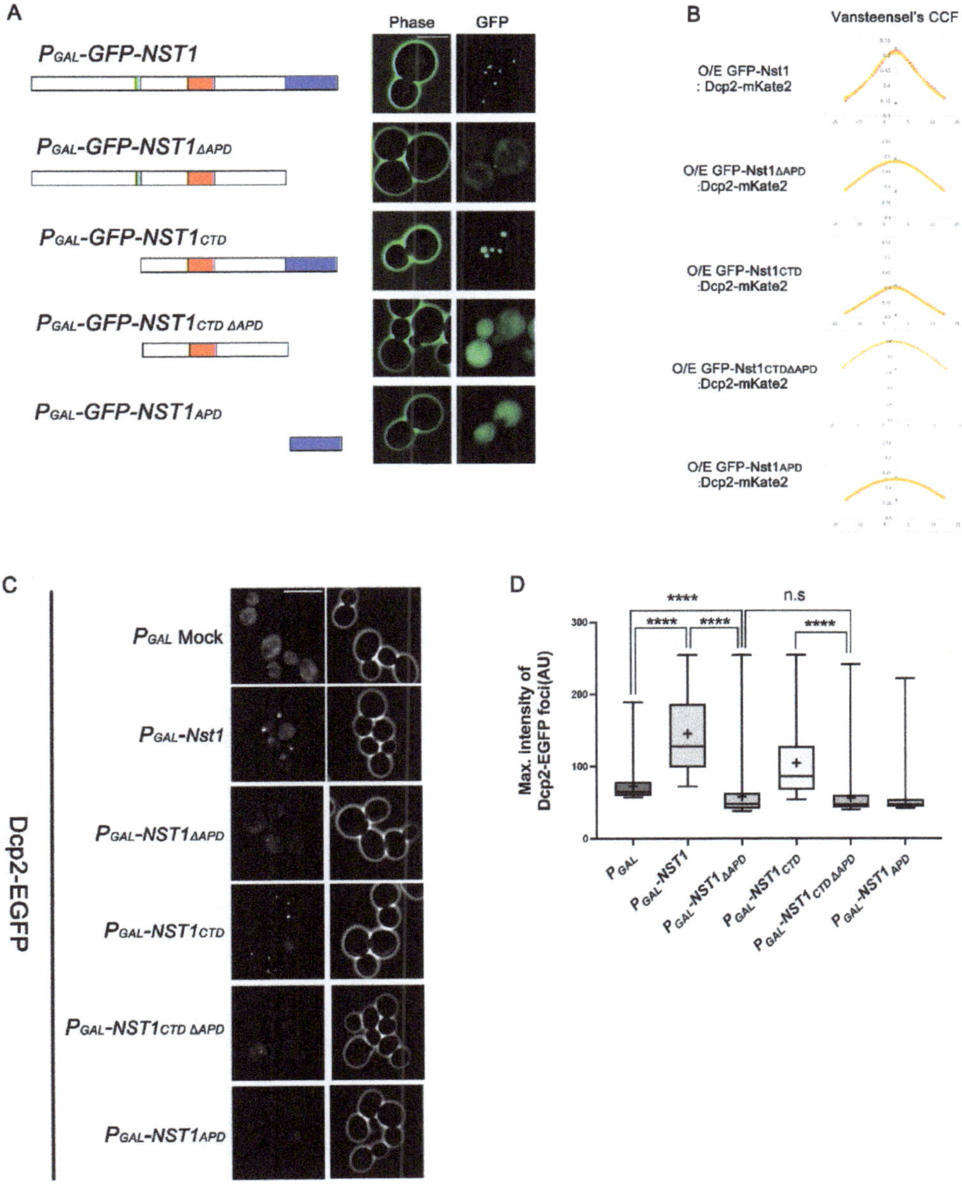

Figure 3. The aggregation-prone domain (APD) in the $Nst1_{CTD}$ is insufficient but crucial for inducing Nst1 self-condensation. (**A**) Fluorescence microscopy of cells overexpressing EGFP-tagged full-length

Nst1, Nst1$_{\Delta APD}$ (residues 1–1015), Nst1$_{CTD}$ (C-terminal Nst1 residues 430–1240), Nst1$_{CTD\Delta APD}$ (residues 431–1015), and Nst1$_{APD}$ (residues 1016–1240). Schematic diagrams of the designed Nst1 domain deletion mutants are shown on the left. Overexpression of each EGFP-tagged Nst1 domain deletion mutant was induced in wild-type cells, then observed. Scale bar: 5 µm. (**B**) The van Steensel's CCFs between each overexpressed Nst1 domain deletion mutant used in (**A**) and the endogenous Dcp2-mKate2 signals were analyzed and presented. Overexpression of each EGFP-tagged Nst1 domain deletion mutant was induced in wild-type cells whose chromosomal *DCP2* was tagged with mKate2. Each Nst1 domain deletion mutant (n = total observed cell number): P_{GAL}-*GFP-NST1* (n = 257), P_{GAL}-*GFP-NST1*$_{\Delta APD}$ (n = 387), P_{GAL}-*GFP-NST1*$_{CTD}$ (n = 161), P_{GAL}-*GFP-NST1*$_{CTD\Delta APD}$ (n = 191), and P_{GAL}-*GFP-NST1*$_{APD}$ (n = 167). All images were analyzed by FIJI (https://imagej.net/Fiji, accessed on 9 August 2020). (**C**,**D**) Each Nst1 domain deletion mutant was overexpressed in the wild-type cells with EGFP-tagged *DCP2* (YSK3485). (**C**) Fluorescence microscopy of endogenous EGFP-Dcp2-tagged cells overexpressing full-length Nst1, Nst1$_{\Delta APD}$, Nst1$_{CTD}$, Nst1 $_{CTD\Delta APD}$, and Nst1$_{APD}$. Scale bar: 10 µm. (**D**) Quantification of the endogenous EGFP-Dcp2 puncta of (**B**). The pixels of the top 0.1% EGFP-Dcp2 signal intensities were segmented for puncta analysis. The maximal intensity of each segmented punctum was plotted. '+' in the boxplot indicates the mean value of maximal intensities of foci. Each Nst1 domain deletion mutant (n = total observed cell number): P_{GAL} vector-only control (n = 218), P_{GAL}-*NST1* (n = 307), P_{GAL}-*NST1*$_{\Delta APD}$ (n = 337), P_{GAL}-*NST1*$_{CTD}$ (n = 300), P_{GAL}-*NST1*$_{CTD\Delta APD}$ (n = 260), and P_{GAL}-*NST1*$_{APD}$ (n = 261). All measurements and analyses were performed by FIJI (https://imagej.net/Fiji, accessed on 31 March 2022). Statistical significance was determined by a Mann–Whitney test (**** $p < 0.0001$).

In monitoring EGFP-Dcp2 in cells overexpressing these domain deletion mutants, Nst1$_{\Delta APD}$ overexpression did not show EGFP-Dcp2 condensate accumulation, while overexpression of the full-length Nst1 induced EGFP-Dcp2 condensation (Figure 3C), as expected from van Steensel's CCF. In the quantitative analysis of the EGFP-Dcp2 condensates, the overexpression of full-length Nst1 and Nst1$_{CTD}$ increased the maximal intensities of EGFP-Dcp2 condensates compared to the vector control, while the overexpression of Nst1$_{\Delta APD}$ and Nst1$_{CTD\Delta APD}$ canceled out the effect (Figure 3D). The maximal intensities of the EGFP-Dcp2 condensates were reduced in both Nst1$_{\Delta APD}$- and Nst1$_{CTD\Delta APD}$-overexpressing cells compared to those of the vector control cells (Figure 3D). Endogenous EGFP-Dcp2 did not appear as puncta in cells overexpressing Nst1$_{\Delta APD}$, suggesting that the APD of Nst1 alone was unable to induce self-aggregation. Endogenous EGFP-Dcp2 expression levels in cells overexpressing each mutant were monitored by Western blotting to confirm that the overexpression of each mutant did not affect EGFP-Dcp2 expression levels (Supplemental Figure S2C). These data demonstrate that the Nst1 APD is the critical region for inducing Nst1 self-condensation but functions in a context-dependent manner.

2.4. The Nst1 PD Is Not a Critical Component in Self-Condensation but Is Responsible for Inducing Dcp2 Condensation

The polyampholyte region of proteins is a representative IDR and is anticipated to be closely related to biomolecular condensation [48,49]. We attempted to demonstrate the function of the PD in Nst1 self-condensation because an obvious polyampholyte region is present in the Nst1 CTD. Considering previous reports on the function of the polyampholyte region in LLPS [48,49], we expected that PD deletion in various Nst1 domain mutants would negatively affect self-condensate generation. We compared the punctum formation of the GFP-tagged PD deletion mutants with that of the GFP-tagged full-length Nst1 and Nst1$_{CTD}$ upon overexpression. Unexpectedly, GFP-tagged Nst1$_{\Delta PD}$ generated condensates similar to wild-type Nst1 when overexpressed, demonstrating that PD does not control Nst1 self-condensation (Figure 4A). Nst1$_{CTD\Delta PD}$ overexpression also formed puncta (Figure 4A). However, the size and intensity of puncta in cells overexpressing GFP-tagged Nst1$_{CTD\Delta PD}$ were reduced, compared with those in cells overexpressing GFP-tagged Nst1$_{CTD}$ (Figure 4A), suggesting that the PD in Nst1 may only partially contribute to Nst1 self-condensation. Both condensates induced by GFP-tagged Nst1$_{\Delta PD}$ and Nst1$_{CTD\Delta PD}$

overexpression were dispersed in the cytoplasm in 1,6-hexanediol-treated cells, indicating their liquid-like properties (Figure 4A).

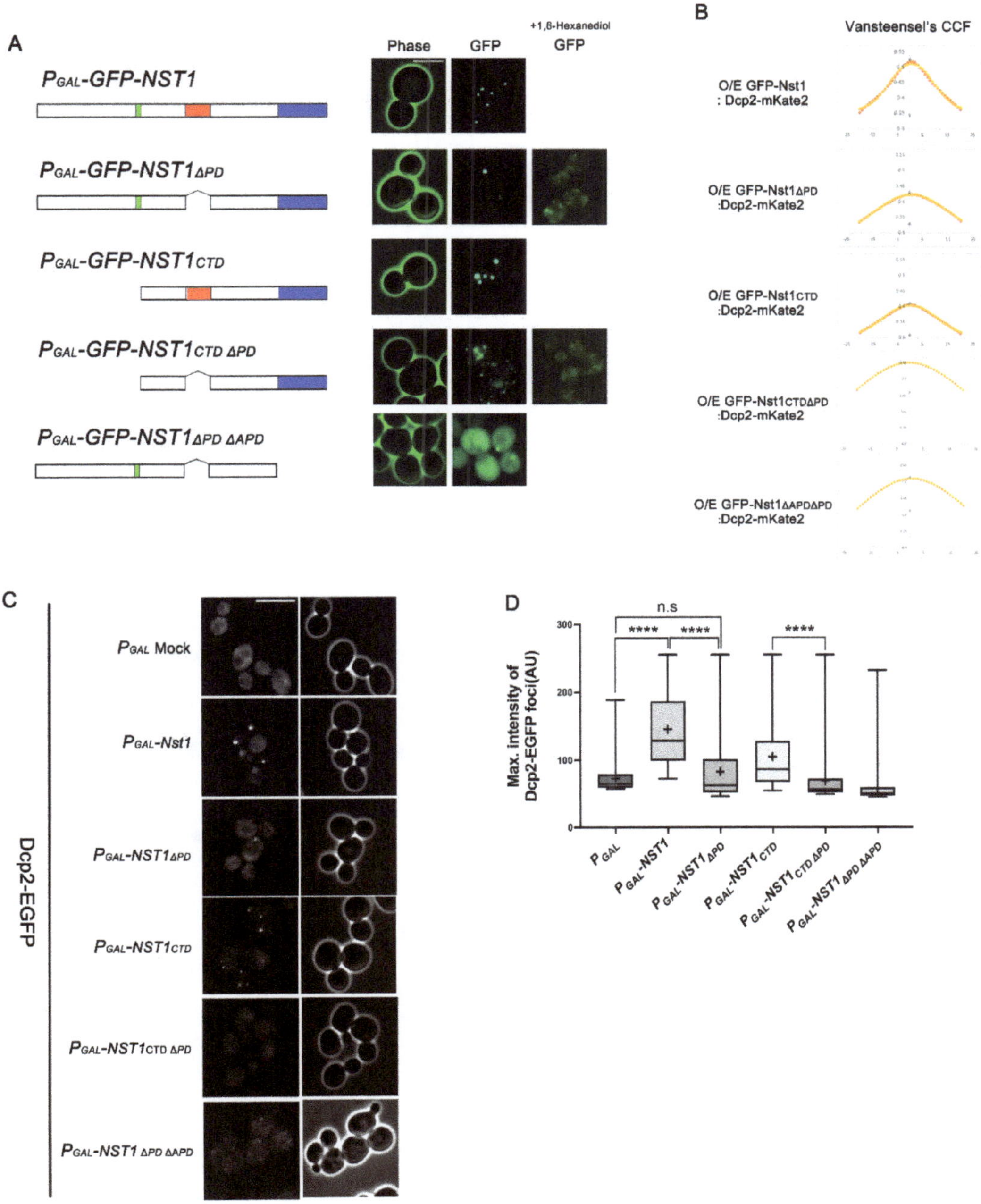

Figure 4. The polyampholyte region is not critical for Nst1 self-condensation but is responsible for inducing Dcp2 condensation. (**A**) Fluorescence microscopy of cells overexpressing EGFP-tagged

full-length Nst1, Nst1$_{\Delta PD}$ (residues 1–630 and 753–1240), Nst1$_{CTD}$ (C-terminal Nst1 residues 430–1240), Nst1$_{CTD\Delta PD}$ (residues 431–630 and 753–1240), and Nst1$_{\Delta PD\Delta APD}$ (residues 1–630 and 753–1015). Overexpression of each EGFP-tagged Nst1 domain deletion mutant was induced in wild-type cells, then observed before and after 1,6-hexanediol treatment. Scale bar: 5 µm. (**B**) The van Steensel's CCFs between each overexpressed Nst1 domain deletion mutant used in (**A**) and the endogenous Dcp2-mKate2 signals were analyzed and presented. Overexpression of each EGFP-tagged Nst1 domain deletion mutant was induced in wild-type cells whose chromosomal *DCP2* was tagged with mKate2. Each Nst1 domain deletion mutant (n = total observed cell number): P_{GAL}-*GFP-NST1* (n = 257), P_{GAL}-*GFP-NST1*$_{\Delta PD}$ (n = 277), P_{GAL}-*GFP-NST1*$_{CTD}$ (n = 161), P_{GAL}-*GFP-NST1*$_{CTD\Delta PD}$ (n = 199), and P_{GAL}-*GFP-NST1*$_{\Delta APD\Delta PD}$ (n = 198). All images were analyzed by FIJI (https://imagej.net/Fiji, accessed on 9 August 2020). (**C**,**D**) Each Nst1 domain deletion mutant was overexpressed in the wild-type cells with EGFP-tagged *DCP2* (YSK3485). Schematic diagrams of the designed Nst1 domain deletion mutants are shown on the left. (**C**) Fluorescence microscopy of endogenous EGFP-Dcp2-tagged cells overexpressing full-length Nst1, Nst1$_{\Delta PD}$ (residues 1–630 and 753–1240), Nst1$_{CTD}$ (C-terminal Nst1 residues 430–1240), Nst1$_{CTD\Delta PD}$ (residues 431–630 and 753–1240), and Nst1$_{\Delta PD\Delta APD}$ (residues 1–630 and 753–1015). Scale bar: 10 µm. (**D**) Quantification of the endogenous EGFP-Dcp2 puncta of (**C**). The pixels of the top 0.1% EGFP-Dcp2 signal intensities were segmented for puncta analysis. The maximal intensity of each segmented punctum was plotted. '+' in the boxplot indicates the mean value of maximal intensities of foci. Each Nst1 domain deletion mutant (n = total observed cell number): P_{GAL} vector-only control (n = 218), P_{GAL}-*NST1* (n = 307), P_{GAL}-*NST1*$_{\Delta PD}$ (n = 284), P_{GAL}-*NST1*$_{CTD}$ (n = 300), P_{GAL}-*NST1*$_{CTD\Delta PD}$ (n = 200), and P_{GAL}-*NST1*$_{\Delta PD\Delta APD}$ (n = 260). All measurements and analyses were performed by FIJI (https://imagej.net/Fiji, accessed on 31 March 2022). Statistical significance was determined by a Mann–Whitney test (**** $p < 0.0001$).

We anticipated that overexpressed Nst1$_{\Delta PD}$ would induce EGFP-Dcp2 condensates and colocalize with the Nst1 overexpression because Nst1 PD deletion did not interrupt Nst1 self-condensation upon overexpression. However, in van Steensel's CCF diagram, the localization of endogenous Dcp2-mKate2 tended to be less correlated with overexpressed GFP-tagged Nst1$_{\Delta PD}$ localization than with overexpressed GFP-tagged full-length Nst1 and Nst1$_{CTD}$ (Figure 4B).

To examine the functional potential of PD deletion in EGFP-Dcp2 condensation, we investigated EGFP-Dcp2 condensation in cells overexpressing Nst1$_{\Delta PD}$. Endogenous EGFP-Dcp2 was monitored in cells overexpressing each Nst1 domain deletion mutant (Figure 4A). Unexpectedly, the intensity of EGFP-Dcp2 puncta hardly increased in cells overexpressing Nst1$_{\Delta PD}$ (Figure 4C,D), although we observed that GFP-tagged Nst1$_{\Delta PD}$ overexpression generated bright puncta via self-condensation (Figure 4A). Endogenous EGFP-Dcp2 expression levels in Nst1 mutant-overexpressing cells monitored by Western blotting indicated that Nst1$_{\Delta PD}$ mutant overexpression did not affect EGFP-Dcp2 expression levels (Supplemental Figure S2D). The intensity of EGFP-Dcp2 puncta in cells overexpressing Nst1$_{\Delta PD}$ was similar to that in cells overexpressing Nst1$_{\Delta APD}$, which did not generate any concentrated EGFP-Dcp2 signals (Figures 3C and 4C). These observations suggest that the PD is less correlated with self-condensation and may play a specific role in recruiting other PB components. Deleting the PD in Nst1$_{CTD}$ also canceled out Nst1$_{CTD}$ overexpression-induced EGFP-Dcp2 accumulation, supporting the role of PD in EGFP-Dcp2 condensation (Figure 4C,D).

The effect of the APD on Nst1 self-condensation was confirmed by Nst1$_{\Delta PD\Delta APD}$ overexpression. We observed that GFP-tagged Nst1$_{\Delta PD\Delta APD}$ was mainly dispersed in the cytoplasm as GFP-tagged Nst1$_{\Delta APD}$, while overexpressed GFP-tagged Nst1$_{\Delta PD}$ was observed as clear puncta (Figure 4A). Van Steensel's CCF between GFP-tagged Nst1$_{\Delta PD\Delta APD}$ and Dcp2-mKate2 also reflected that the double deletion of the PD and APD reduced the intrinsic self-condensation potential of overexpressed Nst1 to be co-localized with the PB marker (Figure 4B). These data confirm that the APD is responsible for Nst1 self-condensation. As expected, endogenous EGFP-Dcp2 condensates did not accumulate in cells overexpressing Nst1$_{\Delta PD \Delta APD}$ (Figure 4C,D).

2.5. Dcp2 Condensation Induced by Nst1 PD Overexpression Is Independent of Free Ribosomal Influx

Observations of endogenous EGFP-Dcp2 in cells overexpressing various Nst1 mutants revealed that the APD and PD are largely responsible for self-condensation and inducing Dcp2 condensation, respectively. Since RNA functions as a scaffold for protein condensation via LLPS [12], we examined whether the PD in Dcp2 condensation is mediated by polysome RNA influx. We investigated EGFP-Dcp2 puncta induced by overexpression of each Nst1 domain deletion mutant after cycloheximide (CHX) treatment. PB formation induced by stress relies on an increase in non-translating mRNA concentration [12]. CHX completely disassembled the endogenous PBs formed during glucose deficiency, which inhibited translation elongation and resulted in a reduction in non-translating RNA [11,12]. Thus, protein-induced PB accumulation can be verified because CHX eliminated RNA-derived PBs.

Nst1, Nst1$_{\Delta PD}$, Nst1$_{CTD}$, and Nst1$_{CTD \Delta PD}$ were overexpressed in cells with EGFP-tagged chromosomal *DCP2*, and EGFP-Dcp2 was observed after treating cells with 100 μg/mL CHX for 10 min [12]. Consistent with our previous report, Nst1 overexpression maintained EGFP-Dcp2 condensates in the presence of CHX, whereas the EGFP-Dcp2 puncta completely disappeared in the vector control (Figure 5A). We also observed that Nst1$_{CTD}$ overexpression maintained EGFP-Dcp2 condensates in the presence of CHX (Figure 5A). The maximal intensity of EGFP-Dcp2 puncta generated by each domain deletion mutant was measured and plotted on the y-axis in the puncta quantification analysis shown in Figure 5A (Figure 5B). Similar to the results shown in Figure 4, the maximal intensity of EGFP-Dcp2 puncta accumulated by Nst1$_{\Delta PD}$ overexpression was significantly decreased compared to that of full-length Nst1 overexpression and was similar to the vector-only control (Figure 5B). The maximal intensity of the EGFP-Dcp2 puncta accumulated by Nst1$_{CTD \Delta PD}$ declined, similar to Nst1$_{\Delta PD}$ (Figure 5B). The ratio of cells with generated EGFP-Dcp2 puncta in the PD deletion mutants (Nst1$_{\Delta PD}$ and Nst1$_{CTD \Delta PD}$) was dramatically decreased compared to that in Nst1 and Nst1$_{CTD}$ (Figure 5C). These observations strongly support the implication that the PD is responsible for inducing the condensation of other PB components.

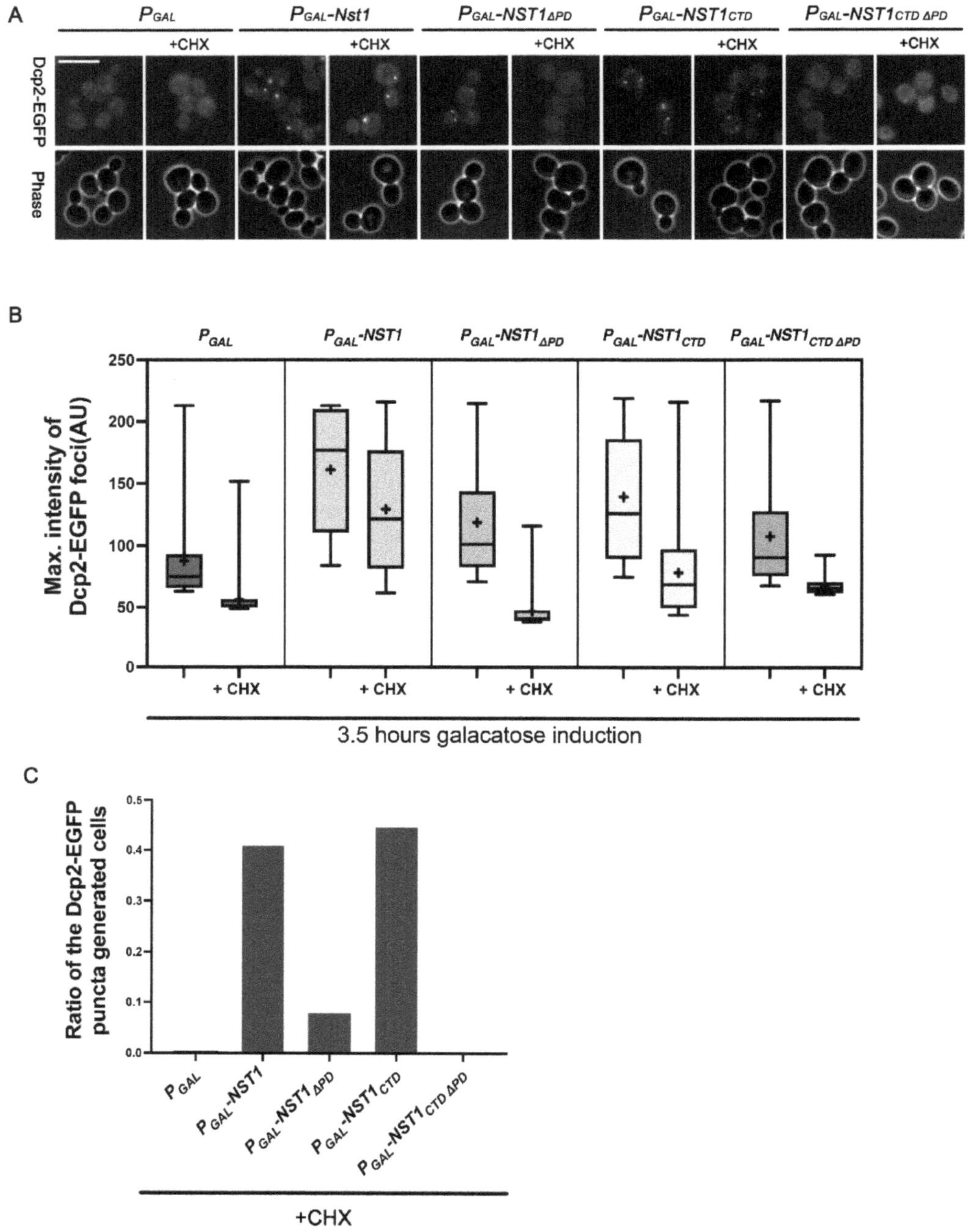

Figure 5. The Nst1 polyampholyte region interacts with a PB component Dcp2 independent of the free ribosomal RNA influx. (**A–C**) In the wild-type strain whose chromosomal *DCP2* was tagged with EGFP, endogenous EGFP-Dcp2 was observed after the overexpression of full-length Nst1, $Nst1_{\Delta PD}$ (residues 1–630 and 753–1240), $Nst1_{CTD}$ (C-terminal Nst1 residues 430–1240), and $Nst1_{CTD\Delta PD}$ (residues 431–630 and 753–1240). In the cells overexpressing each Nst1 domain deletion mutant, endogenous Dcp2 was observed before and after the 10 min 100 µg/mL cycloheximdie (CHX)

treatment. (**A**) Fluorescence microscopy of EGFP-Dcp2 in the cells overexpressing each mutant before and after the 10 min 100 μg/mL CHX treatment. Scale bar: 10 μm. (**B**) Quantification of EGFP-Dcp2 puncta shown in (**A**). The pixels of the top 0.1% signal intensities were segmented and analyzed. The maximal value of each punctum was plotted. '+' in the boxplot indicates the mean value of maximal intensities of foci. All measurements and analyses were performed by FIJI (https://imagej.net/Fiji, accessed on 31 March 2022) (**C**) The ratio of cells producing EGFP-Dcp2 puncta to the total cells by the overexpression of each Nst1 mutant (Nst1 domain deletion mutant (n = total observed cell number): P_{GAL} (vector only, n = 260), P_{GAL}-$NST1$ (n = 312), P_{GAL}-$NST1_{\Delta PD}$ (n = 247), P_{GAL}-$NST1_{CTD}$ (n = 302), and P_{GAL}-$NST1_{CTD\Delta PD}$ (n = 241)) in (**A**).

2.6. The Nst1 PD Serves as a Binding Hub, Mediating the Condensation of other PB Components

Edc3 is a PB scaffold protein in *S. cerevisiae* [11,54]. The Δ*edc3lsm4*ΔC mutant could not induce EGFP-Dcp2 condensates independent of RNA influx, indicating that Edc3 is a critical component in PB generation. In our previous study, EGFP-Dcp2 condensation driven by Nst1 overexpression was suppressed in Δ*edc3 lsm4*ΔC mutant cells, suggesting a functional relationship between Nst1 and Edc3 in condensate formation. The EGFP-tagged *EDC3* strain was transformed with the same Nst1 deletion mutant clones tested in Figure 5 to determine whether the Nst1 PD is also responsible for EGFP-Edc3 condensation. We then treated these cells with 100 μg/mL CHX after galactose induction to examine whether Nst1 PD-mediated Edc3 condensation is independent of polysome RNA influx. Microscopic observations revealed that the puncta of EGFP-Dcp2 and EGFP-Edc3, induced by Nst1 overexpression, behaved analogously. In the presence of CHX, EGFP-Edc3 condensation was highly decreased in cells overexpressing $Nst1_{\Delta PD}$ and $Nst1_{CTD\Delta PD}$ compared to that in cells overexpressing Nst1 and $Nst1_{CTD}$ (Figure 6A). $Nst1_{\Delta PD}$ overexpression did not induce EGFP-Edc3 puncta (Figure 6A), although GFP-tagged $Nst1_{\Delta PD}$ overexpression resulted in its bright puncta (Figure 4B). The pattern of EGFP-Edc3 puncta generated by the overexpression of diverse Nst1 deletion mutants was similar to the pattern of EGFP-Dcp2, both in the maximal intensity and the ratio of puncta-generating cells (Figure 6B). The ratio of cells with generated EGFP-Edc3 puncta in the PD deletion mutants ($Nst1_{\Delta PD}$ and $Nst1_{CTD\Delta PD}$) was dramatically decreased compared to that in Nst1 and $Nst1_{CTD}$ (Figure 6C). Endogenous EGFP-Edc3 expression levels of each Nst1 domain deletion mutant monitored by Western blot analysis showed that altering EGFP-Edc3 expression levels did not induce EGFP-Edc3 puncta reduction in cells overexpressing PD deletion mutants (Supplementary Figure S2F). These analyses strongly suggested that the PD is responsible for recruiting Edc3 as well as Dcp2.

The EGFP-tagged *DHH1* and *XRN1* strains were transformed with Nst1 and $Nst1_{\Delta PD}$ and treated with 100 μg/mL CHX for 10 min after galactose induction to verify whether the PD recruits other PB components. Overexpression of PD deletion mutants ($Nst1_{\Delta PD}$ and $Nst1_{CTD\Delta PD}$) generated fewer Dhh1 and Xrn1 puncta than wild-type CHX-treated Nst1 overexpressing cells (Figure 6D,E). Overall, overexpression of the PD deletion mutants ($Nst1_{\Delta PD}$ and $Nst1_{CTD\Delta PD}$) reduced the condensation of known PB components, suggesting that the Nst1 PD interacts with PB components independent of polysome RNA influx.

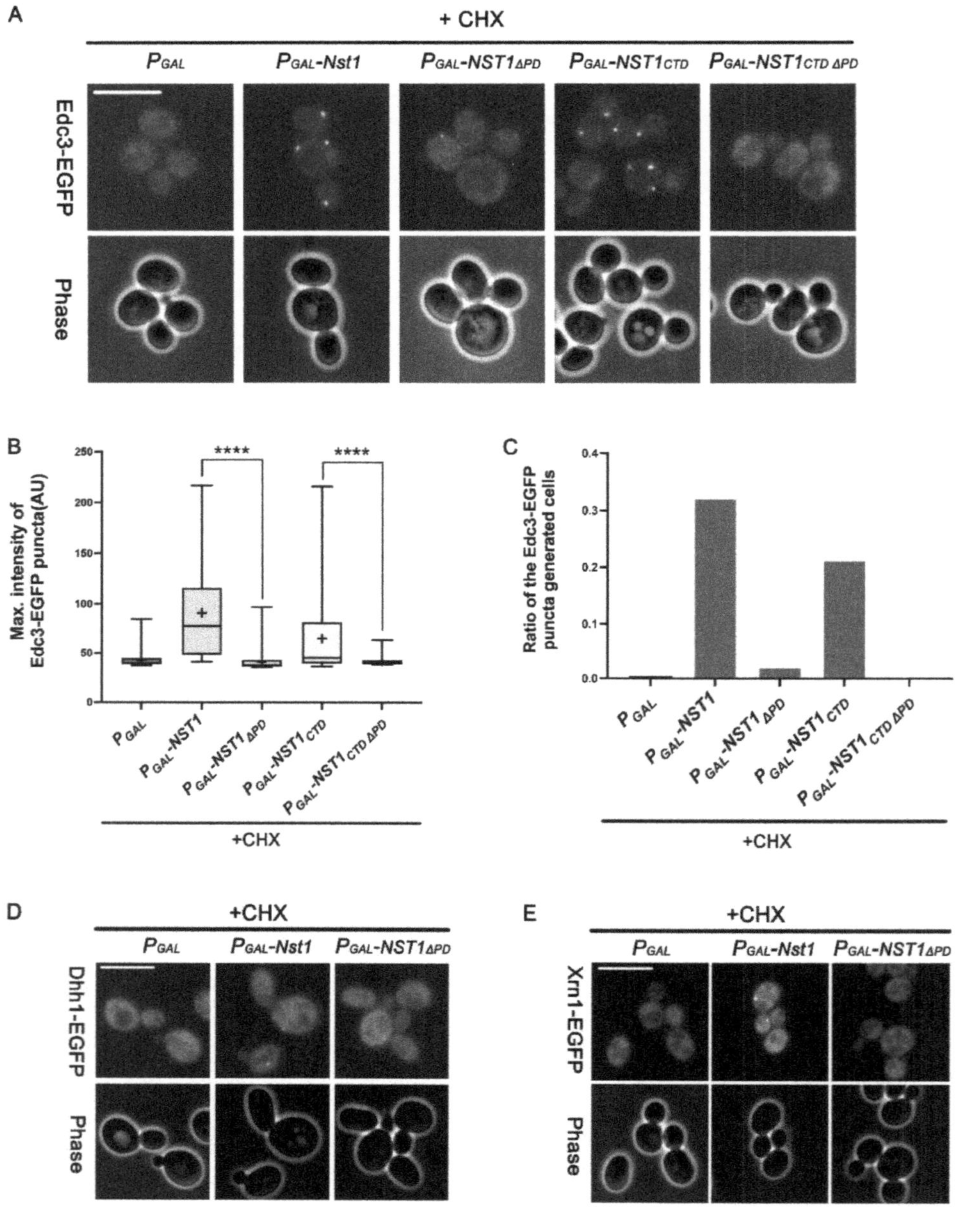

Figure 6. The Nst1 polyampholyte region functions as a binding hub for P-body (PB) components independent of the free ribosomal RNA influx. (**A–C**) In the wild-type strains whose chromosomal EDC3 was tagged with EGFP, the full-length Nst1, Nst1ΔPD (residues 1–630 and 753–1240), Nst1CTD (C-terminal Nst1 residues 430–1240), and Nst1CTDΔPD (residues 431–630 and 753–1240) were overexpressed. Endogenous enhancer of mRNA decapping 3 (Edc3) was observed before and after cells were treated with 100 μg/mL CHX for 10 min. (**A**) Fluorescence microscopy of EGFP-Edc3 in the cells overexpressing each mutant before and after the 10 min 100 μg/mL CHX treatment. Scale bar: 10 μm. (**B**) Quantification of the EGFP-Edc3 puncta shown in (**A**) with CHX. The pixels of the top

0.1% signal intensities were segmented as the EGFP-Edc3 puncta. The EGFP-Edc3 puncta generated were quantified, and the maximal value of each punctum was plotted. '+' in the boxplot indicates the mean value of maximal intensities of foci. All measurements and analyses were performed by FIJI (https://imagej.net/Fiji, accessed on 31 March 2022) Statistical significance was determined by a Mann–Whitney test (**** $p < 0.0001$). **(C)** The ratio of cells producing EGFP-Edc3 puncta to the total cells overexpressing each Nst1 mutant (Nst1 domain deletion mutant in the presence of CHX (n = total observed cell number): PGAL (vector only, n = 233), PGAL-NST1 (n = 320), PGAL-NST1ΔPD (n = 200), PGAL-NST1CTD (n = 244), and PGAL-NST1CTDΔPD (n = 275). **(D,E)** In the wild-type strains with EGFP-tagged chromosomal DHH1 and XRN1, the overexpression of full-length Nst1 and Nst1ΔPD cells was induced, and then cells were treated with 100 µg/mL CHX for 10 min. Fluorescence microscopy of **(D)** EGFP-Dhh1 and **(E)** EGFP-Xrn1 in CHX-treated cells overexpressing full-length Nst1 and Nst1ΔPD. Scale bar: 10 µm. All images were measured and analyzed by FIJI (https://imagej.net/Fiji, accessed on 31 March 2022).

3. Discussion

Understanding the syntax of biomolecular condensation is key to understanding the molecular dynamics of cells. RNA is a powerful scaffold, and the RNA-binding moiety of scaffold proteins is expected to be crucial in biomolecular condensation. However, the protein scaffolds responsible for condensation need to be investigated further. The sequence properties of various scaffold proteins in condensates, such as the low complexity domains (LCDs) of poly-Q or RGG and the polyampholytic region of charged amino acids (lysine or arginine), may be critical for biomolecular condensation [50,55]. Further, scaffold proteins that specifically function in a particular condensation generally have oligomerizing properties and IDRs. A study on the PB component, Lsm4, in budding yeast found that *GAL*-induced Lsm4 overexpression drives self-condensation [27]. Lsm4 is a representative PB component, with a prion-like domain (PrD, poly-Q motif) in its C-terminal region. Although CHX dissipated the stress-responsive endogenous Lsm4-GFP puncta, it did not disperse the bright clear puncta generated by the *GAL*-induced GFP-Lsm4, implying that the physical properties of the puncta induced by overexpressed Lsm4 were not identical to the stress-derived endogenous PBs. However, these observations indicate that Lsm4 has strong self-oligomerizing potential despite the unidentical physical properties of the puncta upon overexpression with native PBs. Similarly, overexpressed GFP-Edc3 appeared as bright clear puncta not dissipated by CHX, supporting previous reports that Edc3 harbors the Yjef-N domain, which induces Edc3 self-oligomerization [56].

We previously found that Nst1 significantly accumulated puncta in the stationary phase. We also reported that GFP-tagged Nst1 overexpression using a *GAL*-inducible promoter yielded condensates of round puncta (Figure 2A) and drove the accumulation of other PB components. These data strongly suggest that Nst1 has the potential to self-condensate and recruit other PB components to condense. CHX did not dissipate the overexpressed GFP-Nst1-generated bright clear puncta, suggesting that Nst1 has a sequence element that induces self-oligomerization similar to Lsm4 and Edc3, although Nst1 does not have a recognizable PrD. Nst1 is a large protein consisting of 1240 amino acids, including diverse sequence elements, a presumptive IDR, and aggregation-prone regions, as predicted by several programs (Figure 1). In this study, we attempted to elucidate the functional sequence elements of Nst1 for its self-condensation and accumulation of other PB component/s by examining GFP-tagged Nst1 domain deletion mutants upon overexpression.

3.1. The Nst1 C-Terminus Is Necessary and Sufficient for Self-Condensation, While the N-Terminus Has an Auxiliary Role in Recruiting other PB Components

Overexpression of GFP-tagged $Nst1_{NTD}$ (residues 1–429) did not generate any self-condensation, whereas CTD (residues 430–1240) overexpression was sufficient for self-condensation, indicating that the oligomerizing domain is present in the Nst1 C-terminus. Nst1 condensation induced EGFP-Dcp2 condensation. The Nst1 domain, serving as a platform to interact with Dcp2, is essential for EGFP-Dcp2 condensation. In the analysis of Dcp2 condensates in cells overexpressing different Nst1 domain deletion mutants, CTD overexpression induced less Dcp2 accumulation than full-length Nst1. GFP-tagged $Nst1_{CTD}$ formed bright clear puncta with a similar intensity to the GFP-tagged full-length Nst1 upon overexpression. Consistently, cells overexpressing the NTD did not seem to produce any EGFP-Dcp2 puncta compared to the cells overexpressing full-length Nst1 (Figure 2C,D), but instead showed reduced EGFP-Dcp2 puncta in comparison with the vector-only control. These observations can be explained by the recent LLPS mechanism suggested by the Brangwynne group, in which node capping could reduce interactor condensation [25]. By functioning as a Dcp2 node capper, the Nst1 NTD may directly or indirectly interact with Dcp2 to cover the Dcp2 node, resulting in Dcp2 condensation inhibition. These data imply that the overexpressed Nst1 NTD does not function in Nst1 self-condensation, but it may support Dcp2 recruitment to PB-associated condensates in full-length Nst1-overexpressing cells.

3.2. The Aggregation-Prone Region May Be Associated with Inducing Nst1 Condensates with Liquid-like Properties

We attempted to identify a specific region in the CTD that is responsible for self-condensation. According to Nst1 sequence-based predictions, the most aggregation-prone region consisted of hydrophobic amino acid residues in the APD. Although the precise link between aggregation propensity and LLPS remains unclear, the degree of aggregation propensity is likely correlated with many types of condensation, such as LLPS [57]. Among the several Nst1 domain deletion mutants constructed, the APD deletion mutant ($Nst1_{\Delta APD}$) was the most powerful suppressor of Nst1 condensation driven by overexpression. Overexpressed GFP-tagged APD mutants, such as $Nst1_{\Delta APD}$, $Nst1_{CTD\Delta APD}$, and $Nst1_{\Delta PD\Delta APD}$, had significantly decreased puncta and were dispersed in the cytoplasm (Figure 3A), suggesting that the APD in the C-terminus is critical for Nst1 self-condensation. Conversely, APD-overexpressing cells did not show any EGFP-Dcp2 condensates, although condensation of EGFP-Dcp2 in cells overexpressing $Nst1_{\Delta APD}$ was alleviated compared to cells overexpressing full-length Nst1 (Figure 3C,D). These data suggest that APD is necessary but insufficient for Nst1 or $Nst1_{CTD}$ self-condensation. The insufficiency of the APD for self-condensation could explain the importance of context and promiscuous interactions in protein condensation by LLPS [26,27,34,53]. These observations imply that although we could obtain clues on the sequence elements involved in Nst1 self-condensation by deleting each element, removing a domain may damage the unique sequence pattern of full-length Nst1 for condensation.

3.3. The Polyampholyte Region May Be Involved in Molecular Condensation as a Platform for Multivalent Protein–Protein Interactions Independent of RNA Influx

Polyampholytes containing a significant proportion (>35%) of positively and negatively charged residues are present in 75% of intrinsically disordered proteins [48,55]. In the analyses of EGFP-Dcp2 puncta formed in cells overexpressing Nst1 domain deletion mutants, the Nst1 PD was responsible for recruiting other PB components. The cells overexpressing the PD deletion mutants showed significantly decreased Dcp2 puncta compared to the full length and CTD of Nst1. However, the overexpressed GFP-tagged PD deletion mutants strongly induced self-condensation, similar to overexpressed GFP-tagged Nst1 and $Nst1_{CTD}$ (Figure 4A). The decreased EGFP-Dcp2 phenotype in the overexpressed PD deletion mutants was more clearly observed after treatment with CHX (Figure 5A,B).

Other PB markers, such as Edc3, Dhh1, and Xrn1, showed similar accumulation patterns to Dcp2 in CHX-treated cells overexpressing PD deletion mutants. These observations clarified that PD does not affect Nst1 oligomerization but significantly contributes to the recruitment of other PB components, namely Dcp2, Dhh1, Xrn1, and Edc3.

To assess the physical interaction of Nst1 and $Nst1_{\Delta PD}$ with PB components upon biomolecular condensation, we first checked the physical interactions of Nst1 with other essential PB constituents via a co-immunoprecipitation (Co-IP) assay of the 6hemagglutinin (HA)-tagged Nst1 with 9Myc-tagged Dcp2/Edc3/Dhh1/Ccr4 in the log phase cells, but could not observe any interaction (data not shown). We also tried to identify whether the interaction between the overexpressed full-length Nst1, $Nst1_{\Delta PD}$, $Nst1_{CTD}$ $Nst1_{CTD\Delta PD}$ and Dcp2 is biochemically detected. Co-IP of the overexpressed Nst1 and Nst1 domain deletion mutants with 9Myc-tagged Dcp2 was performed. Unfortunately, we could not detect any direct physical interaction between them (Supplemental Figure S3A,B), although microscopic examinations provided evidence of Dcp2, Edc3, Dhh1, and Xrn1 condensation and co-localization with Nst1 when Nst1 was overexpressed (Figures 5 and 6). These results imply that the interactions between Nst1 and known PB components were not strong enough to be detected biochemically but were sufficient to induce promiscuous interactions. Considering that polyampholyte regions are among the most frequently occurring IDRs in nature, our findings provide insights into the roles of polyampholyte multivalency in interacting with other PB constituents. Further studies are needed to understand the molecular mechanism by which the Nst1 polyampholyte recruits other PB components to form condensates.

This study on Nst1 domain deletion mutants further improves our understanding of the sequence elements with high aggregation propensity and the polyampholyte region in unstructured proteins involved in the self-condensation and condensation of PB components with liquid-like properties.

4. Materials and Methods

4.1. Yeast strains, Plasmids, and Cultures

Table 1 lists the *S. cerevisiae* strains and genotypes used in this study. The strains were constructed on the BY4741 or w303a wild-type background by integrating templates from the polymerase chain reaction (PCR) toolbox at the 3′ end of each reading frame in each endogenous locus through PCR-based homologous recombination [58]. We used PCR of the integrated locus and Western blotting to verify all the constructed strains.

Table 1. The yeast strains used in this study.

Strain Name	Genotype	Source
YSK3485	*DCP2-EGFP:HIS3MX6 BY4741 MATa his3Δ1 leu2Δ0 met15Δ0 ura3Δ0*	This study
YSK3482	*XRN1-EGFP::HIS3MX6 BY4741 MATa his3Δ1 leu2Δ0 met15Δ0 ura3Δ0*	This study
YSK3484	*DHH1-EGFP:HIS3MX6 BY4741 MATa his3Δ1 leu2Δ0 met15Δ0 ura3Δ0*	This study
YSK3534	*EDC3-EGFP::HIS3MX6 BY4741 MATa his3Δ1 leu2Δ0 met15Δ0 ura3Δ0*	This study
YSK3578	*DCP2-mKate2-sphis5 BY4741 MATa his3Δ1 leu2Δ0 met15Δ0 ura3Δ0*	This study
YSK3483	*BY4741 MATa his3Δ1 leu2Δ0 met15Δ0 ura3Δ0 wild-type*	This study
YSK3592	*DCP2-9MYC::HIS3MX6 BY4741 MATa his3Δ1 leu2Δ0 met15Δ0 ura3Δ0*	This study

All plasmids used in this study were constructed in *pMW20(U)-P_{GAL}-GFP* or *pMW20(U)-P_{GAL}*, as Table 2 describes. Nst1 domain deletion mutant clones were generated using the PCR-mediated deletion method [59] and confirmed by sequencing.

Yeast strains were cultured in YPAD or synthetic complete (SC) media containing 2% glucose. Glucose deprivation was induced in the SC medium without glucose. Yeast cells were cultured at 25 °C to an optical density of 600 nm (OD_{600}) $\leq$ 0.5 for logarithmic phase growth. Cells in the logarithmic phase were primarily cultured in SC-U + 2% glucose media to an OD_{600} of 0.5 and harvested to induce overexpression under the *GAL*. The cells were washed three times with Sc-U + 2% raffinose + 0.1% glucose medium, diluted to half of its concentration, and cultured for an additional 3 h in Sc-U + 2% raffinose + 0.1% glucose. Then, 20% galactose stock was added to the culture to adjust the final galactose concentration to 2%, and the cells were further incubated for 3 h for induction before collection.

Table 2. The plasmids used in this study.

Plasmids
pMW20-P_{GAL}-GFP-NST
pMW20-P_{GAL}-GFP-NSTΔ430–1240
pMW20-P_{GAL}-GFP-NST1Δ1–429
pMW20-P_{GAL}-GFP-NST1Δ1–429 Δ631–752
pMW20-P_{GAL}-GFP-NST1Δ1–429 Δ1016–1240
pMW20-P_{GAL}-GFP-NST1Δ1–1015
pMW20-P_{GAL}-GFP-NST1Δ631–752
pMW20-P_{GAL}-GFP-NST1Δ631–752 Δ1016–1240
pMW20-P_{GAL}-GFP-NST1Δ1016–1240
pMW20-P_{GAL}-GFP-NST1Δ1–1015
pMW20-P_{GAL}-NST
pMW20-P_{GAL}-NSTΔ430–1240
pMW20-P_{GAL}-NST1Δ1–429
pMW20-P_{GAL}-NST1Δ1–429 Δ631–752
pMW20-P_{GAL}-NST1Δ1–429 Δ1016–1240
pMW20-P_{GAL}-NST1Δ631–752
pMW20-P_{GAL}-NST1Δ631–752 Δ1016–1240
pMW20-P_{GAL}-NST1Δ1016–1240
pMW20-P_{GAL}-NST1Δ1–429 Δ631–752

4.2. 1,6-Hexanediol and CHX Treatments and Western Blots

Yeast *GAL* promoter induction was performed as described above. For CHX treatment, *GAL*-induced cells were incubated with 100 μg/mL CHX for 10 min. For 1,6-hexanediol treatment, the *GAL*-induced cells were washed three times with a medium containing 10% 1,6-hexanediol and incubated for 30 min. Western blotting was performed as described by Choi and Song [40] with anti-EGFP antibody (600-101-215 Rockland, Limerick, PA, USA) and anti-Tub1 (T5168, Sigma, St. Louis, MO, USA) positive controls. HRP-conjugated anti-goat (705-035-003, Jackson Immune Research, PA, USA) and anti-mouse (sc-2005, Santa Cruz Biotechnology, Dallas, TX, USA) antibodies were used as secondary antibodies to detect EGFP and anti-Tub1, respectively.

4.3. Nst1 Structure and Domain Predictions Based on the Sequence

Structural prediction of Nst1 was performed using GalaxyWEB [60]. The IUPRED2A, PONDR, and DISOPRED3 [42,43,61,62] algorithms, which predict IDRs, were used to analyze the disorder properties in the Nst1 sequence. AGGRESCAN, Tango, and PASTA 2.0 were used to predict aggregation-prone regions in the Nst1 sequence [45,46,63–65]. Domain deletion mutants of Nst1 were projected in the Das–Pappu phase diagram at http://pappulab.wustl.edu/CIDER/analysis/, (accessed on 6 April 2022).

4.4. Wide-Field Fluorescence Microscopy of Yeast Cells and Image Analysis

Fluorescence-labeled proteins were visualized using an Axioplan2 microscope (Carl Zeiss, Jena, Germany) with a 100× Plan-Neofluar oil immersion objective. Images were acquired using an Axiocam CCD camera and Axio Vision software (Carl Zeiss). The same culture conditions, exposure times, and fluorescence intensities were applied to all the strains observed in this study to compare the degree of puncta intensity. Images were analyzed as described below.

For colocalization analysis to deduce van Steensel's CCF, fluorescent images obtained through green or red channels were analyzed with the plugin JACoP v4.0 analysis tool using FIJI (Image J). CCFs were calculated and presented as a bell-shaped plot.

All images obtained were measured and analyzed with the same optics, filters, and zoom settings throughout the study using FIJI (ImageJ) to quantify PB condensation.

The pixel intensity of each GFP signal in the cells was scaled from 1 to 255 to investigate the intensity of puncta per cell. Pixels of the top 0.05% signal intensities of each strain were segmented to perform particle analysis and determine the individual punctum strength to deduce the maximum intensity value of each punctum. The highest pixel value of each punctum was presented. The total number of cells, the number of cells with puncta, and the ratio of cells with puncta were calculated to analyze the puncta generated.

4.5. Statistical Analysis

Detailed statistics, including the mean values and standard deviations, are indicated in each figure legend. Statistical analyses were performed using GraphPad Prism 6 (GraphPad Software, Inc., La Jolla, CA, USA). A t-test was used to assess statistically significant differences. $p < 0.05$ (*), $p < 0.01$ (**), $p < 0.001$ (***), and $p < 0.0001$ (****) indicate statistical significance compared with the control. $p > 0.05$ indicates statistical non-significance (n.s.).

Supplementary Materials: The following supporting information can be downloaded at: https://www.mdpi.com/article/10.3390/ijms23137380/s1.

Author Contributions: K.S. and Y.-J.C. conceived and designed the study. Y.-J.C. and Y.L. (Yujin Lee) performed the experiments. K.S., Y.-J.C. and Y.-H.L. contributed to the reagents/materials/analytical tools. K.S. and Y.-J.C. drafted the manuscript. Y.-H.L., Y.L. (Yuxi Lin), Y.H. and Y.-J.C. predicted the Nst1 domains. All authors have read and agreed to the published version of the manuscript.

Funding: This research was funded by the National Research Foundation of KOREA (NRF) (No. NRF-2017R1A2B4009785) via the Korean Government (MSIT). Y. Choi was partially supported by an NRF grant from the Korean government (NRF-2020R1A2C1102153). Y.H. Lee was supported by NRF grants funded by the Korean government (NRF-2019R1A2C1004954 and NRF-2022R1A2C1011793) and Korea Basic Science Institute grants (C220000, C230130, and C280320).

Institutional Review Board Statement: Not applicable.

Informed Consent Statement: Not applicable.

Data Availability Statement: Not applicable.

Acknowledgments: The authors deeply appreciate the kind gift of yeast strains from Parker (University of Colorado at Boulder).

Conflicts of Interest: The authors declare that there are no conflict of interest.

References

1. Banani, S.F.; Lee, H.O.; Hyman, A.A.; Rosen, M.K. Biomolecular condensates: Organizers of cellular biochemistry. *Nat. Rev. Mol. Cell Biol.* **2017**, *18*, 285–298. [CrossRef] [PubMed]
2. Gomes, E.; Shorter, J. The molecular language of membraneless organelles. *J. Biol. Chem.* **2019**, *294*, 7115–7127. [CrossRef] [PubMed]
3. Mao, Y.S.; Zhang, B.; Spector, D.L. Biogenesis and function of nuclear bodies. *Trends Genet.* **2011**, *27*, 295–306. [CrossRef] [PubMed]
4. Mitrea, D.M.; Kriwacki, R.W. Phase separation in biology; functional organization of a higher order. *Cell Commun. Signal* **2016**, *14*, 1. [CrossRef]
5. Handwerger, K.E.; Gall, J.G. Subnuclear organelles: New insights into form and function. *Trends Cell Biol.* **2006**, *16*, 19–26. [CrossRef]
6. Decker, C.J.; Parker, R. P-bodies and stress granules: Possible roles in the control of translation and mRNA degradation. *Cold Spring Harb. Perspect. Biol.* **2012**, *4*, a012286. [CrossRef]
7. Shin, Y.; Brangwynne, C.P. Liquid phase condensation in cell physiology and disease. *Science* **2017**, *357*, eaaf4382. [CrossRef]
8. Hyman, A.A.; Weber, C.A.; Julicher, F. Liquid-liquid phase separation in biology. *Annu. Rev. Cell Dev. Biol.* **2014**, *30*, 39–58. [CrossRef]
9. Alberti, S. Phase separation in biology. *Curr. Biol.* **2017**, *27*, R1097–R1102. [CrossRef]

10. Ditlev, J.A.; Case, L.B.; Rosen, M.K. Who's In and Who's Out-Compositional Control of Biomolecular Condensates. *J. Mol. Biol.* **2018**, *430*, 4666–4684. [CrossRef]
11. Teixeira, D.; Parker, R. Analysis of P-body assembly in Saccharomyces cerevisiae. *Mol. Biol. Cell* **2007**, *18*, 2274–2287. [CrossRef] [PubMed]
12. Teixeira, D.; Sheth, U.; Valencia-Sanchez, M.A.; Brengues, M.; Parker, R. Processing bodies require RNA for assembly and contain nontranslating mRNAs. *RNA* **2005**, *11*, 371–382. [CrossRef] [PubMed]
13. Jain, S.; Wheeler, J.R.; Walters, R.W.; Agrawal, A.; Barsic, A.; Parker, R. ATPase-Modulated Stress Granules Contain a Diverse Proteome and Substructure. *Cell* **2016**, *164*, 487–498. [CrossRef] [PubMed]
14. Khong, A.; Matheny, T.; Jain, S.; Mitchell, S.F.; Wheeler, J.R.; Parker, R. The Stress Granule Transcriptome Reveals Principles of mRNA Accumulation in Stress Granules. *Mol. Cell* **2017**, *68*, 808–820.e5. [CrossRef] [PubMed]
15. Youn, J.Y.; Dunham, W.H.; Hong, S.J.; Knight, J.D.R.; Bashkurov, M.; Chen, G.I.; Bagci, H.; Rathod, B.; MacLeod, G.; Eng, S.W.M.; et al. High-Density Proximity Mapping Reveals the Subcellular Organization of mRNA-Associated Granules and Bodies. *Mol. Cell* **2018**, *69*, 517–532.e11. [CrossRef]
16. Markmiller, S.; Soltanieh, S.; Server, K.L.; Mak, R.; Jin, W.; Fang, M.Y.; Luo, E.C.; Krach, F.; Yang, D.; Sen, A.; et al. Context-Dependent and Disease-Specific Diversity in Protein Interactions within Stress Granules. *Cell* **2018**, *172*, 590–604.e13. [CrossRef]
17. Van Damme, E.; Laukens, K.; Dang, T.H.; Van Ostade, X. A manually curated network of the PML nuclear body interactome reveals an important role for PML-NBs in SUMOylation dynamics. *Int. J. Biol. Sci.* **2010**, *6*, 51–67. [CrossRef]
18. Banani, S.F.; Rice, A.M.; Peeples, W.B.; Lin, Y.; Jain, S.; Parker, R.; Rosen, M.K. Compositional Control of Phase-Separated Cellular Bodies. *Cell* **2016**, *166*, 651–663. [CrossRef]
19. Bernardi, R.; Pandolfi, P.P. Structure, dynamics and functions of promyelocytic leukaemia nuclear bodies. *Nat. Rev. Mol. Cell Biol.* **2007**, *8*, 1006–1016. [CrossRef]
20. Ishov, A.M.; Sotnikov, A.G.; Negorev, D.; Vladimirova, O.V.; Neff, N.; Kamitani, T.; Yeh, E.T.; Strauss, J.F., 3rd; Maul, G.G. PML is critical for ND10 formation and recruits the PML-interacting protein daxx to this nuclear structure when modified by SUMO-1. *J. Cell Biol.* **1999**, *147*, 221–234. [CrossRef]
21. Zhong, S.; Muller, S.; Ronchetti, S.; Freemont, P.S.; Dejean, A.; Pandolfi, P.P. Role of SUMO-1-modified PML in nuclear body formation. *Blood* **2000**, *95*, 2748–2752. [CrossRef] [PubMed]
22. Li, Y.; Ma, X.; Chen, Z.; Wu, H.; Wang, P.; Wu, W.; Cheng, N.; Zeng, L.; Zhang, H.; Cai, X.; et al. B1 oligomerization regulates PML nuclear body biogenesis and leukemogenesis. *Nat. Commun.* **2019**, *10*, 3789. [CrossRef] [PubMed]
23. Jensen, K.; Shiels, C.; Freemont, P.S. PML protein isoforms and the RBCC/TRIM motif. *Oncogene* **2001**, *20*, 7223–7233. [CrossRef] [PubMed]
24. Shen, T.H.; Lin, H.K.; Scaglioni, P.P.; Yung, T.M.; Pandolfi, P.P. The Mechanisms of PML-Nuclear Body Formation. *Mol. Cell* **2006**, *24*, 805. [CrossRef]
25. Sanders, D.W.; Kedersha, N.; Lee, D.S.W.; Strom, A.R.; Drake, V.; Riback, J.A.; Bracha, D.; Eeftens, J.M.; Iwanicki, A.; Wang, A.; et al. Competing Protein-RNA Interaction Networks Control Multiphase Intracellular Organization. *Cell* **2020**, *181*, 306–324. [CrossRef]
26. Protter, D.S.W.; Rao, B.S.; Van Treeck, B.; Lin, Y.; Mizoue, L.; Rosen, M.K.; Parker, R. Intrinsically Disordered Regions Can Contribute Promiscuous Interactions to RNP Granule Assembly. *Cell Rep.* **2018**, *22*, 1401–1412. [CrossRef]
27. Kroschwald, S.; Maharana, S.; Mateju, D.; Malinovska, L.; Nuske, E.; Poser, I.; Richter, D.; Alberti, S. Promiscuous interactions and protein disaggregases determine the material state of stress-inducible RNP granules. *eLife* **2015**, *4*, e06807. [CrossRef]
28. Nott, T.J.; Petsalaki, E.; Farber, P.; Jervis, D.; Fussner, E.; Plochowietz, A.; Craggs, T.D.; Bazett-Jones, D.P.; Pawson, T.; Forman-Kay, J.D.; et al. Phase transition of a disordered nuage protein generates environmentally responsive membraneless organelles. *Mol. Cell* **2015**, *57*, 936–947. [CrossRef]
29. Patel, A.; Lee, H.O.; Jawerth, L.; Maharana, S.; Jahnel, M.; Hein, M.Y.; Stoynov, S.; Mahamid, J.; Saha, S.; Franzmann, T.M.; et al. A Liquid-to-Solid Phase Transition of the ALS Protein FUS Accelerated by Disease Mutation. *Cell* **2015**, *162*, 1066–1077. [CrossRef]
30. Lin, Y.; Protter, D.S.; Rosen, M.K.; Parker, R. Formation and Maturation of Phase-Separated Liquid Droplets by RNA-Binding Proteins. *Mol. Cell* **2015**, *60*, 208–219. [CrossRef]
31. Molliex, A.; Temirov, J.; Lee, J.; Coughlin, M.; Kanagaraj, A.P.; Kim, H.J.; Mittag, T.; Taylor, J.P. Phase separation by low complexity domains promotes stress granule assembly and drives pathological fibrillization. *Cell* **2015**, *163*, 123–133. [CrossRef] [PubMed]
32. Murakami, T.; Qamar, S.; Lin, J.Q.; Schierle, G.S.; Rees, E.; Miyashita, A.; Costa, A.R.; Dodd, R.B.; Chan, F.T.; Michel, C.H.; et al. ALS/FTD Mutation-Induced Phase Transition of FUS Liquid Droplets and Reversible Hydrogels into Irreversible Hydrogels Impairs RNP Granule Function. *Neuron* **2015**, *88*, 678–690. [CrossRef] [PubMed]
33. Xiang, S.; Kato, M.; Wu, L.C.; Lin, Y.; Ding, M.; Zhang, Y.; Yu, Y.; McKnight, S.L. The LC Domain of hnRNPA2 Adopts Similar Conformations in Hydrogel Polymers, Liquid-like Droplets, and Nuclei. *Cell* **2015**, *163*, 829–839. [CrossRef]
34. Borcherds, W.; Bremer, A.; Borgia, M.B.; Mittag, T. How do intrinsically disordered protein regions encode a driving force for liquid-liquid phase separation? *Curr. Opin. Struct. Biol.* **2020**, *67*, 41–50. [CrossRef] [PubMed]
35. Elbaum-Garfinkle, S.; Kim, Y.; Szczepaniak, K.; Chen, C.C.; Eckmann, C.R.; Myong, S.; Brangwynne, C.P. The disordered P granule protein LAF-1 drives phase separation into droplets with tunable viscosity and dynamics. *Proc. Natl. Acad. Sci. USA* **2015**, *112*, 7189–7194. [CrossRef] [PubMed]

36. Han, T.W.; Kato, M.; Xie, S.; Wu, L.C.; Mirzaei, H.; Pei, J.; Chen, M.; Xie, Y.; Allen, J.; Xiao, G.; et al. Cell-free formation of RNA granules: Bound RNAs identify features and components of cellular assemblies. *Cell* **2012**, *149*, 768–779. [CrossRef] [PubMed]
37. Kshirsagar, M.; Parker, R. Identification of Edc3p as an enhancer of mRNA decapping in Saccharomyces cerevisiae. *Genetics* **2004**, *166*, 729–739. [CrossRef]
38. Bouveret, E.; Rigaut, G.; Shevchenko, A.; Wilm, M.; Seraphin, B. A Sm-like protein complex that participates in mRNA degradation. *EMBO J.* **2000**, *19*, 1661–1671. [CrossRef]
39. Tharun, S.; He, W.; Mayes, A.E.; Lennertz, P.; Beggs, J.D.; Parker, R. Yeast Sm-like proteins function in mRNA decapping and decay. *Nature* **2000**, *404*, 515–518. [CrossRef]
40. Choi, Y.J.; Song, K. Nst1, Densely Associated to P-Body in the Post-Exponential Phases of Saccharomyces cerevisiae, Shows an Intrinsic Potential of Producing Liquid-Like Condensates of P-Body Components in Cells. *Int. J. Mol. Sci.* **2022**, *23*, 2501. [CrossRef]
41. Meszaros, B.; Erdos, G.; Dosztanyi, Z. IUPred2A: Context-dependent prediction of protein disorder as a function of redox state and protein binding. *Nucleic Acids Res.* **2018**, *46*, W329–W337. [CrossRef] [PubMed]
42. Romero, P.; Obradovic, Z.; Li, X.; Garner, E.C.; Brown, C.J.; Dunker, A.K. Sequence complexity of disordered protein. *Proteins* **2001**, *42*, 38–48. [CrossRef]
43. Jones, D.T.; Cozzetto, D. DISOPRED3: Precise disordered region predictions with annotated protein-binding activity. *Bioinformatics* **2015**, *31*, 857–863. [CrossRef]
44. Dunker, A.K.; Babu, M.M.; Barbar, E.; Blackledge, M.; Bondos, S.E.; Dosztanyi, Z.; Dyson, H.J.; Forman-Kay, J.; Fuxreiter, M.; Gsponer, J.; et al. What's in a name? Why these proteins are intrinsically disordered: Why these proteins are intrinsically disordered. *Intrinsically Disord Proteins* **2013**, *1*, e24157. [CrossRef]
45. Conchillo-Sole, O.; de Groot, N.S.; Aviles, F.X.; Vendrell, J.; Daura, X.; Ventura, S. AGGRESCAN: A server for the prediction and evaluation of "hot spots" of aggregation in polypeptides. *BMC Bioinform.* **2007**, *8*, 65. [CrossRef]
46. Fernandez-Escamilla, A.M.; Rousseau, F.; Schymkowitz, J.; Serrano, L. Prediction of sequence-dependent and mutational effects on the aggregation of peptides and proteins. *Nat. Biotechnol.* **2004**, *22*, 1302–1306. [CrossRef]
47. Walsh, I.; Seno, F.; Tosatto, S.C.; Trovato, A. PASTA 2.0: An improved server for protein aggregation prediction. *Nucleic Acids Res.* **2014**, *42*, W301–W307. [CrossRef] [PubMed]
48. Batchelor, M.; Paci, E. Helical Polyampholyte Sequences Have Unique Thermodynamic Properties. *J. Phys. Chem. B* **2018**, *122*, 11784–11791. [CrossRef]
49. Bianchi, G.; Longhi, S.; Grandori, R.; Brocca, S. Relevance of Electrostatic Charges in Compactness, Aggregation, and Phase Separation of Intrinsically Disordered Proteins. *Int. J. Mol. Sci.* **2020**, *21*, 6208. [CrossRef]
50. Mao, A.H.; Crick, S.L.; Vitalis, A.; Chicoine, C.L.; Pappu, R.V. Net charge per residue modulates conformational ensembles of intrinsically disordered proteins. *Proc. Natl. Acad. Sci. USA* **2010**, *107*, 8183–8188. [CrossRef]
51. Ribbeck, K.; Gorlich, D. The permeability barrier of nuclear pore complexes appears to operate via hydrophobic exclusion. *EMBO J.* **2002**, *21*, 2664–2671. [CrossRef] [PubMed]
52. Patel, S.S.; Belmont, B.J.; Sante, J.M.; Rexach, M.F. Natively unfolded nucleoporins gate protein diffusion across the nuclear pore complex. *Cell* **2007**, *129*, 83–96. [CrossRef] [PubMed]
53. Bolte, S.; Cordelieres, F.P. A guided tour into subcellular colocalization analysis in light microscopy. *J. Microsc-Oxf.* **2006**, *224*, 213–232. [CrossRef]
54. Decker, C.J.; Teixeira, D.; Parker, R. Edc3p and a glutamine/asparagine-rich domain of Lsm4p function in processing body assembly in Saccharomyces cerevisiae. *J. Cell Biol.* **2007**, *179*, 437–449. [CrossRef] [PubMed]
55. Das, R.K.; Pappu, R.V. Conformations of intrinsically disordered proteins are influenced by linear sequence distributions of oppositely charged residues. *Proc. Natl. Acad. Sci. USA* **2013**, *110*, 13392–13397. [CrossRef]
56. Coller, J.; Parker, R. General translational repression by activators of mRNA decapping. *Cell* **2005**, *122*, 875–886. [CrossRef]
57. Pantuso, E.; Mastropietro, T.F.; Briuglia, M.L.; Gerard, C.J.J.; Curcio, E.; Ter Horst, J.H.; Nicoletta, F.P.; Di Profio, G. On the Aggregation and Nucleation Mechanism of the Monoclonal Antibody Anti-CD20 Near Liquid-Liquid Phase Separation (LLPS). *Sci. Rep.* **2020**, *10*, 8902. [CrossRef]
58. Janke, C.; Magiera, M.M.; Rathfelder, N.; Taxis, C.; Reber, S.; Maekawa, H.; Moreno-Borchart, A.; Doenges, G.; Schwob, E.; Schiebel, E.; et al. A versatile toolbox for PCR-based tagging of yeast genes: New fluorescent proteins, more markers and promoter substitution cassettes. *Yeast* **2004**, *21*, 947–962. [CrossRef]
59. Hansson, M.D.; Rzeznicka, K.; Rosenback, M.; Hansson, M.; Sirijovski, N. PCR-mediated deletion of plasmid DNA. *Anal. Biochem.* **2008**, *375*, 373–375. [CrossRef]
60. Ko, J.; Park, H.; Heo, L.; Seok, C. GalaxyWEB server for protein structure prediction and refinement. *Nucleic Acids Res.* **2012**, *40*, W294–W297. [CrossRef]
61. Dosztanyi, Z.; Csizmok, V.; Tompa, P.; Simon, I. IUPred: Web server for the prediction of intrinsically unstructured regions of proteins based on estimated energy content. *Bioinformatics* **2005**, *21*, 3433–3434. [CrossRef] [PubMed]
62. Linding, R.; Russell, R.B.; Neduva, V.; Gibson, T.J. GlobPlot: Exploring protein sequences for globularity and disorder. *Nucleic Acids Res.* **2003**, *31*, 3701–3708. [CrossRef] [PubMed]
63. Frousios, K.K.; Iconomidou, V.A.; Karletidi, C.M.; Hamodrakas, S.J. Amyloidogenic determinants are usually not buried. *BMC Struct. Biol.* **2009**, *9*, 44. [CrossRef] [PubMed]

64. Maurer-Stroh, S.; Debulpaep, M.; Kuemmerer, N.; Lopez de la Paz, M.; Martins, I.C.; Reumers, J.; Morris, K.L.; Copland, A.; Serpell, L.; Serrano, L.; et al. Exploring the sequence determinants of amyloid structure using position-specific scoring matrices. *Nat. Methods* **2010**, *7*, 237–242. [CrossRef] [PubMed]
65. Potenza, E.; Di Domenico, T.; Walsh, I.; Tosatto, S.C. MobiDB 2.0: An improved database of intrinsically disordered and mobile proteins. *Nucleic Acids Res.* **2015**, *43*, D315–D320. [CrossRef]

International Journal of
Molecular Sciences

Article

BIAPSS: A Comprehensive Physicochemical Analyzer of Proteins Undergoing Liquid–Liquid Phase Separation

Aleksandra E. Badaczewska-Dawid [1], Vladimir N. Uversky [2,*] and Davit A. Potoyan [1,3,*]

1 Department of Chemistry, Iowa State University, Ames, IA 50011, USA; abadacz@iastate.edu
2 Department of Molecular Medicine and USF Health Byrd Alzheimer's Research Institute, Morsani College of Medicine, University of South Florida, Tampa, FL 33612, USA
3 Bioinformatics and Computational Biology Program, Iowa State University, Ames, IA 50011, USA
* Correspondence: vuversky@usf.edu (V.N.U.); potoyan@iastate.edu (D.A.P.)

Abstract: The liquid–liquid phase separation (LLPS) of biomolecules is a phenomenon which is nowadays recognized as the driving force for the biogenesis of numerous functional membraneless organelles and cellular bodies. The interplay between the protein primary sequence and phase separation remains poorly understood, despite intensive research. To uncover the sequence-encoded signals of protein capable of undergoing LLPS, we developed a novel web platform named BIAPSS (Bioinformatics Analysis of LLPS Sequences). This web server provides on-the-fly analysis, visualization, and interpretation of the physicochemical and structural features for the superset of curated LLPS proteins.

Keywords: liquid–liquid phase separation; membraneless organelles; intrinsically disordered proteins; proteins with low complexity

Citation: Badaczewska-Dawid, A.E.; Uversky, V.N.; Potoyan, D.A. BIAPSS: A Comprehensive Physicochemical Analyzer of Proteins Undergoing Liquid–Liquid Phase Separation. *Int. J. Mol. Sci.* **2022**, *23*, 6204. https://doi.org/10.3390/ijms23116204

Academic Editor: Vitaly V. Kushnirov

Received: 1 May 2022
Accepted: 27 May 2022
Published: 31 May 2022

1. Introduction

The spatiotemporal organization of biomolecules and biomolecular interactions is essential for the efficient regulation of cellular biochemistry. The underlying biophysical mechanism for membraneless compartmentalization is liquid–liquid phase separation (LLPS). In the past few years, the LLPS of biomolecules has become a unifying physical mechanism for understanding the principles of intracellular compartmentalization, the formation of membraneless organelles (MLOs), and gene regulation [1–14]. In the LLPS process, the relatively well-mixed solution of biomolecules separates into liquid droplets. The ability of proteins to phase separate appears to be encoded primarily in the peculiarities of their primary sequences, which often contain low-complexity regions and intrinsically disordered regions (IDRs) that are enriched in charged and multivalent interaction centers [6–8,10,11,13–19]. While some general sequence trends have emerged, the quantitative aspects of how amino acid sequences encode and decode phase separation still remain largely unknown [20–22]. This is because many different combinations of relevant interactions seem to be contributing to phase separation, without any one being universally necessary [23]. As a consequence (with a few exceptions [24–30]), mostly case-by-case studies of different sequences are performed, with the broader context of many findings, including their statistical significance, remaining unknown.

Following the statistical trends in PubMed, biological LLPS has been gaining widespread attention in the last two decades. The rapidly growing amount of data from both in vitro and in vivo experiments have systematically narrowed the range of the LLPS-promoting conditions [31]. From these studies, we know that the regulatory mechanisms of phase separation appear strongly context-dependent [31]. The key factors include: the physicochemical state of the protein (e.g., posttranslational modifications), the environmental conditions (e.g., temperature, pH), and the concentration of binding partners (e.g., proteins, nucleic acids, carbohydrates, lipids). Many among the recent hypotheses suggest the prevalence of: (i) electrostatics and π-stacking; or (ii) specific sequence decoration in charge or hydrophobicity; and (iii) the role of short sequential (e.g., GARs (glycine–arginine-rich) [32]) or structural (e.g., LARKS (low-complexity amyloid-like reversible kinked segments) [33]) motifs [15]. However, deciphering the interplay between sequence composition and phase separation turns out to be challenging.

In recent years, several databases have emerged that collect LLPS-related protein sequence data and metadata, with prominent examples being PhaSepDB [25], PhaSePro [26], LLPSDB [27,28], and DrLLPS [29]. These databases collect and annotate partially overlapping sets of phase-separating protein sequences, including data on the experimental conditions and significant annotations. In particular, PhaSePro, LLPSDB, and a subset of PhaSepDB contain manually curated proteins, which are recognized for driving the formation of subcellular compartments.

The accumulation of high-quality datasets is certainly a necessary condition for making progress towards uncovering the driving forces of protein phase separation. However, one needs a biophysically motivated computational infrastructure to be able to harness the data from carefully and manually curated sets of phase-separating proteins for revealing the molecular features that determine protein phase separation. We argue that providing the concise but informative patterns of various features, all together horizontally stacked along the protein sequence, could improve the identification of the significant yet nontrivial correlations that contribute to the multivalent interactions. On the basis of these premises, we have developed a novel web platform named BIAPSS: Bioinformatics Analysis of Liquid–Liquid Phase-Separating Protein Sequences (available at https://biapss.chem.iastate.edu/ and last accessed on 31 May 2022). BIAPSS combines a high-throughput interactive deep sequence analysis with a comprehensive pre-parsed bioinformatics database containing a wide array of physicochemical and evolutionary features that are relevant for low-complexity, disordered, and ordered proteins. This platform provides scientists working in the field of biomolecular condensates with a versatile tool for the rapid and on-the-fly deep statistical analysis of LLPS-driver protein sequences.

2. Results and Discussion

2.1. Introduction of BIAPSS

Figure 1 represents the features included in the comprehensive BIAPSS analyzer. These features combine sequence composition and biophysical properties. The composition component is represented by: (i) the amino acid content, including frequencies and patterning (i.e., distribution and enriched regions); and (ii) the sequence complexity, which comprises the detection of low-complexity regions, repeats, short motifs, and the on-the-fly calculation of Shannon entropy. The biophysical component covers the physicochemical and structural properties. Specifically, we provide a set of residue-resolution patterns: polarity, hydrophobicity, aromaticity, charge induced interactions and hydrogen bonding. These properties, correlated with the experimentally confirmed LLPS regions, facilitate the identification of the nature and driving forces of interactions. The structural properties aid in filtering out the interactions involved primarily in stabilizing the structure or in identifying regions prone to disorder-to-order transitions.

Previously, such structural switchers were recognized in low-complexity and internally disordered sequences that function via phase separation [34]. Thus, the collected molecular properties incorporate robust sequence-based predictions for the secondary structure, solvent accessibility, intrinsic disorder, and intramolecular contacts. Finally, the evolutionary context derived from the joint outcome of the HMMER-based analysis [35] and Pfam database search [36] highlights the location of functional domains, including those specialized in nucleic acid recognition. The other highly conserved short motifs or individual positions detected through the analysis of multiple sequence alignments may confirm that evolution deliberately preserves phase separation.

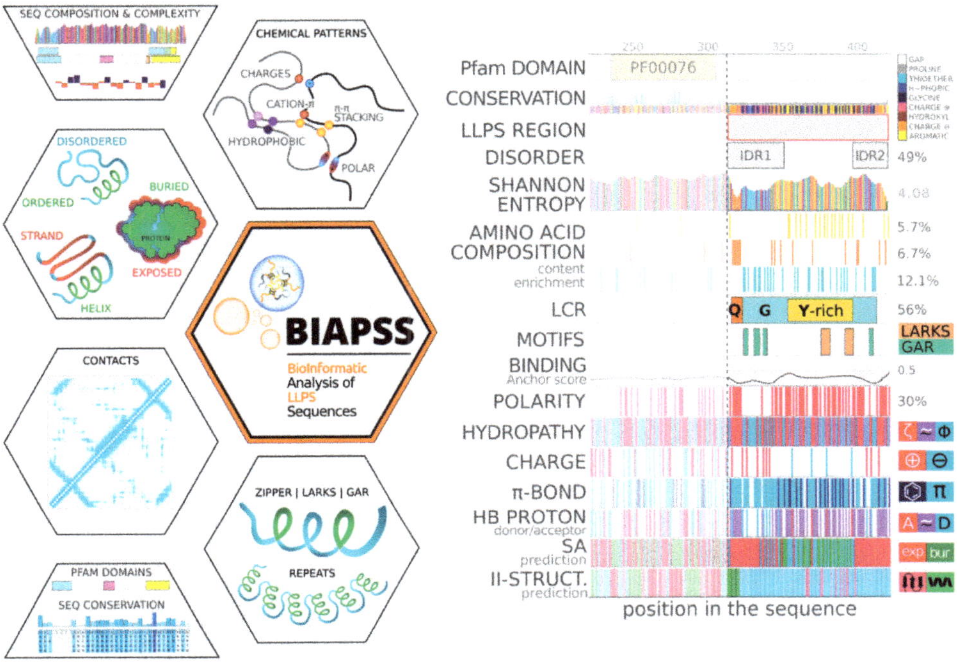

Figure 1. The comprehensive BIAPSS analyzer incorporates the compositional, evolutionary, physicochemical, and structural properties of LLPS proteins. All characteristics can be easily compared and correlated on the horizontally stacked multirow graphs. The interactive exploration helps to filter out sequence signals relevant for phase separation.

The comprehensive approach adopted in BIAPSS (Bioinformatics Analysis of LLPS Sequences) consists of integrating multiple third-party tools and high-performance computing, followed by in-house biostatistical analysis and the extraction of meaningful results (see the Materials and Methods section). The protocol was successfully applied for 501 proteins with experimental evidence of phase separation. Moreover, to the best of our knowledge, the resulting platform represents the broadest database with physicochemically characterized LLPS proteins (see Figure 2). In particular, the pool of entries completely covers the contents of several primary databases of curated LLPS deposits (PhaSePro, PhaSepDB.v1, LLPSDB), which collect annotations and experimental conditions. High interest in the phase-separation phenomenon has already spurred the growth of experimental data repositories. However, the deficiency of the computational infrastructure that targets the integrated biophysical and statistical analysis of phase-separating systems still hampers progress in the field. Therefore, in addition to the open access to the raw yet standardized and well-documented results of our extensive work, we have developed a web-based BIAPSS platform for interactive customized exploration and easy interpretation.

As a user-friendly web server, BIAPSS (https://biapss.chem.iastate.edu/, last accessed on 31 May 2022) is billing itself as a central resource for the systematic and standardized statistical analysis of the biophysical characteristics of the known LLPS sequences.

The web service provides users with:

(i) A database of the superset of experimentally evidenced LLPS-driver protein sequences.
(ii) A repository of precomputed bioinformatics and statistics data.
(iii) Two sets of web applications supporting the interactive analysis and visualization of the physicochemical and biomolecular characteristics of LLPS proteins.

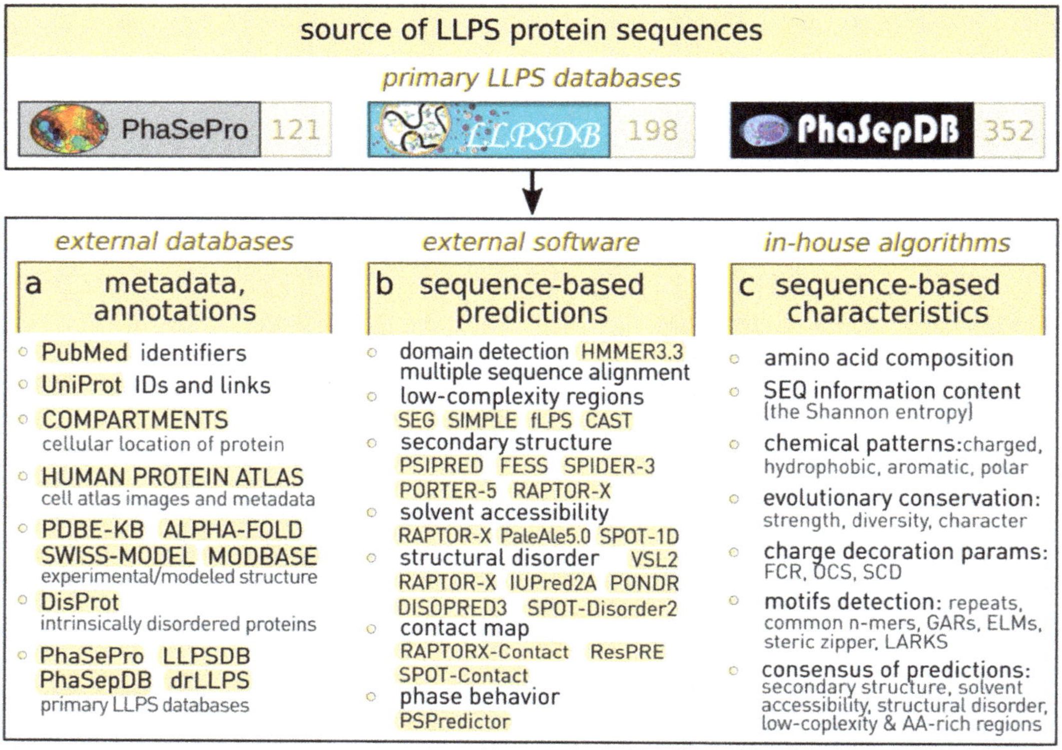

Figure 2. The BIAPSS repository collects the largest dataset of known LLPS proteins that have been identified from carefully curated primary LLPS databases. The computational framework starts from protein sequences downloaded from the UniProt database. The approach builds on three complementary components: (**a**) integration of metadata, annotation, and cross-links from the external databases; (**b**) comprehensive sequence-based bioinformatic analysis of evolutionary and biomolecular properties using state-of-the-art third-party software; and (**c**) meticulous physicochemical and compositional analysis and robust data integration using the in-house algorithms. The BIAPSS interactive web applications enable exploration through the distilled essence of the crafted characteristics.

The applications integrate the results from our comprehensive computational approach. The SingleSEQ module includes a residue-resolution biophysical analyzer for interrogating individual protein sequences. The complementary analyses are organized in nine web applications that toggle between a generalized summary view and details specific to a given characteristic. The latter allows users to correlate regions prone to phase separation with an array of physicochemical attributes, structural properties, detected domains, and various sequential or structural motifs. Many characteristics provided by applications in the SingleSEQ pipeline are qualitative and show a profile or pattern of the feature along the amino acid sequence. Examples include distributions of physicochemical characteristics, such as polarity, hydrophobicity, charge, residues forming hydrogen bonds, and pi-stacking. In these cases, the assignment is binary, and the numerical value is the percentage of residues in the sequence that meet the criterion. The second group of characteristics includes structural features predicted based on the amino acid sequences with top-ranked tools. Here, examples include the secondary structure, the solvent accessibility, the tendency to disorder, and low-complexity regions. The visual representation is developed to assign each position along the amino acid sequence a discrete consensus value (e.g., helical or extended, or coil for the secondary structure). The numerical value is the percentage of residues that meet the given criterion (e.g., % of helical). Figure 1 is a concept image, while, in the interactive graphs, there is a label of what the given value refers to. Furthermore, for those interested in in-depth analyses, the individual applications offer an on-the-fly exploration of the results from the original tools, which typically provide the fractional probabilities for each variant of a feature (e.g., p(helical), p(extended), p(coil)) for each position along the protein sequence.

BIAPSS also includes the MultiSEQ module. One of its aims is to obtain insight into the overall characteristics of the sufficient nonredundant set of LLPS-driver protein sequences. The comparison to the benchmarks of various protein groups enables a statistical inference of specific phase-separating affinities. Finally, BIAPSS incorporates an extensive cross-reference section that links all entries to primary LLPS databases and other external resources, thereby serving as a central navigation hub for the phase-separation community. All the data used by BIAPSS are freely available for download as well-formatted files with detailed descriptions, facilitating rapid implementation in user-defined computational protocols. The long-term plan for BIAPSS is for it to serve as a unifying hub for the experimental and computational community. Thus, it provides a comprehensive set of analytic tools, biophysically featured data, and standardized protocols that facilitate the identification of the sequence signals that drive the LLPS, which altogether can support applications for designing new sequences of biomedical interest.

2.2. Case Study and Tutorial: Fused in Sarcoma (FUS)

To illustrate the practical utility of BIAPSS, we carefully interpreted the results for fused in sarcoma (FUS) (UniProt ID: P35637), which is a widely used model system to study biological phase separation [37]. We provide below the details of the BIAPSS-based analysis, combined with a handy tutorial on the BIAPSS functionalities.

Fused in sarcoma (FUS) is one of the early discovered biological systems that undergoes self-organization by liquid–liquid phase separation (LLPS) [37]. Since then, the protein has been the subject of extensive experimental and computational research to understand the molecular mechanisms and interactions that drive this phenomenon. FUS can be found in the SingleSEQ module of the BIAPSS service by the UniProt identifier (P35637), the gene (FUS), or by using the "RNA-binding" search key (Last accessed on May 30 2022). The summary page contains a high-quality image of the experimentally confirmed cellular location (left panel in Figure 3). Due to its multifunctionality in RNA processing, FUS is mostly observed in the nucleus [38]. In physiological conditions, the low levels of the protein are distributed in the cytoplasm [39], where FUS transports and manages RNA through the dynamic liquid-like subcellular compartments, such as ribonucleoprotein or stress granules [40]. However, the cytoplasmic concentration of FUS significantly increases when noxious mutations lead to aggregation [41].

This progressively aberrant process is manifested by neurodegenerative diseases in humans [41]. Although plenty of accumulated evidence points to the influence of distinct factors on the cellular behavior of FUS, its primary sequence still holds many cues. To frame the physicochemical properties of full-length FUS, we used the analytical approach offered by the SingleSEQ module of BIAPSS.

The average metrics, available in the *Summary* of the SingleSEQ module, indicate that the 526-residue-long sequence of FUS contains over 80% disorder and only 8% order. The solvent-accessibility predictions show the same aspect ratio between exposure and burial. The contents of aromatic, hydrophobic, polar, and charged residues are 10%, 42%, 40%, and 17%, respectively, with a slight excess of positive charge. Such a rough overview described by a set of averages gives some general insight into the protein properties, but it conceals some local distributions that are important for the identification of the preferential interactions.

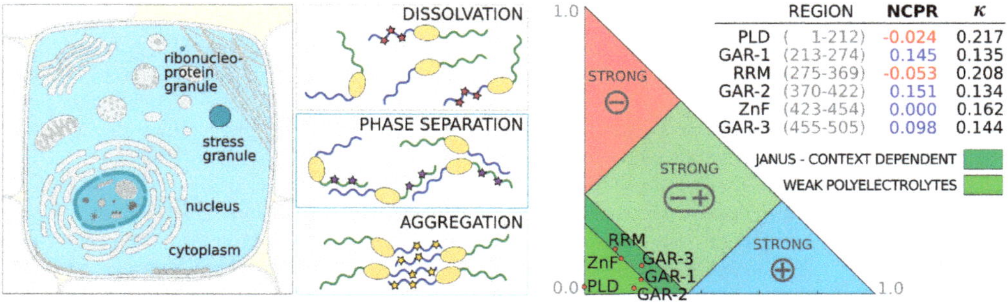

Figure 3. The **left** panel shows the cellular location of FUS (image source: BIAPSS). The protein is predominantly located in the nucleus. The physiological low levels of FUS found in the cytoplasm typically self-organize to membraneless compartments, such as stress granule or ribonucleoprotein granule. The aberrant disease-related aggregates are mostly localized in the cytoplasm. The methylation (purple stars) of C-terminal arginines (green tail) in the wild-type FUS strongly promotes phase separation and gelation. The phosphorylation (magenta stars) of serine and threonine in the N-terminus (blue tail) dissolve liquid-liquid droplets. The tyrosine-to-phenylalanine mutants (yellow stars) in the N-terminus and hypomethylation of arginines in the C-terminus increase aggregation. The **right** panel shows classification of intrinsically disordered ensemble regions (CIDER) [42] for the FUS sequence split into functional segments.

Therefore, we conduct a detailed analysis of the composition and complexity of the FUS sequence, and we present the resulting patterns in Figure 4. Compared to any reference set of proteins (use *Composition and Complexity* app), this one is extremely enriched in glycine, which makes up nearly 1/3 of the full sequence. Another 20% of the amino acid content consists primarily of serine and glutamine. Although the dominant content of these three amino acids suggests the generally low complexity of the sequence, their distribution along the sequence is strongly heterogeneous. Indeed, the calculated low information content of the sequence is mainly localized around protein terminals and clearly corresponds to three fragments with high glycine concentrations (LCR2: residues 164–267; LCR3: residues 370–420; LCR4: residues 454–507). These regions also exclusively accumulate total arginines, which, together with glycine, form a series of RGG repeating motifs that are known to bind RNA specifically [32]. Both serine and glutamine are mostly localized at the N-terminus, being more clearly clustered within LCR1 (1–163). LCR1 additionally gathers 24/35 available tyrosines, and, thus, it has visibly distinct enrichment (SQYG) that is known to occur in prion-like domains (PLD) [43]. By using the *Domains, Motifs, Repeats* application, we also found that the remaining compositionally more complex regions of the C-terminus (I287-L365 and R422-D453) match the PF00076 and PF00641 Pfam domains (i.e., the RNA recognition motif (RRM) and RNA-binding zinc finger (ZnF), respectively). The robust predictions (for details, see Methods) unanimously show that RRM is a well-folded FUS domain, while the other fragments remain disordered.

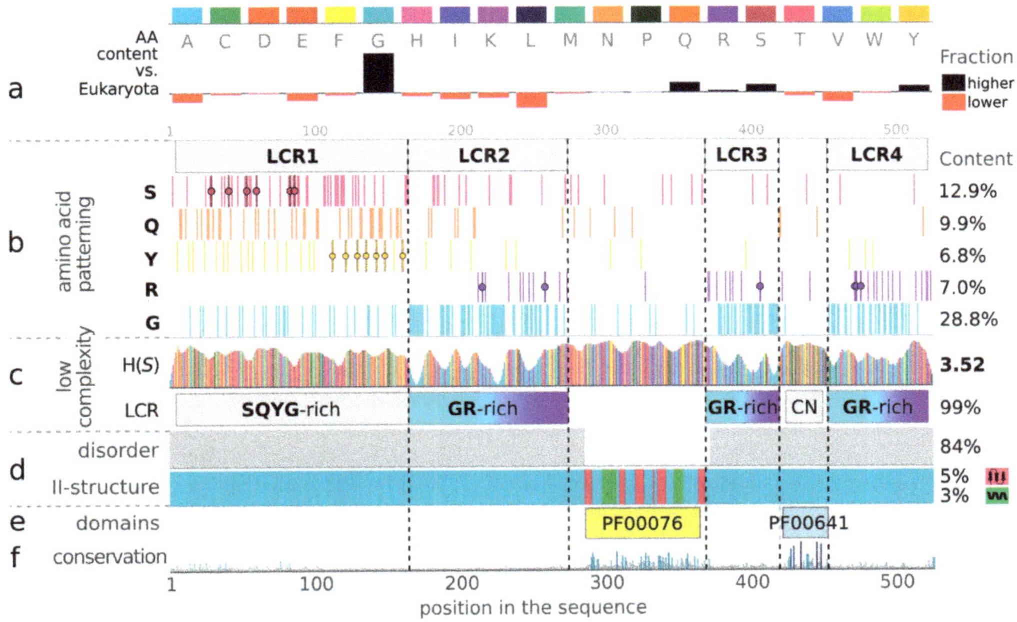

Figure 4. Sequence composition and complexity of FUS. The upper panel (**a**) shows the amino acid (AA) content of the query sequence compared to the Eukaryota dataset (black indicates higher and red the lower content than the reference). The bottom panel shows, the following information for a specific query sequence: (**b**) the patterning of enriched amino acids (S (magenta), Q (orange), Y (yellow), R (purple), G (cyan)); (**c**) low-complexity measures (a color scale corresponding to each amino acid) provided as regions of particular AA enrichments (LCR row), and the sequence information content (H(S) row, Shannon entropy); (**d**) consensus of predicted disorder regions (gray) and secondary-structure assignment (helix in green, strand in magenta, coil in light blue); (**e**) detected Pfam domains; and (**f**) evolutionary conservation derived from the multiple sequence alignment against UniRef50 (blue shades). The amino acid patterning section contains points corresponding to the locations of the most relevant serine phosphorylation sites (residues 30, 42, 54, 61, 84, 87) [37], arginine methylations (residues 216, 259, 407, 472, 473, 476) [44], and tyrosine mutations (residues 113, 122, 130, 136, 143, 149, 161) [44].

The seed MSAs prepared for FUS within the *Sequence Conservation* application further confirm that both domains are evolutionarily conserved members of Pfam families: RRM_1 and zf-RanBP, respectively (see bottom rows in Figure 4). The visual inspection of the amino acid content and the distribution of FUS allows us to identify and isolate specific regions in the protein (Figure 5). Furthermore, we have performed a physicochemical featurization of these segments, using the *Chemical Properties Patterns* app, which reveals the preferred interactions when coupled with biomolecular conditionals that are known from experiments. The recent experimental reports show that the isolated prion-like domain (PLD) (residues 1–214, or even residues 1–163) can undergo self-organization, forming liquid droplets when kept at high protein levels or high salt concentrations [44,45]. This N-terminal fragment is enriched in amino acids, whose side chains are multivalent, as shown in Figure 5. Thus, the dense pattern of polarity comes from the enrichment in S, Q, Y, where Y, Q, and G also provide π-electron centers for π-π-stacking. Most of them are also able to be both donors and acceptors of side-chain protons for hydrogen bonding (HB). In line with this, the intermolecular-interaction profiles derived from simulations of the 120–163 region indicate the most frequent contacts between QQ > QY > YY > SY and other pairs of enriched amino acids [46]. All of these observations suggest that the homotypic phase separation of wild-type PLD monomers is driven by balanced contributions from hydrogen bonding and π-stacking. Indeed, several mutagenesis studies show that Y→A substitution disrupts phase separation by the removal of both components of the interaction, while Y→F mutants are significantly more aggregation-prone, due to the strengthening of the binding via tighter hydrophobic F-π-stacking at the cost of losing HB contributions of polar tyrosine [44,46]. It is also worth noting that the PLD region is completely deficient of positive charge, with a minor net charge per residue of −0.01 (M1-S165: −0.012; and M1-G212: −0.024), which places it within the weak polyelectrolyte region on the CIDER diagram (right panel in Figure 3) [42]. However, an excess of serine and threonine in this region provides an ability to introduce a strongly negative charge through multiple phosphorylations. After phosphorylation, the dominant force becomes electrostatic repulsion, which is known to disrupt both phase separation and aggregation [37]. The central region of the PLD (residues 39–95) was proposed as the core of aberrant fibrils, which, in solid-state form, structured cross-β-sheets [37]. The same structural properties have not been unambiguously confirmed in the condensed phase of liquid–liquid mixing. Undoubtedly, however, our algorithms detected, along this region, structural motifs known as low-complexity amyloid-like reversible folded segments (LARKS) [33]. In our analysis, the most effective predictors of structural properties showed, for these motifs, some tendency towards an extended secondary structure and a slightly increased probability of burial (bottom panel in Figure 5; use *Secondary Structure*, *Solvent Accessibility* and *Structural Disorder* applications of the BIAPSS SingleSEQ module). Interestingly, the prediction in eight-letter notation detected a turn or bend within each of the structural motifs, which explains their flexible nature. These findings, together with the ambiguous experimental results, may suggest some variations in the structural state in the PLD core, and specifically the disorder-to-order transition driven by biomolecular conditionals.

The remaining part of the FUS sequence, referred to as the C-terminus, contains two well-known domains (RRM and ZnF) and three glycine–arginine-rich regions (GARs), which are detected using the *Domains, Motifs, Repeats* application. All components are significant players in binding RNA. Zinc finger supports only the recognition of the specific GGU motif, while the RRM domain and RGG repeats are universal towards a variety of RNAs [47]. Both folded domains of FUS are much less polar than the PLD, as seen from the BIAPSS-based physicochemical features in Figure 5. They also have a lower content of side chains that are able to engage in π-stacking or hydrogen bonding. However, the charged residues are pretty abundant in the composition of RRM and ZnF, which explains the functional role of electrostatic interactions towards the binding of nucleic acids or stabilizing folds via salt bridges [48,49].

All three GARs are the least polar regions of the protein (see Figure 5; use *Chemical Properties Patterns* app of BIAPSS SingleSEQ module). The dense patterning of hydrophobicity arises from glycine excess. The rich π-electron-containing systems, other than aromatic side chains, originate mainly from the abundance of the arginine's guanidino group. Arginine is also a source of excess positive charge at the C-terminus. The experimental studies consistently confirm that the isolated C-terminus does not undergo phase separation [44]. However, liquid–liquid droplets rapidly occur when mixed with N-terminal monomers [44]. Moreover, the LLPS of full-length wild-type FUS is more robust than the heterotypic mixing of the N- and C-terminals and the homotypic self-assembling of N-terminal monomers [44]. This suggests the higher priority of cation-π (R-Y) stacking over π-π (Y-Y) stacking, while both are reinforced by hydrogen bonds. Another experimental study showed that R→K mutants, who no longer have the ability of π-π-stacking but retain charge, can still undergo phase separation. In turn, R→A substitutions prevent phase separation because they lose the π-system, cation, and ability of side-chain hydrogen bonding. Interestingly, the recent report indicates that stacking interactions, including cation-π (e.g., RY, KF) and especially π-π (e.g., YY and RY, and even RQ), are most robust over a wide range of salt concentrations [45]. The hydrophobic contribution from π-electron-containing systems becomes the main force that strengthens the contact in high salt. In these conditions, the screening of usually dominant electrostatic contributions is significant. Surprisingly, changing the partitioning of the different forces makes the interaction of the two positively charged arginines attractive under these conditions [45]. The set of diverse chemical groups in arginine is a unique feature among the other amino acid side chains. With its high reactivity, the need for precise regulation comes, and so arginine can be tuned to a preferred state by posttranslational methylation.

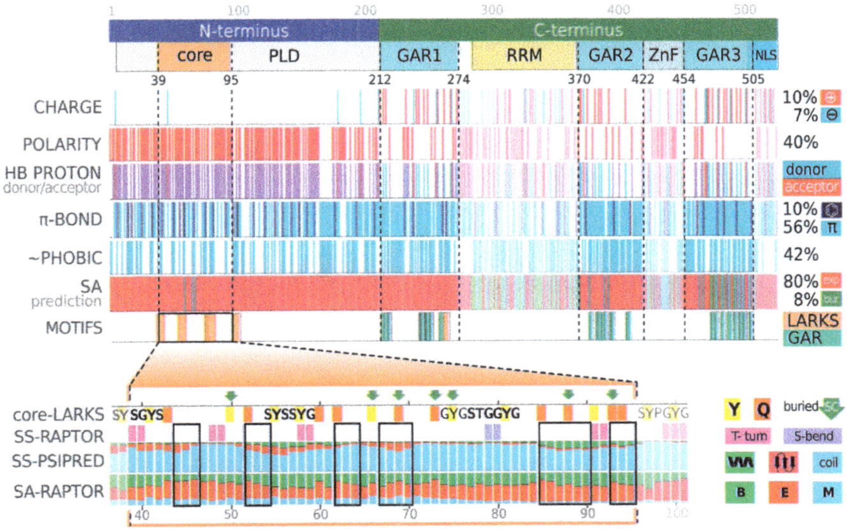

Figure 5. Physicochemical and structural properties of FUS. The various characteristics are shown along the protein-sequence split on the N-terminus (blue) and C-terminus (green). The full-length N-terminus corresponds to a highly polar prion-like domain (PLD in gray). The PLD contains a core region (residues 39–95 in orange), in which multiple LARKS motifs were detected (orange bars in the MOTIFS row) and evidenced to form fibrils in a solid state. The C-terminus contains two well-folded domains detected by Pfam search (RRM in yellow, ZnF in gray), and three glycine–arginine-rich (GAR in cyan) regions. All components of the C-terminus are known to have a functional role in RNA binding. The following rows show the physicochemical patterning along the full-length FUS sequence, including charge (positive (magenta), negative (blue)), polarity, donor (blue)/acceptor (magenta)/both (purple) of side-chain proton for hydrogen bonding, π-electron-containing systems (blue), with separation of aromatic ones (dark blue), hydrophobicity (blue), predicted solvent accessibility (SA) in 3-letter notation (exposed (red), buried (green), medium (blue)). The zoom of the PLD core is shown at the bottom panel, where the SS and SA rows contain the predicted probabilities of secondary structure (helix (green), strand (red), coil (blue)) and solvent accessibility. The green arrows indicate the side chains buried in the fibril core, while the black frames highlight segments that form strands of a cross-β motif [37].

Thus, under physiological conditions, FUS is highly methylated [44]. This limits self-assembly via interactions with tyrosine and promotes a functional role of intermolecular interactions with other proteins and nucleic acids. Therefore, phase separation and the gelation of FUS can increase by the hypomethylation of arginines within RGG-rich regions or the insertion of additional ones into the C-terminus [44]. All of these findings come together to demonstrate the significant role of the arginine side chain in phase separation. Tyrosine and glutamine are similarly relevant.

2.3. FUS, LLPS Regulated in the Context-Dependent Tuning of Preferred Forces

Note that the original results generated for FUS (identifier: P35637) using the BIAPSS web platform are shown in Supplementary Materials. The findings described in the previous section are briefly summarized below. FUS is a predominantly disordered protein (80%), composed of two functional regions: the prion-like domain (residues 1–212) and the RNA-binding C-terminal fragment (residues 213–526). Surprisingly, although the sequence is more than 65% composed of only five amino acids (G >> S >Q > R~Y), their distribution is highly variable. In particular, low sequence complexity occurs mainly in three structurally flexible glycine–arginine-rich regions. They provide hydrophobicity, π-electron-containing centers, and a positive charge of the C-terminus. In contrast, the enrichment of the N-terminus in serine, glutamine, and tyrosine makes it strongly polar but negatively charged (due to a few aspartic acids), with numerous aromatic centers and side chains capable of hydrogen bonding. These very different physicochemical properties of the protein terminals are sensitive to environmental changes and allow for the context-dependent regulation of the protein's cellular behavior. In particular, arginine can be tuned to a preferred state by posttranslational methylation [44], while serine/threonine phosphorylation introduces a highly negative charge [37]. Both modifications prevent the formation of aberrant self-assembly. In the first case, the methyl groups hinder cation-π stacking between arginine and tyrosine, limiting the phase separation driven by contacts between the N- and C-terminals. In the second case, phosphorylation introduces strong electrostatic repulsion between the N-termini. It inhibits homotypic phase separation, driven by the π-stacking of aromatic tyrosines. However, both mechanisms have no effect under high salinity due to the significant screening of electrostatics. In such conditions, even interactions of arginines become attractive due to hydrophobic contributions and π-π stacking [45]. Therefore, these observations demonstrate that the peculiar physicochemical properties of amino acid residues play a significant role in phase separation. The multifunctional chemical groups of amino acids make them reactive and multivalent. These features aid in the context-dependent tuning between preferred modes of interactions. They can work synergistically or alternatively, and their regulation depends on the environmental conditions, the state of posttranslational modifications, and the presence of binding partners.

3. Materials and Methods

3.1. Sequence Complexity and Physicochemical Decoration

3.1.1. Sequence Complexity

Low-complexity regions (LCRs) in proteins are compositionally biased fragments of sequences that often have low amino acid diversity and repeats of short motifs of the sequential or structural kinds. Many reports point to their functional or regulatory roles, frequently also associated with subcellular phase separation [19]. The LCRs of LLPS proteins have been detected by using several state-of-the-art tools, such as SIMPLE [50], CAST [51], fLPS [52], and SEG [53]. The original hits were parsed by in-house algorithms to merge overlapping regions enriched in different amino acids, and only the integrated and unified results have been kept.

Shannon Entropy describes the information content held in data and it is a frequently used measure of protein-sequence complexity. We implemented a module for the on-the-fly calculation of it within BIAPSS services. The typical window length for compositional effects is between 5 and 20. The results can be displayed in:

- Residue-resolution mode (residue option; smoother output):

$$S_i = \frac{1}{N}\sum_{j=1}^{N} S_{(j,N)}$$

where the Shannon entropy ($S_{(i)}$) at sequence position i is a sum of entropies at all windows containing this position, normalized by the window length (N);

- Window-resolution mode (block option):

$$S_{(j,N)} = -\sum_{aa=1}^{AA=20} f_{aa} log_2(f_{aa})$$

where the Shannon entropy ($S_{(j,N)}$) at the *j*-th sequence window of the length (N) is summed over the fractions (f_{aa}) of 20 biogenic amino acids. The value is assigned to the center position within the window. The $S_{(j,N)}$ ranges from 0 (where only one residue is present within the sequence window) to $\log_2(N)$ (all positions are different). Therefore, the lower the Shannon entropy, the less complex the sequence is.

3.1.2. Physicochemical Decoration

To examine the physicochemical properties of LLPS-driving proteins, we identified, along each sequence, the patterns of polarity (Ser, Thr, Tyr, Gln, Asn, Cys, Met), hydrophobicity (Gly, Ala, Val, Ile, Leu, Pro, Phe), and detected π-stacking centers (Arg, Asn, Asp, Gln, Glu, Gly (note that, due to the lack of the side chain, glycine can stack via π-electrons from a peptide bond and hydrogen bonding via backbone carbonyl or amide), including those within aromatic rings (Phe, Tyr, Trp, His). We also provided the charge-distribution split between positively (His, Lys, Arg) and negatively (Glu, Asp) charged residues. For each feature, both the arrangement along the sequence and the fraction of residues are provided.

3.1.3. Electrostatics

It is well established that the electrostatic interactions often affect the solubility and stabilize the binding interface in the liquid–liquid demixing of biomolecules. The recently proposed charge-decoration parameters emerged as a measure of charge distribution along the protein sequence. In addition to the overall charge content, these descriptors are seen as important factors that shape the protein conformations, especially within low-complexity regions [54]. Following these discoveries, we calculated and compared the charge-decoration parameters; namely:

- SCD (sequence charge decoration) is implemented following the formulation by Sawle and Ghosh [55];
- OCS (overall charge symmetry) is implemented following the formulation by Das and Pappu [56];
- FCR (fraction of charged residues) is defined as a sum of the fractions of positive and negative charges.

3.2. HMMER-Based Sequence Conservation and Functional-Domain Detection

The multiple sequence alignment (MSA) and consensus profile were prepared using an efficient HMMER method (*phmmer* + *hmmalign* and *hmmbuild*, respectively), which employs a probabilistic hidden Markov model (HMM) [35], and are significantly more accurate compared to BLAST-based searches. Because some of the LLPS sequences are highly unique (detection of the remote homologs is needed), and because the MSA is reliable if at least several dozen homologous sequences are available, we used sequences selected from various UniProt subsets. Specifically, SwissProt, UniRef50, and UniRef90 differ in the size and increasing sequence identity of entries [57–59]. To identify sequence regions with significant evolutionary conservation, we derived three additional MSA-based parameters: strength, diversity, and character. The MSA strength of the sequence conservation informs on how much the specific position is held by evolution. This measure normalizes results from the *hmmlogo* tool to a discrete range from 0 (poorly conserved) to 5 (highly conserved). The *hmmlogo* computes letter heights along the sequence, depending on the information content of the position. The MSA diversity defines the number of different amino acids detected at a given position in the MSA, and is provided in discrete scale from 0 (highly conserved) to 5 (poorly conserved) (0—one, 1—two, 2—three, 3—four, 4—five or six, 5–7 and more amino acids at the aligned position). The MSA character describes the chemical nature of the most common amino acids at a given position in the multiple sequence alignment. We distinguished the following attributes: polar, charge, aromatic, another π-system, hydrophobic, and other (G or P).

Some LLPS proteins are composed of one or more well-known domains. The identification of these functional regions alongside regions of low complexity or disorder can provide additional insights into the regulatory role of phase separation. Therefore, we have performed a Pfam search for all LLPS proteins, reporting the detected domains and incorporating the original Pfam seed-MSAs for corresponding regions of LLPS sequence (instead of full-length ones) to derive more reliable evolutionary conservation descriptors.

3.3. *Short Sequential and Structural Motifs Specific for LLPS Sequences*

Short linear motifs (SLiMs) are short fragments along the sequence, often situated in the intrinsically disordered regions, generally showing high structural flexibility and evolutionary conservation. We systematically detected various short sequential and structural motifs. The implemented algorithms used the list of grouped motifs' instances, defined by regular expressions, as the keys to search protein sequences prone to phase separation. Among motifs known from the literature as relevant for phase behavior, our analysis includes short structural stretches of protein sequence, such as LARKS [33] and steric zippers [60]; glycine–arginine-rich regions (GARs) [32]; and new sequential repetitive n-mers.

3.4. *Structural Properties Derived from Sequence-Based Predictions*

Bearing in mind the predictive nature of sequence-based methods and, hence, their limited accuracy, comparing several of them and choosing the final consensus has proven to be successful in many approaches. In our study, we comprised predictions from at least three to six widely used tools for each biomolecular characteristic. While almost every method is available as a web server, due to the size and complexity of our analyses, we employed standalone versions. The raw data derived from these standalone tools during the high-performance computing was initially parsed, filtered, and simplified to a uniform CSV format, and deposited in our online repository at https://biapss.chem.iastate.edu/download.html, accessed on 1 April 2022.

3.4.1. Secondary Structure

Protein secondary structure is a regular three-dimensional organization of local fragments along a polypeptide chain. The two most common secondary structural elements are alpha helices and beta sheets. Most of the predictors provide the secondary-structure assignment in 3-letter notation (ss3): H—helix, E—strand, C—coil, while the advanced ones (RaptorX and PORTER-5) also deliver more detailed 8-letter notation (ss8): H—α-helix; G—3_{10}-helix; I—π-helix; E—β-strand; B—β-bridge; T—HB-turn; S—bend; C–loop. In our benchmark study, we employed five well-established secondary-structure predictors: PSIPRED [61], RaptorX-SS8 [62], PORTER-5 [63], SPIDER-3-Single [64], and FELLS [65].

3.4.2. Solvent Accessibility

Solvent accessibility gives some insight into protein structural flexibility, indicating the exposed patches on the protein surface available for interactions with the solvent molecules. Some surface sites have high evolutionary conservation, which is suggestive of functional or structural importance. Since not many structures of phase-separating IDPs are known, the robust prediction of solvent accessibility can help to identify flexible regions prone to conformational changes upon binding. The assignment of solvent accessibility is usually provided in the 3-letter code: B—buried, E—exposed, M—medium. In our benchmark study, we employed three well-established solvent-accessibility predictors: RaptorX-Property [66], PaleAle 5.0 [67], SPOT-1D [68].

3.4.3. Structural Disorder

The sequence-based predictions indicate regions of increased structural flexibility, usually estimating the disorder probability at a given position in the sequence. Detecting highly flexible regions may support the identification of short sequence stretches of multivalent interactions that can be relevant to phase separation. In our benchmark study, we employed seven well-established predictors of structural disorder: RaptorX-Property [66], IUPred2A [69], SPOT-Disorder [70], DISOPRED (v2 and v3) [71], and PONDR® (FIT, VLXT, VSL2) [72]. Most of these methods return the probability of disorder for each position in the sequence. Usually, the residue is considered as ordered when the score is below 0.5. The protein-binding regions in disordered fragments were estimated using the ANCHOR method [73].

3.4.4. Contact Map

Contact-map application provides a more reduced representation of a protein structure using a binary two-dimensional matrix of distances between all possible amino acid-residue pairs. The commonly used definition assumes the threshold 6–10 Å as the distance between the pair of two Cα or Cβ atoms being in contact. The contact number of protein residues limits the number of possible protein conformations and helps encode a three-dimensional structure. In our benchmark study, we employed three state-of-the-art predictors of intramolecular contacts: RaptorX-Contact [74], ResPRE [75], SPOT-Contact [68,76].

3.5. Data Availability

The UniProt IDs of LLPS sequences were collected as a joint superset of deposits from primary LLPS databases (i.e., PhaSePro (https://phasepro.elte.hu/, accessed on 1 April 2022), PhaSepDB.v1 (http://db.phasep.pro/, accessed on 1 April 2022), LLPSDB (http://bio-comp.org.cn/llpsdb/, accessed on 1 April 2022)). Then, protein sequences were taken from the UniProt database, available at https://www.uniprot.org/, accessed on 1 April 2022. The cellular location of the protein was derived via web scraping of primary LLPS databases, UniProt and COMPARTMENTS (https://compartments.jensenlab.org/, accessed on 1 April 2022). The following resources were reviewed for the corresponding entries of experimental or predicted three-dimensional structures: PDBe (https://pdbe-kb.org/, accessed on 1 April 2022), Swiss-Model Repository (https://swissmodel.expasy.org/repository/, accessed on 1 April 2022), ModBase (http://salilab.org/modbase-cgi/, accessed on 1 April 2022), and AlphaFold DB (https://alphafold.ebi.ac.uk/, accessed on 1 April 2022).

The results of our comprehensive analysis, performed on 501 proteins, are available at https://biapss.chem.iastate.edu/download.html, accessed on 1 April 2022. For each file, the details of its content and methods used are comprehensively described. These files are used directly as the input for the web applications of the SingleSEQ and MultiSEQ modules in the BIAPSS platform. The results of the analysis can be explored interactively online, saved as high-quality PNG images, and used directly as figures in the publication.

3.6. Code Availability

3.6.1. Phase-Separation Predictors:

1. PSPredictor, web server version available at http://www.pkumdl.cn:8000/PSPredictor, accessed on 1 April 2022.

3.6.2. Low-Complexity-Region (LCR) Predictors (Sequence-Based):

1. SEG, standalone (1999), available at https://ftp.ncbi.nih.gov/pub/seg/seg/, accessed on 1 April 2022;

2. fLPS, standalone Sep 2017, available at https://github.com/pmharrison/flps, accessed on 1 April 2022;

3. SIMPLE, standalone V6-6.1, available at https://github.com/john-hancock/SIMPLE-V6, accessed on 1 April 2022;

4. CAST2, web server version available at http://structure.biol.ucy.ac.cy/CAST2/index.html, accessed on 1 April 2022.

3.6.3. Multiple Sequence Alignment/Build a Profile/Conservation Logo:

1. HMMER (phmmer, hmmalign, hmmbuild, hmmlogo), standalone 3.3, available at http://hmmer.org/download.html, accessed on 1 April 2022;
2. Pfam, the database search was used to detect functional domains, available at http://pfam.xfam.org/, accessed on 1 April 2022.

3.6.4. Secondary-Structure Prediction (Sequence-Based):

1. PSIPRED, standalone 4.02, available at https://github.com/psipred/psipred, accessed on 1 April 2022;
2. RAPTOR-X, standalone Version ID: Rev: 37223, available upon request at http://raptorx.uchicago.edu/download/, accessed on 1 April 2022;
3. PORTER, standalone v5, available at https://github.com/mircare/Porter5/, accessed on 1 April 2022;
4. SPIDER, standalone v3, available upon request at https://sparks-lab.org/downloads/, accessed on 1 April 2022;
5. FESS, standalone 2.0 (November 2016), available upon request at http://old.protein.bio.unipd.it/download/, accessed on 1 April 2022.

3.6.5. Solvent-Accessibility Prediction (Sequence-Based):

1. RAPTOR-X property, standalone v1.01 (October 2018), available upon request at http://raptorx.uchicago.edu/StructurePropertyPred/predict/, accessed on 1 April 2022;
2. PaleAle, standalone 5.0 (December 2019), available at https://github.com/mircare/Brewery, accessed on 1 April 2022;
3. SPOT-1D, standalone (July 2019), available upon request at https://servers.sparks-lab.org/downloads/, accessed on 1 April 2022.

3.6.6. Structural-Disorder Prediction (Sequence-Based):

1. RAPTOR-X property, standalone v1.01 (October 2018), available upon request at http://raptorx.uchicago.edu/StructurePropertyPred/predict/, accessed on 1 April 2022;
2. UPred2A, standalone (November 2019), available upon request at https://iupred2a.elte.hu/download_new, accessed on 1 April 2022;
3. DISOPRED, standalone v2 and v3.1, available at https://github.com/psipred/disopred, accessed on 1 April 2022;
4. SPOT-Disorder2, standalone (February 2019), available upon request at https://sparks-lab.org/downloads/, accessed on 1 April 2022;
5. VSL2, standalone (November 2019), downloaded from http://www.dabi.temple.edu/disprot/download/VSL2.tar.gz (not available now), accessed on 1 April 2022;
6. PONDR-FIT, web server available at http://original.disprot.org/pondr-fit.php, accessed on 1 April 2022;
7. PONDR-VLXT, web server available at http://www.pondr.com/, accessed on 1 April 2022.

3.6.7. Contact-Map Prediction (Sequence-Based):

1. RAPTOR-X Contact, web server available at http://raptorx.uchicago.edu/ContactMap/, accessed on 1 April 2022;
2. ResPRE, standalone (November 2019), available at https://zhanglab.ccmb.med.umich.edu/ResPRE/download/ResPRE.zip, accessed on 1 April 2022;
3. SPOT-Contact, standalone v3 (June 2007), available upon request at https://sparks-lab.org/downloads/, accessed on 1 April 2022.

The raw data collected from the third-party software were parsed utilizing custom python/bash algorithms to provide the unified format and derive the consensus properties.

The output files are available at https://biapss.chem.iastate.edu/download.html, accessed on 1 April 2022.

4. Conclusions

In conclusion, many proteins undergo liquid–liquid phase separation (LLPS), which drives the biogenesis of various membraneless organelles. The interplay between the protein sequence and the LLPS potential is poorly understood. The BIAPSS web platform, which provides the means for the analysis, visualization, and interpretation of data for LLPS proteins, is designed to uncover the sequence-encoded signals of LLPS proteins. This BIAPSS platform stands out as an efficient and user-friendly visualization framework that facilitates the integration and comparison of the physicochemical and structural features of the vast majority of known phase-separating proteins. With the rapid growth of experimental data on a single-case basis, we expect the increased need for computational infrastructure that consolidates some generalized insights. Hence, we have also developed a feature-rich module for analyzing multiple protein sequences. The interactive interface, with content-rich labels and tooltips, makes data exploration and interpretation easy. Both the web applications and raw datasets are broadly accessible on multiple operating systems and popular browsers. The presented case study of FUS shows that the BIAPSS-inferred biophysical regularities accurately identify regions prone to phase separation and facilitate the design of precise sequence modifications for various applications.

While the current version of BIAPSS enables the convenient and insightful analysis of large nonredundant and high-quality LLPS protein supersets, there are several directions through which our platform could expand to discover unknown LLPS proteins. These include analyzing the physicochemical and structural properties for customized protein sequences and introducing several LLPS indicators trained with machine learning to reveal coupled effects. The new functionality will find applications for flexible sequence redesign to introduce or modulate phase separation. Among many beneficial uses, it could tailor the properties of modern biomaterials or open up new directions in the development of medical therapies.

Supplementary Materials: The following supporting information can be downloaded at: https://www.mdpi.com/article/10.3390/ijms23116204/s1.

Author Contributions: Conceptualization, A.E.B.-D.; methodology, A.E.B.-D.; software, A.E.B.-D.; validation, A.E.B.-D., D.A.P. and V.N.U.; formal analysis, A.E.B.-D., D.A.P., V.N.U.; investigation, A.E.B.-D., D.A.P. and V.N.U.; writing—original draft preparation, A.E.B.-D., D.A.P. and V.N.U.; writing—review and editing, A.E.B.-D., D.A.P. and V.N.U.; visualization, A.E.B.-D.; funding acquisition, A.E.B.-D. and D.A.P. All authors have read and agreed to the published version of the manuscript.

Funding: This research was funded by the Roy J. Carver Charitable Trust through the Iowa State University Bioscience Innovation Postdoctoral Fellowship (to A.E.B.-D.) and the National Institute of General Medical Sciences of the National Institutes of Health [R35GM138243 to D.A.P.].

Institutional Review Board Statement: Not applicable.

Informed Consent Statement: Not applicable.

Data Availability Statement: The data presented in this study are available in this article and in Supplementary Materials.

Conflicts of Interest: The authors declare no conflict of interest. The funders had no role in the design of the study; in the collection, analyses, or interpretation of data; in the writing of the manuscript; or in the decision to publish the results. The content is solely the responsibility of the authors and does not necessarily represent the official views of the National Institutes of Health.

References

1. Brangwynne, C.P.; Eckmann, C.R.; Courson, D.S.; Rybarska, A.; Hoege, C.; Gharakhani, J.; Julicher, F.; Hyman, A.A. Germline P granules are liquid droplets that localize by controlled dissolution/condensation. *Science* **2009**, *324*, 1729–1732. [CrossRef] [PubMed]
2. Brangwynne, C.P.; Mitchison, T.J.; Hyman, A.A. Active liquid-like behavior of nucleoli determines their size and shape in Xenopus laevis oocytes. *Proc. Natl. Acad. Sci. USA* **2011**, *108*, 4334–4339. [CrossRef] [PubMed]
3. Li, P.; Banjade, S.; Cheng, H.C.; Kim, S.; Chen, B.; Guo, L.; Llaguno, M.; Hollingsworth, J.V.; King, D.S.; Banani, S.F.; et al. Phase transitions in the assembly of multivalent signalling proteins. *Nature* **2012**, *483*, 336–340. [CrossRef] [PubMed]
4. Banjade, S.; Rosen, M.K. Phase transitions of multivalent proteins can promote clustering of membrane receptors. *eLife* **2014**, *3*, e04123. [CrossRef]
5. Banjade, S.; Wu, Q.; Mittal, A.; Peeples, W.B.; Pappu, R.V.; Rosen, M.K. Conserved interdomain linker promotes phase separation of the multivalent adaptor protein Nck. *Proc. Natl. Acad. Sci. USA* **2015**, *112*, E6426–E6435. [CrossRef]
6. Uversky, V.N. Intrinsically disordered proteins in overcrowded milieu: Membrane-less organelles, phase separation, and intrinsic disorder. *Curr. Opin. Struct. Biol.* **2017**, *44*, 18–30. [CrossRef]
7. Uversky, V.N. Protein intrinsic disorder-based liquid-liquid phase transitions in biological systems: Complex coacervates and membrane-less organelles. *Adv. Colloid. Interface Sci.* **2017**, *239*, 97–114. [CrossRef]
8. Darling, A.L.; Zaslavsky, B.Y.; Uversky, V.N. Intrinsic disorder-based emergence in cellular biology: Physiological and pathological liquid-liquid phase transitions in cells. *Polymers* **2019**, *11*, 990. [CrossRef]
9. Sawyer, I.A.; Bartek, J.; Dundr, M. Phase separated microenvironments inside the cell nucleus are linked to disease and regulate epigenetic state, transcription and RNA processing. *Semin. Cell Dev. Biol.* **2019**, *90*, 94–103. [CrossRef]
10. Turoverov, K.K.; Kuznetsova, I.M.; Fonin, A.V.; Darling, A.L.; Zaslavsky, B.Y.; Uversky, V.N. Stochasticity of biological soft matter: Emerging concepts in intrinsically disordered proteins and biological phase separation. *Trends Biochem. Sci.* **2019**, *44*, 716–728. [CrossRef]
11. Uversky, V.N.; Finkelstein, A.V. Life in phases: Intra- and inter- molecular phase transitions in protein solutions. *Biomolecules* **2019**, *9*, 842. [CrossRef] [PubMed]
12. Nesterov, S.V.; Ilyinsky, N.S.; Uversky, V.N. Liquid-liquid phase separation as a common organizing principle of intracellular space and biomembranes providing dynamic adaptive responses. *Biochim. Biophys. Acta Mol. Cell. Res.* **2021**, *1868*, 119102. [CrossRef] [PubMed]
13. Uversky, V.N. Recent developments in the field of intrinsically disordered proteins: Intrinsic disorder-based emergence in cellular biology in light of the physiological and pathological liquid-liquid phase transitions. *Annu. Rev. Biophys.* **2021**, *50*, 135–156. [CrossRef] [PubMed]
14. Antifeeva, I.A.; Fonin, A.V.; Fefilova, A.S.; Stepanenko, O.V.; Povarova, O.I.; Silonov, S.A.; Kuznetsova, I.M.; Uversky, V.N.; Turoverov, K.K. Liquid-liquid phase separation as an organizing principle of intracellular space: Overview of the evolution of the cell compartmentalization concept. *Cell. Mol. Life Sci.* **2022**, *79*, 251. [CrossRef]
15. Dignon, G.L.; Best, R.B.; Mittal, J. Biomolecular phase separation: From molecular driving forces to macroscopic properties. *Annu. Rev. Phys. Chem.* **2020**, *71*, 53–75. [CrossRef] [PubMed]
16. Choi, J.M.; Holehouse, A.S.; Pappu, R.V. Physical principles underlying the complex biology of intracellular phase transitions. *Annu. Rev. Biophys.* **2020**, *49*, 107–133. [CrossRef]
17. Wang, J.; Choi, J.M.; Holehouse, A.S.; Lee, H.O.; Zhang, X.; Jahnel, M.; Maharana, S.; Lemaitre, R.; Pozniakovsky, A.; Drechsel, D.; et al. A molecular grammar governing the driving forces for phase separation of prion-like rna binding proteins. *Cell* **2018**, *174*, 688–699.e16. [CrossRef]
18. Uversky, V.N.; Kuznetsova, I.M.; Turoverov, K.K.; Zaslavsky, B. Intrinsically disordered proteins as crucial constituents of cellular aqueous two phase systems and coacervates. *FEBS Lett.* **2015**, *589*, 15–22. [CrossRef]
19. Banani, S.F.; Lee, H.O.; Hyman, A.A.; Rosen, M.K. Biomolecular condensates: Organizers of cellular biochemistry. *Nat. Rev. Mol. Cell. Biol.* **2017**, *18*, 285–298. [CrossRef]
20. Savojardo, C.; Martelli, P.L.; Casadio, R. Protein-protein interaction methods and protein phase separation. *Annu. Rev. Biomed. Data Sci.* **2020**, *3*, 89–112. [CrossRef]
21. Borcherds, W.; Bremer, A.; Borgia, M.B.; Mittag, T. How do intrinsically disordered protein regions encode a driving force for liquid-liquid phase separation? *Curr. Opin. Struct. Biol.* **2021**, *67*, 41–50. [CrossRef] [PubMed]
22. Zaslavsky, B.Y.; Ferreira, L.A.; Uversky, V.N. Driving forces of liquid-liquid phase separation in biological systems. *Biomolecules* **2019**, *9*, 473. [CrossRef] [PubMed]
23. Tsang, B.; Pritisanac, I.; Scherer, S.W.; Moses, A.M.; Forman-Kay, J.D. Phase separation as a missing mechanism for interpretation of disease mutations. *Cell* **2020**, *183*, 1742–1756. [CrossRef] [PubMed]
24. Saar, K.L.; Morgunov, A.S.; Qi, R.; Arter, W.E.; Krainer, G.; Knowles, T.P. Machine learning models for predicting protein condensate formation from sequence determinants and embeddings. *bioRxiv* **2020**. [CrossRef]
25. You, K.; Huang, Q.; Yu, C.; Shen, B.; Sevilla, C.; Shi, M.; Hermjakob, H.; Chen, Y.; Li, T. PhaSepDB: A database of liquid-liquid phase separation related proteins. *Nucleic Acids Res.* **2020**, *48*, D354–D359. [CrossRef]
26. Meszaros, B.; Erdos, G.; Szabo, B.; Schad, E.; Tantos, A.; Abukhairan, R.; Horvath, T.; Murvai, N.; Kovacs, O.P.; Kovacs, M.; et al. PhaSePro: The database of proteins driving liquid-liquid phase separation. *Nucleic Acids Res.* **2020**, *48*, D360–D367. [CrossRef]

27. Li, Q.; Peng, X.; Li, Y.; Tang, W.; Zhu, J.; Huang, J.; Qi, Y.; Zhang, Z. LLPSDB: A database of proteins undergoing liquid-liquid phase separation in vitro. *Nucleic Acids Res.* **2020**, *48*, D320–D327. [CrossRef]
28. Wang, X.; Zhou, X.; Yan, Q.; Liao, S.; Tang, W.; Xu, P.; Gao, Y.; Li, Q.; Dou, Z.; Yang, W.; et al. LLPSDB v2.0: An updated database of proteins undergoing liquid-liquid phase separation in vitro. *Bioinformatics* **2022**, *38*, 2010–2014. [CrossRef]
29. Ning, W.; Guo, Y.; Lin, S.; Mei, B.; Wu, Y.; Jiang, P.; Tan, X.; Zhang, W.; Chen, G.; Peng, D.; et al. DrLLPS: A data resource of liquid-liquid phase separation in eukaryotes. *Nucleic Acids Res.* **2020**, *48*, D288–D295. [CrossRef]
30. Li, Q.; Wang, X.; Dou, Z.; Yang, W.; Huang, B.; Lou, J.; Zhang, Z. Protein databases related to liquid-liquid phase separation. *Int. J. Mol. Sci.* **2020**, *21*, 6796. [CrossRef]
31. Alberti, S.; Gladfelter, A.; Mittag, T. Considerations and challenges in studying liquid-liquid phase separation and biomolecular condensates. *Cell* **2019**, *176*, 419–434. [CrossRef] [PubMed]
32. Chong, P.A.; Vernon, R.M.; Forman-Kay, J.D. RGG/RG motif regions in RNA binding and phase separation. *J. Mol. Biol.* **2018**, *430*, 4650–4665. [CrossRef] [PubMed]
33. Hughes, M.P.; Sawaya, M.R.; Boyer, D.R.; Goldschmidt, L.; Rodriguez, J.A.; Cascio, D.; Chong, L.; Gonen, T.; Eisenberg, D.S. Atomic structures of low-complexity protein segments reveal kinked beta sheets that assemble networks. *Science* **2018**, *359*, 698–701. [CrossRef] [PubMed]
34. Do Amaral, M.J.; Araujo, T.S.; Diaz, N.C.; Accornero, F.; Polycarpo, C.R.; Cordeiro, Y.; Cabral, K.M.S.; Almeida, M.S. Phase separation and disorder-to-order transition of human brain expressed X-linked 3 (hBEX3) in the presence of small fragments of tRNA. *J. Mol. Biol.* **2020**, *432*, 2319–2348. [CrossRef] [PubMed]
35. Mistry, J.; Finn, R.D.; Eddy, S.R.; Bateman, A.; Punta, M. Challenges in homology search: HMMER3 and convergent evolution of coiled-coil regions. *Nucleic Acids Res.* **2013**, *41*, e121. [CrossRef] [PubMed]
36. Mistry, J.; Chuguransky, S.; Williams, L.; Qureshi, M.; Salazar, G.A.; Sonnhammer, E.L.L.; Tosatto, S.C.E.; Paladin, L.; Raj, S.; Richardson, L.J.; et al. Pfam: The protein families database in 2021. *Nucleic Acids Res.* **2021**, *49*, D412–D419. [CrossRef]
37. Murray, D.T.; Kato, M.; Lin, Y.; Thurber, K.R.; Hung, I.; McKnight, S.L.; Tycko, R. Structure of FUS protein fibrils and its relevance to self-Assembly and phase separation of low-complexity domains. *Cell* **2017**, *171*, 615–627.e16. [CrossRef]
38. Birsa, N.; Ule, A.M.; Garone, M.G.; Tsang, B.; Mattedi, F.; Chong, P.A.; Humphrey, J.; Jarvis, S.; Pisiren, M.; Wilkins, O.G.; et al. FUS-ALS mutants alter FMRP phase separation equilibrium and impair protein translation. *Sci. Adv.* **2021**, *7*, eabf8660. [CrossRef]
39. Birsa, N.; Bentham, M.P.; Fratta, P. Cytoplasmic functions of TDP-43 and FUS and their role in ALS. *Semin. Cell Dev. Biol.* **2020**, *99*, 193–201. [CrossRef]
40. Protter, D.S.W.; Parker, R. Principles and properties of stress granules. *Trends Cell Biol.* **2016**, *26*, 668–679. [CrossRef]
41. Emmanouilidis, L.; Esteban-Hofer, L.; Damberger, F.F.; de Vries, T.; Nguyen, C.K.X.; Ibanez, L.F.; Mergenthal, S.; Klotzsch, E.; Yulikov, M.; Jeschke, G.; et al. NMR and EPR reveal a compaction of the RNA-binding protein FUS upon droplet formation. *Nat. Chem. Biol.* **2021**, *17*, 608–614. [CrossRef] [PubMed]
42. Holehouse, A.S.; Das, R.K.; Ahad, J.N.; Richardson, M.O.; Pappu, R.V. CIDER: Resources to analyze sequence-ensemble relationships of intrinsically disordered proteins. *Biophys. J.* **2017**, *112*, 16–21. [CrossRef] [PubMed]
43. King, O.D.; Gitler, A.D.; Shorter, J. The tip of the iceberg: RNA-binding proteins with prion-like domains in neurodegenerative disease. *Brain Res.* **2012**, *1462*, 61–80. [CrossRef]
44. Qamar, S.; Wang, G.; Randle, S.J.; Ruggeri, F.S.; Varela, J.A.; Lin, J.Q.; Phillips, E.C.; Miyashita, A.; Williams, D.; Strohl, F.; et al. FUS phase separation is modulated by a molecular chaperone and methylation of arginine cation-pi interactions. *Cell* **2018**, *173*, 720–734. [CrossRef] [PubMed]
45. Krainer, G.; Welsh, T.J.; Joseph, J.A.; Espinosa, J.R.; Wittmann, S.; de Csillery, E.; Sridhar, A.; Toprakcioglu, Z.; Gudiskyte, G.; Czekalska, M.A.; et al. Reentrant liquid condensate phase of proteins is stabilized by hydrophobic and non-ionic interactions. *Nat. Commun.* **2021**, *12*, 1085. [CrossRef]
46. Murthy, A.C.; Dignon, G.L.; Kan, Y.; Zerze, G.H.; Parekh, S.H.; Mittal, J.; Fawzi, N.L. Molecular interactions underlying liquid-liquid phase separation of the FUS low-complexity domain. *Nat. Struct. Mol. Biol.* **2019**, *26*, 637–648. [CrossRef]
47. Loughlin, F.E.; Lukavsky, P.J.; Kazeeva, T.; Reber, S.; Hock, E.M.; Colombo, M.; Von Schroetter, C.; Pauli, P.; Clery, A.; Muhlemann, O.; et al. The solution structure of FUS bound to RNA reveals a bipartite mode of RNA recognition with both sequence and shape specificity. *Mol. Cell* **2019**, *73*, 490–504.e496. [CrossRef]
48. Maris, C.; Dominguez, C.; Allain, F.H. The RNA recognition motif, a plastic RNA-binding platform to regulate post-transcriptional gene expression. *FEBS J.* **2005**, *272*, 2118–2131. [CrossRef]
49. Shazman, S.; Mandel-Gutfreund, Y. Classifying RNA-binding proteins based on electrostatic properties. *PLoS Comput. Biol.* **2008**, *4*, e1000146. [CrossRef]
50. Alba, M.M.; Laskowski, R.A.; Hancock, J.M. Detecting cryptically simple protein sequences using the SIMPLE algorithm. *Bioinformatics* **2002**, *18*, 672–678. [CrossRef]
51. Promponas, V.J.; Enright, A.J.; Tsoka, S.; Kreil, D.P.; Leroy, C.; Hamodrakas, S.; Sander, C.; Ouzounis, C.A. CAST: An iterative algorithm for the complexity analysis of sequence tracts. Complexity analysis of sequence tracts. *Bioinformatics* **2000**, *16*, 915–922. [CrossRef] [PubMed]
52. Harrison, P.M. fLPS: Fast discovery of compositional biases for the protein universe. *BMC Bioinform.* **2017**, *18*, 476. [CrossRef]
53. Wootton, J.C.; Federhen, S. Statistics of local complexity in amino acid sequences and sequence databases. *Comput. Chem.* **1993**, *17*, 149–163. [CrossRef]

54. Bianchi, G.; Longhi, S.; Grandori, R.; Brocca, S. Relevance of electrostatic charges in compactness, aggregation, and phase separation of intrinsically disordered proteins. *Int. J. Mol. Sci.* **2020**, *21*, 6208. [CrossRef] [PubMed]
55. Sawle, L.; Ghosh, K. A theoretical method to compute sequence dependent configurational properties in charged polymers and proteins. *J. Chem. Phys.* **2015**, *143*, 085101. [CrossRef]
56. Das, R.K.; Pappu, R.V. Conformations of intrinsically disordered proteins are influenced by linear sequence distributions of oppositely charged residues. *Proc. Natl. Acad. Sci. USA* **2013**, *110*, 13392–13397. [CrossRef]
57. UniProt, C. UniProt: The universal protein knowledgebase in 2021. *Nucleic Acids Res.* **2021**, *49*, D480–D489. [CrossRef]
58. The UniProt Consortium. UniProt: The universal protein knowledgebase. *Nucleic Acids Res.* **2017**, *45*, D158–D169. [CrossRef]
59. Apweiler, R.; Bairoch, A.; Wu, C.H.; Barker, W.C.; Boeckmann, B.; Ferro, S.; Gasteiger, E.; Huang, H.; Lopez, R.; Magrane, M.; et al. UniProt: The Universal Protein knowledgebase. *Nucleic Acids Res.* **2004**, *32*, D115–D119. [CrossRef]
60. Riek, R. The three-dimensional structures of amyloids. *Cold Spring Harb. Perspect. Biol.* **2017**, *9*, a023572. [CrossRef]
61. Buchan, D.W.A.; Jones, D.T. The PSIPRED protein analysis workbench: 20 years on. *Nucleic Acids Res.* **2019**, *47*, W402–W407. [CrossRef]
62. Wang, Z.; Zhao, F.; Peng, J.; Xu, J. Protein 8-class secondary structure prediction using conditional neural fields. *Proteomics* **2011**, *11*, 3786–3792. [CrossRef] [PubMed]
63. Torrisi, M.; Kaleel, M.; Pollastri, G. Deeper profiles and cascaded recurrent and convolutional neural networks for state-of-the-art protein secondary structure prediction. *Sci. Rep.* **2019**, *9*, 12374. [CrossRef] [PubMed]
64. Heffernan, R.; Paliwal, K.; Lyons, J.; Singh, J.; Yang, Y.; Zhou, Y. Single-sequence-based prediction of protein secondary structures and solvent accessibility by deep whole-sequence learning. *J. Comput. Chem.* **2018**, *39*, 2210–2216. [CrossRef] [PubMed]
65. Piovesan, D.; Walsh, I.; Minervini, G.; Tosatto, S.C.E. FELLS: Fast estimator of latent local structure. *Bioinformatics* **2017**, *33*, 1889–1891. [CrossRef]
66. Wang, S.; Li, W.; Liu, S.; Xu, J. RaptorX-Property: A web server for protein structure property prediction. *Nucleic Acids Res.* **2016**, *44*, W430–W435. [CrossRef] [PubMed]
67. Kaleel, M.; Torrisi, M.; Mooney, C.; Pollastri, G. PaleAle 5.0: Prediction of protein relative solvent accessibility by deep learning. *Amino Acids* **2019**, *51*, 1289–1296. [CrossRef]
68. Hanson, J.; Paliwal, K.; Litfin, T.; Yang, Y.; Zhou, Y. Improving prediction of protein secondary structure, backbone angles, solvent accessibility and contact numbers by using predicted contact maps and an ensemble of recurrent and residual convolutional neural networks. *Bioinformatics* **2019**, *35*, 2403–2410. [CrossRef]
69. Meszaros, B.; Erdos, G.; Dosztanyi, Z. IUPred2A: Context-dependent prediction of protein disorder as a function of redox state and protein binding. *Nucleic Acids Res.* **2018**, *46*, W329–W337. [CrossRef]
70. Hanson, J.; Paliwal, K.K.; Litfin, T.; Zhou, Y. SPOT-Disorder2: Improved protein intrinsic disorder prediction by ensembled deep learning. *Genom. Proteom. Bioinform.* **2019**, *17*, 645–656. [CrossRef]
71. Jones, D.T.; Cozzetto, D. DISOPRED3: Precise disordered region predictions with annotated protein-binding activity. *Bioinformatics* **2015**, *31*, 857–863. [CrossRef] [PubMed]
72. Xue, B.; Dunbrack, R.L.; Williams, R.W.; Dunker, A.K.; Uversky, V.N. PONDR-FIT: A meta-predictor of intrinsically disordered amino acids. *Biochim. Biophys. Acta* **2010**, *1804*, 996–1010. [CrossRef] [PubMed]
73. Dosztanyi, Z.; Meszaros, B.; Simon, I. ANCHOR: Web server for predicting protein binding regions in disordered proteins. *Bioinformatics* **2009**, *25*, 2745–2746. [CrossRef] [PubMed]
74. Wang, S.; Sun, S.; Li, Z.; Zhang, R.; Xu, J. Accurate de novo prediction of protein contact map by ultra-deep learning model. *PLoS Comput. Biol.* **2017**, *13*, e1005324. [CrossRef]
75. Li, Y.; Hu, J.; Zhang, C.; Yu, D.J.; Zhang, Y. ResPRE: High-accuracy protein contact prediction by coupling precision matrix with deep residual neural networks. *Bioinformatics* **2019**, *35*, 4647–4655. [CrossRef]
76. Singh, J.; Litfin, T.; Singh, J.; Paliwal, K.; Zhou, Y. SPOT-Contact-LM: Improving single-sequence-based prediction of protein contact map using a transformer language model. *Bioinformatics* **2022**, *38*, 1888–1894. [CrossRef]

International Journal of
Molecular Sciences

MDPI

Review

Phase-Separated Subcellular Compartmentation and Related Human Diseases

Lin Zhang [1,†], Shubo Wang [1,†], Wenmeng Wang [1], Jinming Shi [1], Daniel B. Stovall [2], Dangdang Li [1,*] and Guangchao Sui [1,*]

[1] College of Life Science, Northeast Forestry University, Harbin 150040, China; linzhang@nefu.edu.cn (L.Z.); shubowang@nefu.edu.cn (S.W.); wangwenmeng@nefu.edu.cn (W.W.); jmshi@nefu.edu.cn (J.S.)
[2] College of Arts and Sciences, Winthrop University, Rock Hill, SC 29733, USA; stovalld@winthrop.edu
* Correspondence: lidd@nefu.edu.cn (D.L.); gcsui@nefu.edu.cn (G.S.); Tel.: +86-451-82191081 (D.L. & G.S.)
† These authors contributed equally to this work.

Abstract: In live cells, proteins and nucleic acids can associate together through multivalent interactions, and form relatively isolated phases that undertake designated biological functions and activities. In the past decade, liquid–liquid phase separation (LLPS) has gradually been recognized as a general mechanism for the intracellular organization of biomolecules. LLPS regulates the assembly and composition of dozens of membraneless organelles and condensates in cells. Due to the altered physiological conditions or genetic mutations, phase-separated condensates may undergo aberrant formation, maturation or gelation that contributes to the onset and progression of various diseases, including neurodegenerative disorders and cancers. In this review, we summarize the properties of different membraneless organelles and condensates, and discuss multiple phase separation-regulated biological processes. Based on the dysregulation and mutations of several key regulatory proteins and signaling pathways, we also exemplify how aberrantly regulated LLPS may contribute to human diseases.

Keywords: liquid–liquid phase separation (LLPS); membraneless organelles; phase-separated condensates; human diseases

Citation: Zhang, L.; Wang, S.; Wang, W.; Shi, J.; Stovall, D.B.; Li, D.; Sui, G. Phase-Separated Subcellular Compartmentation and Related Human Diseases. *Int. J. Mol. Sci.* **2022**, *23*, 5491. https://doi.org/10.3390/ijms23105491

Academic Editor: Vladimir N. Uversky

Received: 22 April 2022
Accepted: 13 May 2022
Published: 14 May 2022

1. Introduction

To organize complex biochemical reactions in a cellular environment, cells create compartments, or organelles. A compartment needs a boundary to separate it from the surroundings, and the components within it are mostly able to freely diffuse, so that biological processes can take place inside [1]. Many compartments, such as the endoplasmic reticulum and Golgi apparatus, are organelles surrounded by lipid bilayer membranes. However, many other cellular compartments are not restricted by any membrane, such as nucleoli, Cajal bodies, PML nuclear bodies, stress granules and germ granules [2–6]. In a cell, these compartments harbor a variety of biomolecules with specific functions in a spatiotemporally controlled manner to ensure undisturbed biological processes and fulfill designated cellular functions [7]. In the past decade, accumulating studies suggest a physical process, known as phase separation, that can drive the assembly of these membraneless compartments. The concept that liquid—liquid phase separation (LLPS) may be generally involved in many cellular processes has been gradually uncovered and increasingly appreciated.

Phase separation is a common phenomenon in physics and chemistry: two liquids do not compatibly dissolve in a homogeneous liquid phase, resulting in a distinct phase–phase separation state. In other words, a uniformly mixed and supersaturated solution without further dispersion will spontaneously separate into a dense phase and a dilute phase that can stably coexist. The droplets or condensates produced by LLPS are different from ordinary droplets. For example, droplets composed of proteins and RNAs are not completely uniform, such as nucleoli with three layers regulating different stages of ribosomal biogenesis, but show the characteristics of liquid flow [8]. LLPS is quickly accepted as a key and general mechanism underlying the creation of biomolecular condensates that can promote the formation of membraneless organelles to regulate various cellular functions and activities [9]. However, phase separation is highly sensitive to altered physical and chemical conditions. For example, many protein condensates are regulated by environmental factors that determine the strength and valency of intermolecular interactions, including temperature, pH, salt concentration, component concentration and composition [10]. A molecule may need to reach a threshold concentration to initiate LLPS, and even a small difference in temperature and protein, nucleic acid or salt concentration can lead to distinct outcomes [11]. Moreover, the presence of crowding molecules, such as polyethylene glycol (PEG), dextran and ficoll, can greatly enhance the process of LLPS [12]. In compositional studies of different membraneless organelles, proteins and nucleic acids may utilize multivalent interactions to form phase-separated condensates with designated physical and chemical properties different from the originally uniform cellular environment. Many key regulatory proteins have been reported to undergo phase separation, of which the dysregulation has been etiologically associated with the onset and progression of many diseases, such as amyotrophic lateral sclerosis (ALS), Alzheimer's disease, Huntington's disease and different cancers [13,14]. In the current review, we summarize recent studies of phase-separation-mediated compartmentation, and discuss how aberrantly regulated LLPS causes human diseases, especially neurodegenerative disorders and cancers.

2. Biomolecular Condensates

Biomolecular condensates are commonly present in live cells, and they troubled scientists for many decades as they attempted to elucidate their formation and functions. Phase separation provides a mechanism for the formation of these condensates that separate or isolate different molecules with related activities in defined compartments. It has also been proposed that the ability to undergo LLPS may be a universal property of proteins and nucleic acids under specific circumstances [15–17].

2.1. The Molecular Features of Biomolecular Aggregates

Many studies indicate that phase separation requires the establishment of an interactive network through multivalent protein molecules that are composed of multiple modular interactive domains and/or contain disordered regions [18]. The interactions include charge–charge, cation–π, π–π stacking and hydrogen bonds, involving both side chains and backbones of the proteins. For example, Nephrin, Nck and Neural Wiskott–Aldrich syndrome protein (N-WASP) can be assembled into a highly ordered and multivalent protein complex through the interactions between phosphorylated tyrosines of Nephrin and SH2 domains of Nck, and between SH3 domains of Nck and proline-rich motifs of N-WASP [19–21].

The phase separation phenomenon has unique physical characteristics, including fluidity, fusion and fluorescence recovery after photobleaching when fused with a fluorescent protein. Meanwhile, the formation of droplets is generally both concentration- and valence-dependent. Intrinsically disordered regions (IDRs) are featured characteristics of many proteins with LLPS capability, and are often both necessary and sufficient for the formation of phase-separated condensates. IDRs usually have low complexity and contain homo-polymeric repeats of specific amino acids, such as glycine, serine, proline and glutamine, with strong self-sustaining aggregation potentials [22,23]. Recently, we reported that histidine clusters could decide the phase separation of several proteins, including YY1, HOXA1, FOXG1B, ZIC3 and HNF6 [24]. Several algorithms have been developed to help researchers predict IDRs in a protein [25,26]. However, not all highly scored sequences based on the prediction software could necessarily form phase-separated condensates [27]. Meanwhile, IDR mutations are causally related to various human diseases, such as cardiovascular disorders, cancers and neurodegenerative diseases [27,28]. Vacic et al. investigated about 100,000 annotated missense disease mutations and discovered that 21.7% of them were located in the IDRs [29]. Among these mutations, 20% led to disorder-to-order transitions, such as increased α-helical propensity, significantly higher than those of annotated polymorphisms and neutral evolutionary substitutions [29].

A classic example is the correlation between fused in sarcoma (FUS) mutations and neurodegenerative diseases, including ALS, essential tremor and rare forms of frontotemporal lobar degeneration [30]. FUS protein contains a prion-like domain that is intrinsically disordered and can form liquid compartments in both the nucleus and cytoplasm [31]. Multiple FUS mutants exhibit significantly reduced mobility and eventually cause prion-like propagation of proteinaceous aggregates in neurons and glial support cells, characteristic of ALS [32]. Another example is the MutL Homolog 1 (MLH1) protein that is essential in DNA mismatch repair. The residue V384 located in the disordered segment of MLH1 is the most common site of mutations. The mutant MLH1 (V384D) is associated with increased susceptibility to colorectal cancer and is prevalent in HER2-positive luminal B breast cancer [33,34]. Phase separation is also involved in the antiviral immune response against the novel severe acute respiratory syndrome (SARS) coronavirus 2. The nucleocapsid protein of SARS2 may undergo LLPS with RNA and subsequently reduced Lys63-linked polyubiquitination and aggregation of mitochondrial antiviral-signaling protein (MAVS), which suppresses the innate antiviral immune response [35].

2.2. Materials Properties of Phase-Separated Condensates

LLPS contributes to the assembly of different membraneless organelles with different functional commitments in cells [36]. Whether a macromolecule can undergo phase separation depends on its concentration and property, as well as environmental conditions, such as pH, temperature, salt type and concentration. Meanwhile, phase-separated condensates formed under a particular physiological circumstance are accessible to various, but also selective, molecules in cells. The condensation process through the LLPS mechanism is generally reversible with a mobile liquid-like dense phase, and constant exchanges between the dense and light phases. However, the phase-separated condensates are subject to further transitions, such as gelation to form hydrogel that is virtually irreversible under physiological conditions. Whether LLPS condensates remain in a liquid and mobile state or become gelatinous and even solidified are physiologically or pathologically relevant [1,22,37,38]. We have illustrated previously reported membraneless organelles with their subcellular localization and functions in Figure 1. Meanwhile, we also summarized their sizes, components, functions and related diseases in Table 1. Here, we discuss the formation, compositions and other properties of several membraneless organelles and key regulatory protein-mediated condensates in the context of human diseases.

Table 1. Membraneless organelles and condensates assembled through the LLPS mechanism.

Localization	Name	Alias	Size (nm)	Components	Functions	Diseases	References
Cytoplasm	P-body	GW-body, RNA processing body, decapping body	100–300	K63, TRAF6, Tob1, TUT4, NoBody, LSM1, GW182, DDX3, DDX6, XRN1, etc.	mRNA degradation, post-transcriptional gene silencing, response to stress, storage of translationally repressed mRNAs	viral infection, neurodegenerative diseases, autoimmune diseases.	[39,40]
	Stress granule	—	1000–2000	RBPs, non-RBPs, TDRD3, TDP43, G3BP1, eIF3, eIF4G, PABPC1, etc.	translational regulation, response to stresses, antiviral defense, response to stresses, store mRNA and proteins	amyotrophic lateral sclerosis, frontotemporal lobar degeneration, cancer, viral infection, inflammatory diseases	[5,41]
	Germ granule	P-granule, chromatoid body, polar granule	250–4000	MEG-3, PGL, RNA, etc.	post-transcriptional regulation, regulation of Germ cell development and function, cell division	Germ cell development	[42]

Table 1. *Cont.*

Localization	Name	Alias	Size (nm)	Components	Functions	Diseases	References
	Synaptic density	Postsynaptic density	500	PSD-95, GKAP, Shank, Homer, etc.	responsible for signal processing	neuropsychiatric diseases	[43]
	RNA transport granule	Neuronal RNA granule	500–1000	Sam68, RNG105, SMN, etc.	mRNA storage and transport	neurodegenerative diseases	[44]
	Balbiani Body	Balbiani's vesicle, the yolk body of Balbiani, yolk nucleus	50–250,000	RNA, mitochondria, Golgi, endoplasmic reticulum, etc.	store RNA, proteins and mitochondria	—	[45]
	Sec body	—	1000	COPII components, Sec16, etc.	response to the nutrient stress of amino acid starvation, protect ERES components from degradation	—	[46]
	U-body	Uridine-rich snRNP body	500	SnRNP, SMN, etc.	storage and assembly of snRNPs	spinal muscular atrophy	[47]
	PSG	—	500	proteasomes, free ubiquitin, etc.	protein-specific degradation, store proteasome	aging and age-related disease	[48]
	Signaling puncta	Dvl puncta	500–1000	Dvl-2, etc.	signal transduction	—	[49]
	Metabolic granule	G-body	1000–5000	glycolytic enzymes, etc.	glycolysis and storage	—	[50]
	STAT3 cytoplasmic body	STAT3 sequestering endosomes	—	STAT3	prolongation of signaling and/or cross talk	hepatoma	[51,52]
	TIS granule	—	1000–5000	TIS11B, membrane protein-encoding mRNAs	3′UTR-dependent nurturing of nascent proteins	—	[53]
Nuclear membrane	Nuclear pore complex	—	40–100	nucleoporins, NDC1, GP210, POM121 etc.	facilitate nucleocytoplasmic transport, chromatin organization	neurological disorders and the aging brain, viral infections and immunity, the development and progression of cancers	[54]
Nucleus	Nucleolus	—	1000–10,000	Nucleolin, rRNA, rDNA, etc.	ribosome biogenesis	Werner syndrome, Bloom syndrome, Treacher Collins syndrome, dyskeratosis congenita syndrome, Rothmund–Thomson syndrome	[55]
	HLB	—	1000	NPAT, FLASH, SLBP, $p220^{NPAT}$, NELF, symplekin, etc.	processing of the histone pre-mRNAs, histone gene transcription	breast cancer	[56]
	DNA damage foci	—	500	γH2AX, ATM, 53BP1, RAD51, etc.	response to DNA damage	neurodegenerative diseases	[57]
	PML body	PML oncogenic domain, nuclear dot, Kremer body,	250–500	UBC9, RNF4, SP100, P53, DAXX, SUMO, PML, RNF168, etc.	transcription regulation, apoptosis signaling, epigenetic gene silencing, sequester partner proteins, SUMOylation sites	Acute Promyelocytic Leukemia, liver fibrosis	[58]
	Nuclear stress body	Peroxisome granule (PG)	300–3000	HSF1, HAP, SAM68, etc.	response to stress, control of gene expression and RNA splicing activities	metabolic syndrome	[59]
	Cajal body	accessory body	100–2000	RNA, snRNPs, scaRNAs, Coilin, SMN, etc.	pre-mRNA and pre-rRNA processing	amyotrophic lateral sclerosis, spinal muscular atrophy	[56]
	PcG body	—	200–1500	PRC1, PRC2, EZH2, etc.	transcriptional repression	malignant lymphomas, epithelial tumors	[60]
	CNB	—	1000–3000	CBP, SUMO-1, etc.	response to DNA damage, protein SUMOylation	—	[36]
	Paraspeckle	—	500–1000	CTN-RNA, PSP1, p54nrb, NEAT1, NONO, etc.	regulate gene expression, RNA processing	breast cancer, hepatocellular carcinoma, viral infection, neurodegenerative diseases	[61]

Table 1. *Cont.*

Localization	Name	Alias	Size (nm)	Components	Functions	Diseases	References
	PNC	—	250–4000	CUGBP, KSRP, polymerase III, Nucleolin, PTB, SRP RNA, etc.	transcriptional regulation, RNA metabolism	breast cancer, ovarian cancer	[62]
	Nuclear gem	Gemini of Cajal body, Gemini of coiled body	100–2000	SMN, etc.	mRNA processing	spinal muscular atrophy	[63]
	OPT domain body	53P1-OPT domain	1000–1500	Nascent mRNA, transcription factors, etc.	transcriptional regulation, response to the replication stress	—	[64]
	STAT3 nuclear body	—	—	STAT3, CREB binding protein (CBP), acetylated histone H4	activation of target genes	hepatoma	[51]
Nucleolus	Amyloid body	A-body	500–2000	Amyloid beta peptides, etc.	store proteins	neurodegenerative diseases	[65]

PSG: Proteasome storage granule; HLB: Histone locus body; PML: Promyelocytic Leukemia; PcG: Polycomb group; CNB: SUMO-1 nuclear body; PNC: Perinucleolar compartment; OPT: OCT1/PTF/transcription.

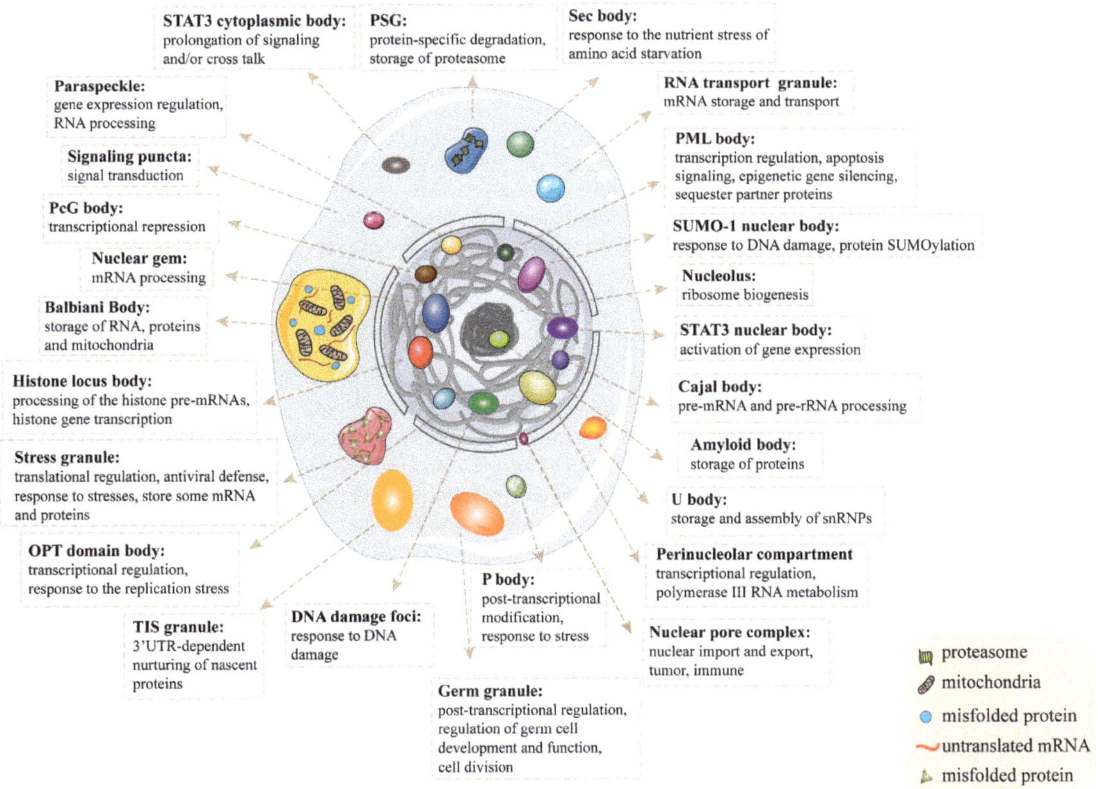

Figure 1. Schematic diagram of membraneless organelles and their functions in a eukaryotic cell.

2.2.1. Stress Granules

Both stress granules and processing bodies (P bodies) are composed of RNA and protein molecules that drive the phase separation of these membraneless organelles. Stress granule formation is exclusively induced by stress signals imposed on the cells, while P bodies can be constitutively visible in many cell types, but their size and number may increase in response to stress [40]. Stress granules contain translation-initiation molecules, and P bodies harbor factors regulating mRNA degradation, but they share many common proteins related to RNA metabolism. Mechanistically, in response to certain stresses, translation initiation can be stalled and ribosomes will disassociate from mRNA, which is the so-called ribosome run-off phenomenon. The released mRNA binds to RNA-binding proteins (RBPs) that promote stress granule formation. The mechanism of constitutive presence of P bodies remains unclear, but the stress-induced retardation of translation preinitiation directly contributes to their increased size and number [40].

Dysregulation of stress granules and P bodies is causally related to different diseases. Stress granules are considered an adaptive response of cells to acute stress, and their formation, composition and life span are associated with cancers, heart diseases, neurodegenerative disorders, inflammatory diseases and viral infections [66]. The oncogenic process consists of hypoxia, ER stress and osmotic alterations that all constitute the signals to induce stress granule formation. Meanwhile, chemotherapeutic challenges can also induce the assembly of stress granules, which contributes to the development of chemoresistance and metastasis of cancer cells [66]. Thus, drugs, such as 15d-PGJ2 targeting the eukaryotic initiation factor 4A-I (eIF4A1) in the stress granules, can inhibit proliferation and induce apoptosis of leukemic and colorectal cancer cells [67]. Several neurodegenerative diseases are caused by dysregulated stress granules that generally exhibit increased formation or reduced dissociation of stress granules compared to in the cognate normal cells. In particular, genetic mutations of certain RNA-binding proteins may impair stress granule assembly and composition leading to neurodegenerative diseases. For example, an FMRP mutant with defective stress granule assembly represents an etiologic cause of the Fragile X syndrome with mild-to-moderate intellectual disability [68]. Mutations in other stress granule-associated RNA-binding proteins are also discovered in Alzheimer's disease patients [69]. In addition, the neurons of Alzheimer's disease patients exhibited pathological aggregates by the nucleation of the proteins in stress granules, such as TIA1/R and G3BP1 [69]. During viral infection, many viruses can use a special viral protease to cleave essential stress granule proteins, which can circumvent the cellular defense against viral infection [70,71].

2.2.2. P Bodies

As another type of cytoplasmic ribonucleoprotein granules, P bodies are relatively understudied for their relevance to human diseases, although current evidence strongly suggests their involvement in neurodegenerative disorders, viral infection and autoimmune diseases. Mutations of DDX6 disrupt P body assembly, which is causally linked to intellectual developmental disorders with impaired language and dysmorphic facies [72]. In response to infection by RNA viruses, the number and stability of P bodies may change, and their components may be recruited to viral replication centers, although the underlying mechanisms remain unclear [73]. In addition, autoantibodies against P body components have been reported to contribute to autoimmune diseases [74,75].

2.2.3. Nucleolus

The nucleolus is an important membraneless organelle consisting of ribonucleoproteins and RNAs is assembled in multilayers through the LLPS mechanism [55]. In the past century, the roles of the nucleolus in hosting RNA polymerase I-mediated transcription, ribosomal RNA (rRNA) modification and processing, and rRNA complex assembly have been gradually recognized. A nucleolus of a mammalian cell may contain several functional modules, each of which constitutes three subcompartments or layers. From the inner to periphery, the three layers include the fibrillar center, the dense fibrillar component and the granular component, responsible for different steps of ribosomal biogenesis [55]. The nucleolus is separated from other compartments of the nucleus; however, due to the membraneless status, the nucleolus harbors various contents that dynamically exchange with the remaining nuclear components. Therefore, nucleoli are important organelles for transient sequestration of crucial factors involved in various biological functions, including the responses to genotoxic and oxidative stress, heat shock, starvation, oncogenic insults and viral infection [55,76]. These stresses may affect the shape, size and number of nucleoli, and the diseased states can markedly alter nucleolar morphology. Interestingly, despite the relatively isolated compartment and spatially distinct layers of each nucleolus, spontaneous coalescence may occur when two nucleoli have intimate contact, resembling droplet fusion during LLPS. Meanwhile, many nucleolar proteins contain IDRs, which are especially enriched by positively charged arginine and lysine residues [77,78].

Dysregulation of the nucleolus may aberrantly change nucleolar morphology, size and number per nucleus, and is tightly linked to various diseases. Excessive production of ribosomes by nucleoli may drive oncogenic transformation. On the other hand, defective activity of ribosome biogenesis may cause a shortage of properly formed ribosomes, and even cause aberrant nucleolar hardening, leading to reduced rRNA and ribonucleoprotein processing. These kinds of ribosomopathies may eventually cause different diseases, such as muscle atrophy and X-linked subtype of dyskeratosis congenita [79,80]. A hexanucleotide repeat GGGGCC (or G4C2) is present in an intron of the *C9ORF72* in chromosome 9, and its expansion can reach up to thousands of copies in ALS patients. Mechanistically, the expanded G4C2 sequence can generate arginine-containing toxic dipeptide repeats that promiscuously interact with the IDRs of RNA-binding proteins to form protein aggregates, and thus impair the dynamics of membraneless organelles, such us nucleoli, leading to the diseases [81]. In addition, the material state of the nucleolus is relevant to aging or longevity. Studies using *C. elegans* as a model revealed that both reduced rRNA production and knockdown of fibrillarin were associated with smaller nucleolar size and extended life span of the worm [82].

2.2.4. Examples of Regulatory Proteins with LLPS Potential

Besides the reported membraneless organelles, many intrinsically disordered proteins, especially those with nucleic acid binding affinity, can form isolated compartments through the LLPS mechanism, and their dysregulation may undergo liquid-to-solid transitions, leading to various diseases [19,83].

The prion-like domains (PrLDs) have relatively low complexity, and are enriched in glycine and uncharged polar amino acids [84]. The PrLDs have been identified in about 240 human proteins, especially many RNA-binding proteins, such as FUS, EWSR1, TDF-43 and TAF15 that are etiologically related to several neurodegenerative diseases, including frontotemporal dementia and ALS.

The RNA-binding protein FUS has 526 amino acids and belongs to the FET (FUS, EWSR1 and TAF15) family. *FUS* was originally discovered to fuse with the *CHOP* gene, and the fusion oncoprotein promotes the development of round cell liposarcoma and myeloid leukemia [85]. In addition to an RNA-binding motif, FUS contains a highly conserved C-terminal nuclear localization signal (NLS) that may harbor various mutations discovered in patients [86]. The EWSR1 protein has a transcriptional activation domain at the N-terminus, and regulates gene expression, cell signaling, RNA processing and RNA transport. The chromosomal translocation between the *EWSR1* and *FLI* genes can produce an oncogenic fusion gene that accounts for about 90% of Ewing sarcomas [87].

Since the N-terminus of FUS contains the IDR, the FUS-CHOP fusion created more intensified nuclear puncta than FUS and CHOP alone, with incorporation of BRD4, a bona fide marker of super-enhancers. Similarly, LLPS is considered as a driving force for the *EWSR1-FLI* fusion gene to regulate transcription and initiate cell transformation [88].

2.3. *Regulation of Condensate Assembly*

The assembly and biophysical properties of LLPS condensates are precisely regulated by chaperone proteins, enzymes for post-translational modifications (PTMs) and other cellular factors [89].

2.3.1. Effects of PTMs on Protein Phase Separation

Different PTMs, such as phosphorylation, acetylation, arginine methylation and SUMOylation that regulate protein–protein or protein–nucleic acid interaction strengths, are well-recognized key regulatory factors of phase separation. Furthermore, PTMs are engaged in the assembly and disassembly of condensates, as well as the regulation of their material properties. As a rapid and reversible process, phosphorylation is one of the most well-characterized PTMs modulating biomolecular phase transitions [90,91]. For example, in Alzheimer's disease, phosphorylation of Tau, a microtubule-associated protein, alters the charge distribution to promote its electrostatic interactions, leading to the formation of Tau aggregates [92]. Additionally, phosphorylation hinders tubulin assembly within Tau condensates. Previous studies indicated that neuronal loss and memory impairment were causally related to the presence of highly phosphorylated soluble Tau protein [93].

Phosphorylation of α-synuclein (α-syn) at Tyr39 (pY39) is enriched in patients with Parkinson's disease, and plays an important role in regulating the liquid–solid phase transition of α-syn [94]. pY39 can accelerate α-syn aggregation and inhibit its degradation through autophagy and proteasome pathways in cortical neurons. In general, α-syn phosphorylation may alter its fibril structure and exacerbate pathogenesis of Parkinson's disease [94,95]. As discussed above, FUS is a protein tightly related to neuronal degeneration diseases. FUS phosphorylation at its IDR could disrupt its phase separation and cytoplasmic aggregation, which reduces FUS-associated cytotoxicity [96], suggesting that FUS is a potential therapeutic target in the treatment of neurodegenerative diseases. In addition, the interactions between tyrosines in the IDR and arginines in the C-terminal regions of the FUS protein are crucial to its phase separation. The methylation of these arginines disrupts these interactions, leading to reduced FUS phase separation; however, hypomethylation of these arginines strongly promotes FUS phase separation and gelation, leading to the formation of immobile hydrogels stabilized by intermolecular β-sheets. The loss of FUS mobility causes impairment of neuron terminals and leads to the disease manifestation of frontotemporal lobar degeneration [97].

Polycomb repressive complexes (PRCs) are important regulators for gene repression during embryonic development and oncogenic progression [98]. In *C. elegans*, a polycomb protein SOP-2 functions as the counterpart of the human PRC1 complex to regulate *HOX* gene expression [99]. Qu et al. reported that SOP-2 contained an IDR and could form phase-separated droplets. Importantly, sumoylation at K453 and K594 SOP-2 could allow it to produce droplets with increased sizes and abundancy, and slightly improved internal mobility compared to the droplets formed by the unmodified protein [100]. Sumo-conjugation is likely essential for both phase separation and transcriptional regulation of SOP-2, because its sumoylation is required for both its localization into nuclear bodies and physiological repression of the *HOX* genes [101].

Phase separation-mediated formation of membraneless organelles is cell-cycle-dependent. Most membraneless organelles are dissolved when the nuclear envelope breaks down during mitosis, but are reformed as mitosis is completed. The kinase activity of DYRK3 plays an important role in dissolving several types of membraneless organelles during mitosis [102]. In fact, DYRK3 has been demonstrated to cause the dissolution of stress granules upon stress relief [103], and this activity is dependent on DYRK3's association with HSP90. In the absence of the heat-shock protein, the inactive DYRK3 either stays in stress granules or undergoes degradation [104].

2.3.2. Effects of Chaperones on Protein Phase Separation

Molecular chaperones play a key role in the assembly of phase-separated condensates. The historically recognized functions of chaperones are their abilities to promote correct protein folding and subsequently prevent protein aggregation into nonfunctional structures. A number of recent studies have revealed the activity of molecular chaperones, including several heat shock proteins, to regulate phase separation [105].

Chaperones regulate protein–protein interplay and assist in protein folding through directly interacting with them in an energy-consuming manner [97,106]. Molecular chaperones, including many heat shock proteins, are extensively involved in the maintenance of intracellular protein homeostasis. Previous studies indicate the presence of different heat shock proteins in a variety of membraneless organelles, such as HSP40, HSP70, HSP90, etc. Gu et al. reported that classes I and II of the HSP40 proteins could undergo phase separation due to their contents of flexible regions enriched with glycine and tyrosine [107]. DNAJB1, a member of the class II HSP40 proteins, could form condensates in nuclear bodies. In response to stress, DNAJB1 can translocate into stress granules. Interestingly, when cophase-separated with FUS, DNAJB1 can prevent FUS from forming amyloid fibrils in vitro and reduce aberrant FUS aggregation in cells [107]. As discussed above, hypomethylation of arginines in the C-terminus of FUS facilitates its phase separation and gelation. However, transportin 1 can serve as a chaperone protein of FUS to reduce its granule formation without affecting its methylation status, and eventually rescue attenuated protein synthesis caused by FUS aggregation in axon terminals [97].

As a canonical small chaperone, HSP27 localizes in stress granules. Due to the interaction with the IDR of FUS, HSP27 can reduce its LLPS. In addition, stress can induce HSP27 phosphorylation that subsequently promotes its co-phase separation with FUS. The presence of HSP27 can prevent FUS from forming amyloid fibrillar aggregates, and thus preserve its liquid phase [106]. Consistently, when mice of an Alzheimer's disease model were crossed with human *HSP27* transgenic mice, overexpressed HSP27 could rescue multiple neurodegenerative defects of the disease, including impaired spatial learning, increased neuronal excitability, reduced long-term potentiation, and widespread amyloid deposition in the brains [108].

As a histone chaperone, CAF-1 has LLPS properties and can form nuclear bodies through recruiting histone modifiers and other chaperones, which contributes to the establishment and maintenance of HIV-1 latency. Therefore, disruption of phase-separated nuclear bodies of CAF-1 can potentially reactivate latent HIV-1 to eradicate the viral reservoir caused by its latency [109].

2.4. Functions of Phase-Separation Condensates

LLPS have been reported to be involved in various biological processes and regulations. We summarize the LLPS-associated functions into the following four categories.

2.4.1. Regulation of Biological Reactions

In cells, the coordinated processes of biochemical reactions benefit from both membrane-restricted and membraneless organelles. The membraneless particles or condensates formed by LLPS are rich in selective proteins and nucleic acids, increasing their local concentrations and subsequently accelerating biochemical reactions [38].

Strulson et al. mimicked the intracellular compartmentalization by partitioning RNA in an aqueous two-phase system established by PEG and dextran. The RNA molecules could show up to 3000-fold enrichment in the dextran-rich phase, and compartmentalization could enhance the rate of ribozyme cleavage by 70-fold [110]. The histone locus body (HLB) is an evolutionarily conserved nuclear body with enriched protein and RNA factors required for histone gene transcription and pre-mRNA processing [111]. In this liquid-like compartment, many factors, such as FLASH and U7 snRNP, essential and constitutive components in HLB, exhibit greatly increased concentrations over the levels in the exterior cellular environment [112].

MicroRNAs (miRNAs) can promote mRNA degradation and/or block translation through targeting the 3′-UTRs. In this regulation, the formation of a miRNA-induced silencing complex (miRISC) consisting of multiple proteins is crucial to the miRNA-mediated gene repression. AGO2 and TNRC6B are the core components of the miRISC. The glycine/tryptophan (GW)-rich domain of TNRC6B is an intrinsically disordered region that promotes phase separation through multivalent interactions with three tryptophan-binding pockets in the PIWI domain of AGO2 [113]. The phase-separation process can enrich both AGO2 and TNRC6B in the condensates, and sequester RNAs to be degraded, which accelerates AGO2-mediated deadenylation of target RNAs.

In addition to the compartmentalizing phenomena discussed above, many other LLPS-mediated membraneless organelles, such as Cajal bodies, nucleoli and PML bodies, can concentrate proteins and nucleic acids involved in different designated biological processes in a confined space, which can enhance both reaction rates and efficiency [114].

The LLPS may also provide a platform that allows nascent proteins to quickly associate with their functional partners, which may determine their activities and destinies. Ma et al. reported the membraneless TIS granules formed by an RNA-binding protein TIS11B, which could partially cover of the cytoplasmic side of the rough endoplasmic reticulum (ER) [53]. The integration of these TIS granules and the ER can generate subcellular compartments, termed as TIS granule-ER, or TIGER, that constructs a biophysically and biochemically distinct environment from the cytoplasm. The TIS granules can promote the association between the SET protein and membrane proteins to be translated, such as CD47 and PD-L1, through a mechanism that the 3′-UTRs of the mRNAs of the membrane proteins facilitate the interaction between SET and CD47 or PD-L1. As a result of the SET-binding, the cell surface expression of the CD47 or PD-L1 can be significantly enhanced, which determines the cell identity. This discovery revealed an exciting notion that protein functions can be regulated by the lengths of the 3′-UTRs. In other words, proteins with the same amino acid sequence but encoded by mRNA isoforms with alternative 3′-UTR lengths may have different functions or subcellular localizations [115]. Therefore, 3′-UTRs may act as a medium or scaffold to nurture nascent proteins, and qualitatively change their properties and fates. Noteworthily, it has been reported that over 50% of protein-coding genes can generate mRNA isoforms with alternative 3′-UTRs [116]. Whether the nurturing niche provided by the TIS granules or TIGER compartments can be generalized to the regulation of the functions, localizations or fates of other proteins, in addition to CD47 and PD-L1, is a very intriguing question and deserves future exploration.

2.4.2. Regulation of Gene Expression

RNA polymerase II (Pol II) is responsible for the transcription of mRNAs and many noncoding RNAs, such as lncRNA and microRNAs. RNA Pol II has a highly conserved C-terminal domain (CTD) that contains 52 repeats of the YSPTSPS heptapeptide essential to polymerase activity [117]. The hyperphosphorylation of the CTD mediated by CDK9 can stimulate target gene transcription. As the kinase component of the positive transcription elongation factor b (P-TEFb), CDK9 can release the paused Pol II at a promoter periphery and facilitate its entry to the gene body, to achieve transcriptional elongation. CDK9 also regulates transcription termination through phosphorylating a Pol II-associated protein, SPT5, and promoting its interaction with the poly(A) site [118,119].

Lu et al. reported that a phase-separation mechanism is also critical for CTD hyperphosphorylation that activates RNA Pol II [117]. Despite the inclusion of a low-complexity region, the isolated CTD of RNA Pol II does not undergo phase separation by itself. However, the CTD can be trapped by the phase-separated condensates formed by the IDR of cyclin T1 that interacts with CDK7. Through this interaction, cyclin T1 compartmentalizes CKD7 and the CTD in restricted condensates to facilitate the hyperphosphorylation reaction of RNA Pol II. Additionally, the CTD can also bind to the low-complexity domains of transactivating proteins FUS, TAF15 and hnRNPA2 to form nuclear granules that promote transcription [120,121].

In the past few years, a rapid surge of studies has demonstrated that many transcription factors and coactivators are able to undergo phase separation that can help them create dynamic hubs, clusters or condensates to regulate target gene expression (Figure 2). Some of these condensates can be assembled into super-enhancers with many tandemly adjacent enhancers, each of which is typically 50 to 1500 base pairs in length [122]. The transcription factors, such as OCT4 and GCN4, harbor IDRs in their transactivation domains that can undergo phase separation to form clustered enhancers or super-enhancers and activate gene expression [122]. Meanwhile, many coactivators, such as BRD4, MED1 and p300, act as key components of the enhancer complexes that drive the expression of the master genes to determine cell identity or promote oncogenesis [11]. As we recently reported, the transcription factor YY1 has an IDR featured with an 11-histidine cluster. Deletion of the histidine cluster or replacing it with 11 alanines abolishes YY1's ability to form nuclear puncta and even deprive its dominant nuclear localization. Through the phase-separation mechanism, YY1 compartmentalizes many coactivators, including p300, BRD4, MED1 and CDK9, to assemble clustered enhancers that activate *FOXM1* gene expression and contribute to mammary tumor formation in a mouse model [24].

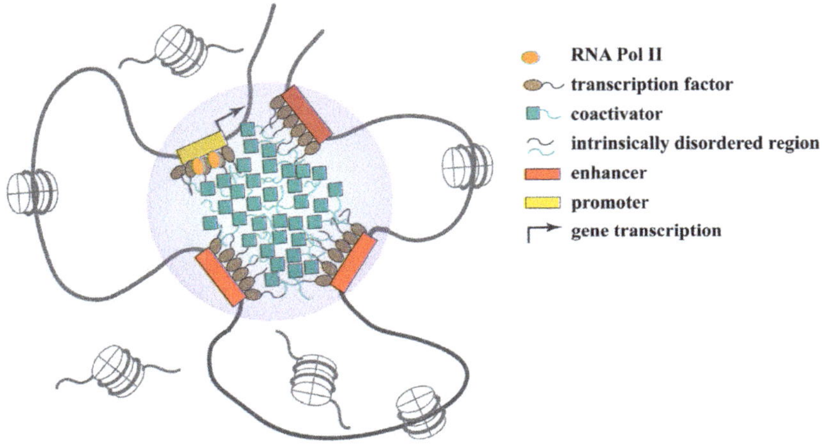

Figure 2. Regulation of gene expression by the phase-separation mechanism.

Another example is the coactivator YAP that can cause chromatin reorganization to activate its target genes. In this regulation, YAP forms phase-separated condensates to compartmentalize the transcription factor TEAD1 and other coactivators, such as TAZ. The YAP condensates in the nucleus consist of super-enhancers with an accessible chromatin structure [123].

Signal transducer and activator of transcription 3 (STAT3) is a transcription factor regulating the expression of a variety of genes involved in different biological processes. As a key regulator in the anti-cancer immune response, STAT3 can be activated by various cytokines. Aberrant activation of STAT3 has been observed in many cancers, which serves as a bona fide target in cancer therapies [124]. Early studies indicated that tyrosine phosphorylation of STAT3 stimulated by interleukin 6 could cause its translocation into nucleus where STAT3 was activated, bound to the enhancer elements of target genes, and formed nuclear bodies. Thus, it was proposed that the STAT3 nuclear bodies could either be directly involved in activated gene transcription or serve as reservoirs of activated STAT3 [125]. Recent studies revealed that the biomolecular condensates formed by activated STAT3 exhibited LLPS properties, suggesting that the phase-separation mechanism contributes to STAT3-mediated gene activation [51,126].

2.4.3. Regulation of Viral Infection

Many studies have demonstrated the regulatory roles of LLPS in both the viral life cycle and virus–host interactions [17,127]. Viral proteins with IDRs can promote the formation of membraneless compartments used for the replication of viruses. These compartments are enriched with specific proteins and nucleic acids, and serve as "viral factories" for the replication, assembly and trafficking of viruses. The LLPS condensates are selective for the inclusion or exclusion of components to allow optimal viral production, and may also avoid the defense of the host immune system. For example, cells infected by negative-strand RNA viruses, such as rabies virus (RABV), rotavirus, vesicular stomatitis virus (VSV), Ebola virus, measles virus, influenza A virus and respiratory syncytial virus (RSV), may form cytoplasmic LLPS condensates that allow all the ribonucleoparticle (RNP) components and viral RNAs to be synthesized inside and assembled into viral particles [128–134]. A report by Fouquet et al. revealed that the phosphoprotein P, essential for viral transcription and replication of RABV, could shuttle between the cytosol and the Negri bodies formed by the virus, leading to the recruitment of focal adhesion kinase (FAK) and HSP70, two cellular proteins with proviral activities [135].

Viral protein-mediated LLPS can interfere with the functions of host cells through two mechanisms, either regulating the expression of cellular genes or modulating the activities of cellular proteins. The oncogenic effects of Epstein–Barr virus (EBV) can be used as an example of the first mechanism. EBV is a human virus with potent activities to induce malignant transformation of infected cells through the activation of both viral oncogenes and cellular proto-oncogenes [136,137]. EBNA2 and EBNALP are two EBV-encoded transcription factors that form nuclear puncta using their IDRs, leading to the formation of super-enhancers on the promoters of the oncogenes MYC and RUNX3 to promote their transcription and subsequent oncogenesis [138]. In contrast, the functional interplays between viral and cellular proteins in the context of LLPS have been relatively understudied [17,127]. The formation of fibrillar aggregates by viral proteins may exert various effects on host cells, including inhibition of key cellular processes, such as such as necroptosis, and sequestration cellular transcription factors to block host cell RNA synthesis [17,139].

LLPS-related mechanisms not only mediate the impairments of infected cells caused by viruses, but also contribute to the defense system of host cells against viral infection. Human myxovirus-resistance protein A (MxA) is a cytoplasmic dynamin-family large GTPase with a molecular weight of about 70 kDa, and can be induced by 50- to 100-fold when cells are treated by type I and III interferons [140]. MxA associates with the endoplasmic reticulum and Golgi apparatus, and exhibits antiviral activity against several RNA and DNA viruses. A study by Davis et al. demonstrated that MxA formed metastable membraneless cytoplasmic spherical or irregular bodies, filaments, or reticula with variable sizes. Importantly, in VSV-infected cells, the nucleocapsid protein of the virus could blend with the MxA condensates in cells showing a concomitant antiviral phenotype [141]. Similarly, Mx1, the murine ortholog of human MxA, could also form nuclear condensates when being transfected into human cells. Interestingly, 20–30% of transfected cells also formed cytoplasmic giantin-based filaments, and these cells, but not the ones with only nuclear bodies, showed antiviral activity against VSV [142]. The mechanism underlying the antiviral effects of the cytoplasmic filaments formed by Mx1 remains unclear.

2.4.4. Sequestration and Storage of Molecules

Cellular condensates work as compartments to selectively sequester biomolecules and stock them, which serves as an approach of resource conservation. For example, each proteasome consists of a catalytic core particle (CP) and a regulatory particle (RP). With yeast as a model, the proteasome holo-enzyme constituted by the CP and RP mostly stays in the nucleus in proliferative cells; however, in the quiescent state, they are transported into the cytoplasm and sequestered as protein condensates called proteasome storage granules (PSGs) [143]. The functions of PSGs include protecting yeast cells against stress and maintaining their fitness during aging [144]. When the cells exit quiescence, the PSGs will be disassembled and the proteasome will reenter the nucleus [145]. Furthermore, P bodies and stress granules are also able to sequester highly expressed mRNAs. Whether the stored mRNAs undergo translation or decay by individual cells in future can generate different phenotypes and improve their ability to withstand stress [146]. Meanwhile, P bodies and stress granules can also serve as protein quality control compartments that help cells to sequester misfolded proteins from the other cellular milieu [147].

Cellular condensates can also confiscate proteins to temporally curb their functions. For example, the death domain-associated protein (DAXX) is a chaperone of the histone H3.3 variant, and recruits HDACs to repress basal transcription [148]. Due to the interaction with PML, DAXX can be sequestered into the PML bodies to block its activity in repressing transcription, and sumoylation of PML is prerequisite for this process [148,149].

The nucleolus is a reputed storage apparatus in the nucleus and can sequester many regulatory proteins in response to different signals [150]. Many proteins involved in cell cycle progression, apoptosis and oncogenesis can be sequestered in nucleoli through different mechanisms. As a ubiquitination E3 ligase, MDM2 can be confined in nucleoli through its interaction with p14ARF or ATP molecules, which leads to p53 activation [151, 152]. Another E3 ligase, VHL, can also be sequestered in the nucleolus in response to reduced extracellular pH. This can prevent the ubiquitination and degradation of its substrate HIF in the presence of oxygen, and allow it to activate its target genes [153]. Other important regulatory proteins with reported nucleolar sequestration include MYC, hTERT and CDC14 [154–156].

3. The Phase Separation of Proteins in Diseases

Accumulating evidence suggests that aberrant assembly of condensates is associated with cancers [157]. Below, we employ several examples to discuss how dysregulated phase separation of key regulatory proteins may contribute to neurodegenerative diseases and cancers.

3.1. LLPS and Neurodegenerative Diseases

3.1.1. FUS

As a multifunctional DNA- and RNA-binding protein, FUS has been reportedly involved in transcription regulation, RNA splicing, RNA transport and DNA damage repair [158]. The FUS protein has an N terminal PrLD that is intrinsically disordered and critical to its phase-separated condensation [120,159]. The RNA-recognition motif (RRM) of FUS can bind to RNA molecules that promote FUS phase separation. Two domains are involved in FUS nuclear localization. First, the three RGG (arginine-glycine-glycine) repeats, designated as the RGG3 domain, can transport FUS from cytoplasm to nucleus. Second, the C-terminal proline tyrosine (PY) domain is a PY-NLS that can also promote FUS's nuclear transportation, but it needs the assistance of the nuclear import receptor transportin, also known as karyopherin β2, to cross the nuclear pore complex [160,161]. Most FUS mutants showed impaired binding to the receptor transportin, leading to their increased cytoplasmic retention. The defective nuclear import of the FUS mutants causes their cytoplasmic aggregation in neuronal and sometimes glial cells, linked to disease pathogenesis, such as ALS [162].

ALS patient-derived mutations of G156E and R244C, located in or adjacent to the prion-like domain of the FUS protein, could convert its droplets to fibrous structures, which eventually form amyloid-like fibrillar aggregates and subsequently contribute to the protein misfolding diseases [31,163]. While the fusion of two adjacent wild-type FUS droplets could occur in seconds, the event would take many hours for the FUS(G156E) mutant [31]. Interestingly, the fibrillar aggregates of FUS(G156E) could act as seeds to efficiently induce the aggregation of wt FUS [163]. Both wt FUS and the G156E mutant could produce similar condensates in cells; however, in a rat model, FUS(G156E) mutant could create intranuclear inclusions in hippocampal neurons with cytotoxicity, likely due to the defects in regulating translation and RNA splicing [163].

It has been reported that the methylation of the arginine in front of the PY-NLS reduced FUS binding to the receptor transportin, and thus caused its cytoplasmic accumulation [164]. Interestingly, the arginine methylation of FUS also decreases its phase separation and stress granule association. Therefore, the NLS mutations of FUS in ALS patients not only weaken transportin-mediated nuclear import, but also abolish its arginine methylation, which promotes phase separation and stress granule formation of FUS [165]. Besides methylation, the PrLD of FUS can be phosphorylated by DNA-PK. The phosphorylated FUS protein exhibits reduced FUS phase separation and subsequently decreased aggregation tendency, which can ameliorate FUS-associated cytotoxicity [96].

3.1.2. Tau

In 1975, Weingarten et al. isolated Tau as a protein essential for microtubule assembly [166], and the subsequent studies indicated this microtubule-associated protein as a regulator of axonal outgrowth and transport in neurons. Tau aggregation leads to the formation of intracellular fibrillary deposits that have been recognized as a hallmark of various neurodegenerative diseases, including Alzheimer's disease, frontotemporal dementia and Parkinson disease, with a common name of tauopathies [167,168]. The intrinsically disordered property and phase-separation potential of Tau can be attributed to its high content of proline and glycine, and many polar and charged amino acids. The LLPS propensity of Tau is primarily controlled by the proline-rich domain in its middle region, which also contains many phosphorylation sites. Tau is a protein that harbors different posttranslational modifications, including phosphorylation, acetylation, glycosylation, glycation and ubiquitination [169]. Some of these modifications have been demonstrated to impact the LLPS of Tau through altering its net charge, conformation and interactions with other molecules. Hyperphosphorylation of Tau can promote the maturation of its condensates into insoluble amyloid-like fibrils contributing to the diseases [170]. Lysine residues are crucial for the LLPS of Tau, and thus their acetylation mediated by p300 and CBP can reduce its interaction with RNA and reverse its condensation [171]. Despite the repressive effects of acetylation on LLPS-mediated aggregation, acetylated Tau is associated with neurotoxicity because it shows dampened interaction with tubulin and impaired ability to promote the growth of microtubule filaments [172].

3.1.3. TDP-43

TDP-43 was initially identified as a protein binding to a regulatory element in the long terminal repeat of HIV-1 and blockint the assembly of its transcription complex [173]. Other studies also revealed TDP-43 as an essential DNA/RNA-binding protein regulating RNA splicing [174]. Among ALS patients, 90–95% are sporadic, with mutations in the genes *C9ORF72*, *SOD1*, *FUS*, etc. Strikingly, about 97% of these ALS patients and 45% of FTLD patients exhibited TDP-43 aggregation, implicating its pathogenic role in causing the motor neuron diseases [175]. TDP-43 is one of the PrLD-containing proteins that are prone to aggregation. Either pre-mRNA alternative splicing or aberrant proteolytic cleavage of the full-length TDP-43 can generate the PrLD fragment, suggesting its high potential in forming aggregates [176,177].

Posttranslational modifications play a regulatory role in TDP-43 condensation. Despite the predominantly nuclear presence, TDP-43 phosphorylation is associated with its cytoplasmic translocation, which can drive early pathology of the diseases [178]. Hyperphosphorylated TDP-43 tends to aggregate and generate inclusion bodies in the brains and spinal cords of the patients. Actually, phosphorylation of S409 and S410 has been considered a signature for ALS pathological analysis [179]. TDP-43 acetylation reduces its RNA-binding affinity and promotes accumulation of insoluble, hyper-phosphorylated TDP-43, which resembles the pathological inclusions observed in ALS and FTLD [180]. Additionally, ubiquitination of TDP-43 by its E3 ligase Parkin does not show clear degradation-orientated effects, but instead causes its cytoplasmic accumulation to form insoluble aggregates [181]. In addition, TDP-43 aggregation is associated with its C-terminal domain consisting of a prion-like glutamine/asparagine-rich domain and glycine-rich region that drives LLPS [175,182].

3.2. LLPS and Cancers

3.2.1. SHP2

Src homology region 2 domain-containing phosphatase-2 (SHP2) is a non-receptor protein tyrosine phosphatase (PTP), encoded by the *PTPN11* gene. SHP2 contains two SH2 domains, a central PTP catalytic domain and a C-terminal tail. The two SH2 domains, C-SH2 and N-SH2, serve as phospho-tyrosine-binding regions to interact with the substrates [183]. As a ubiquitously expressed protein, SHP2 regulates many signaling pathways involved in mitogenic activation, metabolic control, and transcription regulation [184]. Germline mutations of SHP2 accounts for 50% of Noonan syndrome and 90% of LEOPARD syndrome (i.e., Noonan syndrome with multiple lentigines) [185,186] cases. Somatic SHP2 mutations are significantly associated with different human malignancies [187].

The intramolecular interaction between the N-SH2 and PTP domains serves as a "molecular switch" to block the phosphatase activity of SHP2. This switch can be turned on by the N-SH2 domain binding to specific phospho-tyrosine sequences of upstream growth factor receptors and/or scaffold proteins, leading to SHP2 activation. Mutations of SHP2 may either abolish the autoinhibitory switch or impair its PTP activity, which cause either Noonan syndrome or LEOPARD syndrome, respectively [188].

E76 is the most frequently mutated site of SHP2 in human cancers and the mutations disrupt the inhibition of PTP domain by the N-SH2, while R498 mutations in SHP2's PTP domain are also commonly observed and associated with LEOPARD syndrome [189]. Interestingly, a recent study demonstrated that two disease-associated mutant proteins, SHP2(E76K) and SHP2(R498L), showed significantly increased tendency of droplet formation compared to the wild-type SHP2. Consistently, the two mutants also formed nuclear puncta in cells, but wild-type SHP2 did not [190]. However, unlike most previously reported proteins with LLPS capability, the SHP2 protein does not contain any IDR or repetitive multivalent modular domain. Interestingly, the catalytic PTP domain is also responsible for the phase separation of the SHP2 mutants. The mutations of N-SH2 enhance the PTP activity and subsequently promote ERK1/2 activation [190].

3.2.2. YAP and TAZ

As downstream effectors of the Hippo signaling pathway, YAP (Yes-associated protein) and TAZ regulate many biological processes including cell proliferation, apoptosis and differentiation [191]. As transcription coactivators, unphosphorylated YAP/TAZ complexes can be translocated to the nucleus, and bind to the TEAD transcription factors that regulate the expression of several genes involved in cell proliferation and survival, such as MYC and BIRC5 [192]. In recent years, both YAP and TAZ have been demonstrated to undergo LLPS that plays an essential role in activating the expression of their target genes, subsequently promoting oncogenesis. The phase-separated condensates can help YAP and TAZ to compartmentalize transcription machinery, including BRD4, MED1, CDK9 and TEAD [193]. Noticeably, in the Hippo signaling pathway, LATS1/2 phosphorylate YAP at S172 and TAZ at S89 to increase their cytoplasmic retention, which can both inhibit the LLPS of YAP and TAZ and reduce their activity as coactivators [10,123,194]. It has been demonstrated that the Hippo pathway can be frequently inactivated through nonmutational mechanisms during oncogenesis [195], which may explain the consistent hyperactivation of the YAP and TAZ in various cancers [194].

4. Conclusions and Future Perspectives

In a cell, many different types of membraneless organelles or condensates existand provide relatively defined but still dynamic compartments for various biological reactions or material sequestration to occur in an undisturbed fashion. Biomolecules, biochemical reactions and various biological regulations are not present or happen in a chaotic or random manner in the complex cellular milieu. The concept of phase separation that regulates the formation of these compartments is likely a general mechanism to restrict biomolecules into particular compartments for designated biological activities. The questions concerning how large biomolecules, especially proteins and RNAs, are self-organized and undergo LLPS, and how their phase-separation capability can be linked and contribute to their specific activities, have intrigued many researchers and attracted increasing interest. Through the research endeavors over the past two decades, we have gained extensive knowledge regarding the molecular features, assembly requirements and material properties of these membraneless organelles or condensates. We have also obtained many insights in phase-separation-regulated biological processes, including biological reactions, resource storage or sequestration, and gene expression. Importantly, dysregulated LLPS of different proteins due to mutations or aberrant posttranslational modifications are causal causes of various human diseases, such as many neurodegenerative disorders and cancer. Despite the knowledge obtained from the reported studies related to LLPS in normal and diseased cellular conditions, many questions remain to be answered and fertile areas need to be explored. First, although IDRs are likely prerequisite elements for protein phase separation, there are still reported exceptions. Thus, how amino acid sequence and/or composition can precisely determine the LLPS properties of a protein needs to be further defined. Second, the sequences, secondary structures or other properties of nucleic acids involved in phase separation are still largely unexplored. Third, with neurodegenerative diseases as an example, the reasons accounting for the occurrence of phase-separated insoluble aggregates only observed in specific cell types deserve special investigation. Finally, we have only just begun to explore therapeutic applications that take advantage of the phase-separation mechanism for disease treatment.

Author Contributions: L.Z., S.W., D.L. and G.S. wrote the majority of the manuscript and prepared the figures and table. W.W. participated in writing a few sections of the manuscript, and extensively modified the figures. J.S. and D.B.S. contributed to preparing the scheme of the manuscript and extensively revised and worked on finalizing the manuscript. All authors have read and agreed to the published version of the manuscript.

Funding: This research was funded by the Fundamental Research Funds for the Central Universities (2572021BD03) to D.L., and the National Natural Science Foundation of China (81872293) to G.S.

Institutional Review Board Statement: Not applicable.

Informed Consent Statement: Not applicable.

Data Availability Statement: The data presented in this study are available on request from the corresponding author.

Conflicts of Interest: The authors declare no conflict of interest.

References

1. Hyman, A.A.; Weber, C.A.; Jülicher, F. Liquid-liquid phase separation in biology. *Annu. Rev. Cell Dev. Biol.* **2014**, *30*, 39–58. [CrossRef] [PubMed]
2. Liu, X.; Liu, X.; Wang, H.; Dou, Z.; Ruan, K.; Hill, D.L.; Li, L.; Shi, Y.; Yao, X. Phase separation drives decision making in cell division. *J. Biol. Chem.* **2020**, *295*, 13419–13431. [CrossRef] [PubMed]
3. Gomes, E.; Shorter, J. The molecular language of membraneless organelles. *J. Biol. Chem.* **2019**, *294*, 7115–7127. [CrossRef] [PubMed]
4. Kurihara, M.; Kato, K.; Sanbo, C.; Shigenobu, S.; Ohkawa, Y.; Fuchigami, T.; Miyanari, Y. Genomic Profiling by ALaP-Seq Reveals Transcriptional Regulation by PML Bodies through DNMT3A Exclusion. *Mol. Cell* **2020**, *78*, 493–505.e498. [CrossRef]
5. Wolozin, B.; Ivanov, P. Stress granules and neurodegeneration. *Nat. Rev. Neurosci.* **2019**, *20*, 649–666. [CrossRef]
6. Dodson, A.E.; Kennedy, S. Phase Separation in Germ Cells and Development. *Dev. Cell* **2020**, *55*, 4–17. [CrossRef]
7. Banani, S.F.; Lee, H.O.; Hyman, A.A.; Rosen, M.K. Biomolecular condensates: Organizers of cellular biochemistry. *Nat. Rev. Mol. Cell Biol.* **2017**, *18*, 285–298. [CrossRef]
8. Woodruff, J.B.; Hyman, A.A.; Boke, E. Organization and Function of Non-dynamic Biomolecular Condensates. *Trends Biochem. Sci.* **2018**, *43*, 81–94. [CrossRef]
9. Case, L.B.; Zhang, X.; Ditlev, J.A.; Rosen, M.K. Stoichiometry controls activity of phase-separated clusters of actin signaling proteins. *Science* **2019**, *363*, 1093–1097. [CrossRef]
10. Lu, Y.; Wu, T.; Gutman, O.; Lu, H.; Zhou, Q.; Henis, Y.I.; Luo, K. Phase separation of TAZ compartmentalizes the transcription machinery to promote gene expression. *Nat. Cell Biol.* **2020**, *22*, 453–464. [CrossRef]

11. Sabari, B.R.; Dall'Agnese, A.; Boija, A.; Klein, I.A.; Coffey, E.L.; Shrinivas, K.; Abraham, B.J.; Hannett, N.M.; Zamudio, A.V.; Manteiga, J.C.; et al. Coactivator condensation at super-enhancers links phase separation and gene control. *Science* **2018**, *361*, eaar3958. [CrossRef] [PubMed]
12. Richter, K.; Nessling, M.; Lichter, P. Macromolecular crowding and its potential impact on nuclear function. *Biochim. Biophys. Acta* **2008**, *1783*, 2100–2107. [CrossRef] [PubMed]
13. Kanekura, K.; Kuroda, M. How can we interpret the relationship between liquid-liquid phase separation and amyotrophic lateral sclerosis? *Lab. Investig.* 2022; *online ahead of print*.
14. Ambadipudi, S.; Biernat, J.; Riedel, D.; Mandelkow, E.; Zweckstetter, M. Liquid-liquid phase separation of the microtubule-binding repeats of the Alzheimer-related protein Tau. *Nat. Commun.* **2017**, *8*, 275. [CrossRef] [PubMed]
15. Chong, S.; Dugast-Darzacq, C.; Liu, Z.; Dong, P.; Dailey, G.M.; Cattoglio, C.; Heckert, A.; Banala, S.; Lavis, L.; Darzacq, X.; et al. Imaging dynamic and selective low-complexity domain interactions that control gene transcription. *Science* **2018**, *361*, eaar2555. [CrossRef]
16. Altmeyer, M.; Neelsen, K.J.; Teloni, F.; Pozdnyakova, I.; Pellegrino, S.; Grøfte, M.; Rask, M.D.; Streicher, W.; Jungmichel, S.; Nielsen, M.L.; et al. Liquid demixing of intrinsically disordered proteins is seeded by poly(ADP-ribose). *Nat. Commun.* **2015**, *6*, 8088. [CrossRef]
17. Brocca, S.; Grandori, R.; Longhi, S.; Uversky, V. Liquid-Liquid Phase Separation by Intrinsically Disordered Protein Regions of Viruses: Roles in Viral Life Cycle and Control of Virus-Host Interactions. *Int. J. Mol. Sci.* **2020**, *21*, 9045. [CrossRef]
18. Shen, B.; Chen, Z.; Yu, C.; Chen, T.; Shi, M.; Li, T. Computational Screening of Phase-separating Proteins. *Genom. Proteom. Bioinform.* **2021**, *19*, 13–24. [CrossRef]
19. Zhang, H.; Ji, X.; Li, P.; Liu, C.; Lou, J.; Wang, Z.; Wen, W.; Xiao, Y.; Zhang, M.; Zhu, X. Liquid-liquid phase separation in biology: Mechanisms, physiological functions and human diseases. *Sci. China Life Sci.* **2020**, *63*, 953–985. [CrossRef]
20. Peran, I.; Mittag, T. Molecular structure in biomolecular condensates. *Curr. Opin. Struct. Biol.* **2020**, *60*, 17–26. [CrossRef]
21. Alberti, S.; Dormann, D. Liquid-Liquid Phase Separation in Disease. *Annu. Rev. Genet.* **2019**, *53*, 171–194. [CrossRef]
22. Alberti, S.; Gladfelter, A.; Mittag, T. Considerations and Challenges in Studying Liquid-Liquid Phase Separation and Biomolecular Condensates. *Cell* **2019**, *176*, 419–434. [CrossRef] [PubMed]
23. Basu, S.; Mackowiak, S.D.; Niskanen, H.; Knezevic, D.; Asimi, V.; Grosswendt, S.; Geertsema, H.; Ali, S.; Jerkovic, I.; Ewers, H.; et al. Unblending of Transcriptional Condensates in Human Repeat Expansion Disease. *Cell* **2020**, *181*, 1062–1079.e30. [PubMed]
24. Wang, W.; Qiao, S.; Li, G.; Cheng, J.; Yang, C.; Zhong, C.; Stovall, D.B.; Shi, J.; Teng, C.; Li, D.; et al. A histidine cluster determines YY1-compartmentalized coactivators and chromatin elements in phase-separated enhancer clusters. *Nucleic Acids Res.* **2022**; *online ahead of print*.
25. Dosztanyi, Z.; Csizmok, V.; Tompa, P.; Simon, I. IUPred: Web server for the prediction of intrinsically unstructured regions of proteins based on estimated energy content. *Bioinformatics* **2005**, *21*, 3433–3434. [CrossRef] [PubMed]
26. Uversky, V.N. Analyzing IDPs in Interactomes. *Methods Mol. Biol.* **2020**, *2141*, 895–945.
27. Tsang, B.; Pritisanac, I.; Scherer, S.W.; Moses, A.M.; Forman-Kay, J.D. Phase Separation as a Missing Mechanism for Interpretation of Disease Mutations. *Cell* **2020**, *183*, 1742–1756. [CrossRef]
28. Vacic, V.; Iakoucheva, L.M. Disease mutations in disordered regions–Exception to the rule? *Mol. Biosyst.* **2012**, *8*, 27–32. [CrossRef]
29. Vacic, V.; Markwick, P.R.; Oldfield, C.J.; Zhao, X.; Haynes, C.; Uversky, V.N.; Iakoucheva, L.M. Disease-associated mutations disrupt functionally important regions of intrinsic protein disorder. *PLoS Comput. Biol.* **2012**, *8*, e1002709. [CrossRef]
30. Deng, H.; Gao, K.; Jankovic, J. The role of FUS gene variants in neurodegenerative diseases. *Nat. Rev. Neurol.* **2014**, *10*, 337–348. [CrossRef]
31. Patel, A.; Lee, H.O.; Jawerth, L.; Maharana, S.; Jahnel, M.; Hein, M.Y.; Stoynov, S.; Mahamid, J.; Saha, S.; Franzmann, T.M.; et al. A Liquid-to-Solid Phase Transition of the ALS Protein FUS Accelerated by Disease Mutation. *Cell* **2015**, *162*, 1066–1077. [CrossRef]
32. McAlary, L.; Plotkin, S.S.; Yerbury, J.J.; Cashman, N.R. Prion-Like Propagation of Protein Misfolding and Aggregation in Amyotrophic Lateral Sclerosis. *Front. Mol. Neurosci.* **2019**, *12*, 262. [CrossRef]
33. Ohsawa, T.; Sahara, T.; Muramatsu, S.; Nishimura, Y.; Yathuoka, T.; Tanaka, Y.; Yamaguchi, K.; Ishida, H.; Akagi, K. Colorectal cancer susceptibility associated with the hMLH1 V384D variant. *Mol. Med. Rep.* **2009**, *2*, 887–891. [PubMed]
34. Lee, S.E.; Lee, H.S.; Kim, K.Y.; Park, J.H.; Roh, H.; Park, H.Y.; Kim, W.S. High prevalence of the MLH1 V384D germline mutation in patients with HER2-positive luminal B breast cancer. *Sci. Rep.* **2019**, *9*, 10966. [CrossRef] [PubMed]
35. Wang, S.; Dai, T.; Qin, Z.; Pan, T.; Chu, F.; Lou, L.; Zhang, L.; Yang, B.; Huang, H.; Lu, H.; et al. Targeting liquid-liquid phase separation of SARS-CoV-2 nucleocapsid protein promotes innate antiviral immunity by elevating MAVS activity. *Nat. Cell Biol.* **2021**, *23*, 718–732. [CrossRef] [PubMed]
36. Uversky, V.N. Intrinsically disordered proteins in overcrowded milieu: Membrane-less organelles, phase separation, and intrinsic disorder. *Curr. Opin. Struct. Biol.* **2017**, *44*, 18–30. [CrossRef]
37. Alberti, S.; Hyman, A.A. Are aberrant phase transitions a driver of cellular aging? *Bioessays* **2016**, *38*, 959–968. [CrossRef]
38. Shin, Y.; Brangwynne, C.P. Liquid phase condensation in cell physiology and disease. *Science* **2017**, *357*, eaaf4382. [CrossRef]
39. Luo, Y.; Na, Z.; Slavoff, S.A. P-Bodies: Composition, Properties, and Functions. *Biochemistry* **2018**, *57*, 2424–2431. [CrossRef]
40. Riggs, C.L.; Kedersha, N.; Ivanov, P.; Anderson, P. Mammalian stress granules and P bodies at a glance. *J. Cell Sci.* **2020**, *133*, jcs242487. [CrossRef]
41. Zhang, Y.; Gu, J.; Sun, Q. Aberrant Stress Granule Dynamics and Aggrephagy in ALS Pathogenesis. *Cells* **2021**, *10*, 2247. [CrossRef]

42. Voronina, E.; Seydoux, G.; Sassone-Corsi, P.; Nagamori, I. RNA granules in germ cells. *Cold Spring Harb. Perspect. Biol.* **2011**, *3*, a002774. [CrossRef]
43. Zeng, M.; Chen, X.; Guan, D.; Xu, J.; Wu, H.; Tong, P.; Zhang, M. Reconstituted Postsynaptic Density as a Molecular Platform for Understanding Synapse Formation and Plasticity. *Cell* **2018**, *174*, 1172–1187.e16. [CrossRef] [PubMed]
44. Formicola, N.; Vijayakumar, J.; Besse, F. Neuronal ribonucleoprotein granules: Dynamic sensors of localized signals. *Traffic* **2019**, *20*, 639–649. [CrossRef] [PubMed]
45. Jamieson-Lucy, A.; Mullins, M.C. The vertebrate Balbiani body, germ plasm, and oocyte polarity. *Curr. Top. Dev. Biol.* **2019**, *135*, 1–34. [PubMed]
46. Aguilera-Gomez, A.; Rabouille, C. Membrane-bound organelles versus membrane-less compartments and their control of anabolic pathways in Drosophila. *Dev. Biol.* **2017**, *428*, 310–317. [CrossRef] [PubMed]
47. Buckingham, M.; Liu, J.L. U bodies respond to nutrient stress in Drosophila. *Exp. Cell Res.* **2011**, *317*, 2835–2844. [CrossRef]
48. Li, J.; Hochstrasser, M. Microautophagy regulates proteasome homeostasis. *Curr. Genet.* **2020**, *66*, 683–687. [CrossRef]
49. Smalley, M.J.; Signoret, N.; Robertson, D.; Tilley, A.; Hann, A.; Ewan, K.; Ding, Y.; Paterson, H.; Dale, T.C. Dishevelled (Dvl-2) activates canonical Wnt signalling in the absence of cytoplasmic puncta. *J. Cell Sci.* **2005**, *118*, 5279–5289. [CrossRef]
50. Wunder, T.; Mueller-Cajar, O. Biomolecular condensates in photosynthesis and metabolism. *Curr. Opin. Plant Biol.* **2020**, *58*, 1–7. [CrossRef]
51. Sehgal, P.B. Biomolecular condensates in cancer cell biology: Interleukin-6-induced cytoplasmic and nuclear STAT3/PY-STAT3 condensates in hepatoma cells. *Contemp. Oncol.* **2019**, *23*, 16–22. [CrossRef]
52. Xu, F.; Mukhopadhyay, S.; Sehgal, P.B. Live cell imaging of interleukin-6-induced targeting of "transcription factor" STAT3 to sequestering endosomes in the cytoplasm. *Am. J. Physiol. Cell Physiol.* **2007**, *293*, C1374–C1382. [CrossRef]
53. Ma, W.; Mayr, C. A Membraneless Organelle Associated with the Endoplasmic Reticulum Enables 3′UTR-Mediated Protein-Protein Interactions. *Cell* **2018**, *175*, 1492–1506.e19. [CrossRef] [PubMed]
54. Beck, M.; Hurt, E. The nuclear pore complex: Understanding its function through structural insight. *Nat. Rev. Mol. Cell Biol.* **2017**, *18*, 73–89. [CrossRef] [PubMed]
55. Lafontaine, D.L.J.; Riback, J.A.; Bascetin, R.; Brangwynne, C.P. The nucleolus as a multiphase liquid condensate. *Nat. Rev. Mol. Cell Biol.* **2021**, *22*, 165–182. [CrossRef] [PubMed]
56. Nizami, Z.; Deryusheva, S.; Gall, J.G. The Cajal body and histone locus body. *Cold Spring Harb. Perspect. Biol.* **2010**, *2*, a000653. [CrossRef] [PubMed]
57. Pessina, F.; Gioia, U.; Brandi, O.; Farina, S.; Ceccon, M.; Francia, S.; d'Adda di Fagagna, F. DNA Damage Triggers a New Phase in Neurodegeneration. *Trends Genet.* **2021**, *37*, 337–354. [CrossRef] [PubMed]
58. Lallemand-Breitenbach, V.; de Thé, H. PML nuclear bodies: From architecture to function. *Curr. Opin. Cell Biol.* **2018**, *52*, 154–161. [CrossRef] [PubMed]
59. Biamonti, G. Nuclear stress bodies: A heterochromatin affair? *Nat. Rev. Mol. Cell Biol.* **2004**, *5*, 493–498. [CrossRef]
60. Smigova, J.; Juda, P.; Cmarko, D.; Raska, I. Fine structure of the "PcG body" in human U-2 OS cells established by correlative light-electron microscopy. *Nucleus* **2011**, *2*, 219–228. [CrossRef]
61. McCluggage, F.; Fox, A.H. Paraspeckle nuclear condensates: Global sensors of cell stress? *Bioessays* **2021**, *43*, e2000245. [CrossRef]
62. Pollock, C.; Huang, S. The perinucleolar compartment. *Cold Spring Harb. Perspect. Biol.* **2010**, *2*, a000679. [CrossRef]
63. Matera, A.G.; Frey, M.R. Coiled bodies and gems: Janus or gemini? *Am. J. Hum. Genet.* **1998**, *63*, 317–321. [CrossRef] [PubMed]
64. Harrigan, J.A.; Belotserkovskaya, R.; Coates, J.; Dimitrova, D.S.; Polo, S.E.; Bradshaw, C.R.; Fraser, P.; Jackson, S.P. Replication stress induces 53BP1-containing OPT domains in G1 cells. *J. Cell Biol* **2011**, *193*, 97–108. [CrossRef] [PubMed]
65. Veerabhadrappa, B.; Delaby, C.; Hirtz, C.; Vialaret, J.; Alcolea, D.; Lleo, A.; Fortea, J.; Santosh, M.S.; Choubey, S.; Lehmann, S. Detection of amyloid beta peptides in body fluids for the diagnosis of alzheimer's disease: Where do we stand? *Crit. Rev. Clin. Lab. Sci.* **2020**, *57*, 99–113. [CrossRef] [PubMed]
66. Wang, J.; Gan, Y.; Cao, J.; Dong, X.; Ouyang, W. Pathophysiology of stress granules: An emerging link to diseases (Review). *Int. J. Mol. Med.* **2022**, *49*, 44. [CrossRef]
67. Kim, W.J.; Kim, J.H.; Jang, S.K. Anti-inflammatory lipid mediator 15d-PGJ2 inhibits translation through inactivation of eIF4A. *EMBO J.* **2007**, *26*, 5020–5032. [CrossRef]
68. Di Marco, B.; Dell'Albani, P.; D'Antoni, S.; Spatuzza, M.; Bonaccorso, C.M.; Musumeci, S.A.; Drago, F.; Bardoni, B.; Catania, M.V. Fragile X mental retardation protein (FMRP) and metabotropic glutamate receptor subtype 5 (mGlu5) control stress granule formation in astrocytes. *Neurobiol. Dis.* **2021**, *154*, 105338. [CrossRef]
69. Ash, P.E.; Vanderweyde, T.E.; Youmans, K.L.; Apicco, D.J.; Wolozin, B. Pathological stress granules in Alzheimer's disease. *Brain Res.* **2014**, *1584*, 52–58. [CrossRef]
70. Yang, X.; Hu, Z.; Fan, S.; Zhang, Q.; Zhong, Y.; Guo, D.; Qin, Y.; Chen, M. Picornavirus 2A protease regulates stress granule formation to facilitate viral translation. *PLoS Pathog.* **2018**, *14*, e1006901. [CrossRef]
71. Zhang, Q.; Sharma, N.R.; Zheng, Z.M.; Chen, M. Viral Regulation of RNA Granules in Infected Cells. *Virol. Sin.* **2019**, *34*, 175–191. [CrossRef]
72. Balak, C.; Benard, M.; Schaefer, E.; Iqbal, S.; Ramsey, K.; Ernoult-Lange, M.; Mattioli, F.; Llaci, L.; Geoffroy, V.; Courel, M.; et al. Rare De Novo Missense Variants in RNA Helicase DDX6 Cause Intellectual Disability and Dysmorphic Features and Lead to P-Body Defects and RNA Dysregulation. *Am. J. Hum. Genet.* **2019**, *105*, 509–525. [CrossRef]

73. Liu, L.; Weiss, E.; Panas, M.D.; Gotte, B.; Sellberg, S.; Thaa, B.; McInerney, G.M. RNA processing bodies are disassembled during Old World alphavirus infection. *J. Gen. Virol.* **2019**, *100*, 1375–1389. [CrossRef] [PubMed]
74. Bhanji, R.A.; Eystathioy, T.; Chan, E.K.; Bloch, D.B.; Fritzler, M.J. Clinical and serological features of patients with autoantibodies to GW/P bodies. *Clin. Immunol.* **2007**, *125*, 247–256. [CrossRef] [PubMed]
75. Johnson, M.E.; Grassetti, A.V.; Taroni, J.N.; Lyons, S.M.; Schweppe, D.; Gordon, J.K.; Spiera, R.F.; Lafyatis, R.; Anderson, P.J.; Gerber, S.A.; et al. Stress granules and RNA processing bodies are novel autoantibody targets in systemic sclerosis. *Arthritis Res. Ther.* **2016**, *18*, 27. [CrossRef] [PubMed]
76. Nunez Villacis, L.; Wong, M.S.; Ferguson, L.L.; Hein, N.; George, A.J.; Hannan, K.M. New Roles for the Nucleolus in Health and Disease. *Bioessays* **2018**, *40*, e1700233. [CrossRef]
77. Scott, M.S.; Boisvert, F.M.; McDowall, M.D.; Lamond, A.I.; Barton, G.J. Characterization and prediction of protein nucleolar localization sequences. *Nucleic Acids Res.* **2010**, *38*, 7388–7399. [CrossRef]
78. Frege, T.; Uversky, V.N. Intrinsically disordered proteins in the nucleus of human cells. *Biochem. Biophys. Rep.* **2015**, *1*, 33–51. [CrossRef]
79. Windner, S.E.; Manhart, A.; Brown, A.; Mogilner, A.; Baylies, M.K. Nuclear Scaling Is Coordinated among Individual Nuclei in Multinucleated Muscle Fibers. *Dev. Cell* **2019**, *49*, 48–62.e43. [CrossRef]
80. Walne, A.J.; Vulliamy, T.; Marrone, A.; Beswick, R.; Kirwan, M.; Masunari, Y.; Al-Qurashi, F.H.; Aljurf, M.; Dokal, I. Genetic heterogeneity in autosomal recessive dyskeratosis congenita with one subtype due to mutations in the telomerase-associated protein NOP10. *Hum. Mol. Genet.* **2007**, *16*, 1619–1629. [CrossRef]
81. Lee, K.H.; Zhang, P.; Kim, H.J.; Mitrea, D.M.; Sarkar, M.; Freibaum, B.D.; Cika, J.; Coughlin, M.; Messing, J.; Molliex, A.; et al. C9orf72 Dipeptide Repeats Impair the Assembly, Dynamics, and Function of Membrane-Less Organelles. *Cell* **2016**, *167*, 774–788.e717. [CrossRef]
82. Tiku, V.; Jain, C.; Raz, Y.; Nakamura, S.; Heestand, B.; Liu, W.; Spath, M.; Suchiman, H.E.D.; Muller, R.U.; Slagboom, P.E.; et al. Small nucleoli are a cellular hallmark of longevity. *Nat. Commun.* **2017**, *8*, 16083. [CrossRef]
83. Wang, J.; Choi, J.M.; Holehouse, A.S.; Lee, H.O.; Zhang, X.; Jahnel, M.; Maharana, S.; Lemaitre, R.; Pozniakovsky, A.; Drechsel, D.; et al. A Molecular Grammar Governing the Driving Forces for Phase Separation of Prion-like RNA Binding Proteins. *Cell* **2018**, *174*, 688–699.e16. [CrossRef]
84. March, Z.M.; King, O.D.; Shorter, J. Prion-like domains as epigenetic regulators, scaffolds for subcellular organization, and drivers of neurodegenerative disease. *Brain Res.* **2016**, *1647*, 9–18. [CrossRef] [PubMed]
85. Chen, M.; Xu, E.S.; Leisenring, N.H.; Cardona, D.M.; Luo, L.; Ma, Y.; Ventura, A.; Kirsch, D.G. The Fusion Oncogene FUS-CHOP Drives Sarcomagenesis of High-Grade Spindle Cell Sarcomas in Mice. *Sarcoma* **2019**, *2019*, 1340261. [CrossRef] [PubMed]
86. Nolan, M.; Talbot, K.; Ansorge, O. Pathogenesis of FUS-associated ALS and FTD: Insights from rodent models. *Acta Neuropathol. Commun.* **2016**, *4*, 99. [CrossRef] [PubMed]
87. Grunewald, T.G.P.; Cidre-Aranaz, F.; Surdez, D.; Tomazou, E.M.; de Alava, E.; Kovar, H.; Sorensen, P.H.; Delattre, O.; Dirksen, U. Ewing sarcoma. *Nat. Rev. Dis. Primers* **2018**, *4*, 5. [CrossRef]
88. Ahmed, N.S.; Harrell, L.M.; Wieland, D.R.; Lay, M.A.; Thompson, V.F.; Schwartz, J.C. Fusion protein EWS-FLI1 is incorporated into a protein granule in cells. *RNA* **2021**, *27*, 920–932. [CrossRef]
89. Snead, W.T.; Gladfelter, A.S. The Control Centers of Biomolecular Phase Separation: How Membrane Surfaces, PTMs, and Active Processes Regulate Condensation. *Mol. Cell* **2019**, *76*, 295–305. [CrossRef]
90. Aumiller, W.M., Jr.; Keating, C.D. Experimental models for dynamic compartmentalization of biomolecules in liquid organelles: Reversible formation and partitioning in aqueous biphasic systems. *Adv. Colloid Interface Sci.* **2017**, *239*, 75–87. [CrossRef]
91. Wang, Z.; Zhang, H. Phase Separation, Transition, and Autophagic Degradation of Proteins in Development and Pathogenesis. *Trends Cell Biol.* **2019**, *29*, 417–427. [CrossRef]
92. Wegmann, S.; Eftekharzadeh, B.; Tepper, K.; Zoltowska, K.M.; Bennett, R.E.; Dujardin, S.; Laskowski, P.R.; MacKenzie, D.; Kamath, T.; Commins, C.; et al. Tau protein liquid-liquid phase separation can initiate tau aggregation. *EMBO J.* **2018**, *37*, e98049. [CrossRef]
93. Tepper, K.; Biernat, J.; Kumar, S.; Wegmann, S.; Timm, T.; Hubschmann, S.; Redecke, L.; Mandelkow, E.M.; Muller, D.J.; Mandelkow, E. Oligomer formation of tau protein hyperphosphorylated in cells. *J. Biol. Chem.* **2014**, *289*, 34389–34407. [CrossRef] [PubMed]
94. Zhao, K.; Lim, Y.J.; Liu, Z.; Long, H.; Sun, Y.; Hu, J.J.; Zhao, C.; Tao, Y.; Zhang, X.; Li, D.; et al. Parkinson's disease-related phosphorylation at Tyr39 rearranges alpha-synuclein amyloid fibril structure revealed by cryo-EM. *Proc. Natl. Acad. Sci. USA* **2020**, *117*, 20305–20315. [CrossRef] [PubMed]
95. Brahmachari, S.; Ge, P.; Lee, S.H.; Kim, D.; Karuppagounder, S.S.; Kumar, M.; Mao, X.; Shin, J.H.; Lee, Y.; Pletnikova, O.; et al. Activation of tyrosine kinase c-Abl contributes to alpha-synuclein-induced neurodegeneration. *J. Clin. Investig.* **2016**, *126*, 2970–2988. [CrossRef] [PubMed]
96. Monahan, Z.; Ryan, V.H.; Janke, A.M.; Burke, K.A.; Rhoads, S.N.; Zerze, G.H.; O'Meally, R.; Dignon, G.L.; Conicella, A.E.; Zheng, W.; et al. Phosphorylation of the FUS low-complexity domain disrupts phase separation, aggregation, and toxicity. *EMBO J.* **2017**, *36*, 2951–2967. [CrossRef]

97. Qamar, S.; Wang, G.; Randle, S.J.; Ruggeri, F.S.; Varela, J.A.; Lin, J.Q.; Phillips, E.C.; Miyashita, A.; Williams, D.; Strohl, F.; et al. FUS Phase Separation Is Modulated by a Molecular Chaperone and Methylation of Arginine Cation-pi Interactions. *Cell* **2018**, *173*, 720–734.e15. [CrossRef]
98. Piunti, A.; Shilatifard, A. The roles of Polycomb repressive complexes in mammalian development and cancer. *Nat. Rev. Mol. Cell Biol.* **2021**, *22*, 326–345. [CrossRef]
99. Zhang, H.; Azevedo, R.B.; Lints, R.; Doyle, C.; Teng, Y.; Haber, D.; Emmons, S.W. Global regulation of Hox gene expression in C. elegans by a SAM domain protein. *Dev. Cell* **2003**, *4*, 903–915. [CrossRef]
100. Qu, W.; Wang, Z.; Zhang, H. Phase separation of the C. elegans Polycomb protein SOP-2 is modulated by RNA and sumoylation. *Protein Cell* **2020**, *11*, 202–207. [CrossRef]
101. Zhang, H.; Smolen, G.A.; Palmer, R.; Christoforou, A.; van den Heuvel, S.; Haber, D.A. SUMO modification is required for in vivo Hox gene regulation by the Caenorhabditis elegans Polycomb group protein SOP-2. *Nat. Genet.* **2004**, *36*, 507–511. [CrossRef]
102. Rai, A.K.; Chen, J.X.; Selbach, M.; Pelkmans, L. Kinase-controlled phase transition of membraneless organelles in mitosis. *Nature* **2018**, *559*, 211–216. [CrossRef]
103. Wippich, F.; Bodenmiller, B.; Trajkovska, M.G.; Wanka, S.; Aebersold, R.; Pelkmans, L. Dual specificity kinase DYRK3 couples stress granule condensation/dissolution to mTORC1 signaling. *Cell* **2013**, *152*, 791–805. [CrossRef] [PubMed]
104. Mediani, L.; Antoniani, F.; Galli, V.; Vinet, J.; Carra, A.D.; Bigi, I.; Tripathy, V.; Tiago, T.; Cimino, M.; Leo, G.; et al. Hsp90-mediated regulation of DYRK3 couples stress granule disassembly and growth via mTORC1 signaling. *EMBO Rep.* **2021**, *22*, e51740. [CrossRef] [PubMed]
105. Hervas, R.; Oroz, J. Mechanistic Insights into the Role of Molecular Chaperones in Protein Misfolding Diseases: From Molecular Recognition to Amyloid Disassembly. *Int. J. Mol. Sci.* **2020**, *21*, 9186. [CrossRef] [PubMed]
106. Liu, Z.; Zhang, S.; Gu, J.; Tong, Y.; Li, Y.; Gui, X.; Long, H.; Wang, C.; Zhao, C.; Lu, J.; et al. Hsp27 chaperones FUS phase separation under the modulation of stress-induced phosphorylation. *Nat. Struct. Mol. Biol.* **2020**, *27*, 363–372. [CrossRef] [PubMed]
107. Gu, J.; Liu, Z.; Zhang, S.; Li, Y.; Xia, W.; Wang, C.; Xiang, H.; Liu, Z.; Tan, L.; Fang, Y.; et al. Hsp40 proteins phase separate to chaperone the assembly and maintenance of membraneless organelles. *Proc. Natl. Acad. Sci. USA* **2020**, *117*, 31123–31133. [CrossRef] [PubMed]
108. Toth, M.E.; Szegedi, V.; Varga, E.; Juhasz, G.; Horvath, J.; Borbely, E.; Csibrany, B.; Alfoldi, R.; Lenart, N.; Penke, B.; et al. Overexpression of Hsp27 ameliorates symptoms of Alzheimer's disease in APP/PS1 mice. *Cell Stress Chaperones* **2013**, *18*, 759–771. [CrossRef] [PubMed]
109. Ma, X.; Chen, T.; Peng, Z.; Wang, Z.; Liu, J.; Yang, T.; Wu, L.; Liu, G.; Zhou, M.; Tong, M.; et al. Histone chaperone CAF-1 promotes HIV-1 latency by leading the formation of phase-separated suppressive nuclear bodies. *EMBO J.* **2021**, *40*, e106632. [CrossRef]
110. Strulson, C.A.; Molden, R.C.; Keating, C.D.; Bevilacqua, P.C. RNA catalysis through compartmentalization. *Nat. Chem.* **2012**, *4*, 941–946. [CrossRef]
111. Duronio, R.J.; Marzluff, W.F. Coordinating cell cycle-regulated histone gene expression through assembly and function of the Histone Locus Body. *RNA Biol.* **2017**, *14*, 726–738. [CrossRef]
112. Tatomer, D.C.; Terzo, E.; Curry, K.P.; Salzler, H.; Sabath, I.; Zapotoczny, G.; McKay, D.J.; Dominski, Z.; Marzluff, W.F.; Duronio, R.J. Concentrating pre-mRNA processing factors in the histone locus body facilitates efficient histone mRNA biogenesis. *J. Cell Biol.* **2016**, *213*, 557–570. [CrossRef]
113. Kim, K.M.; Lee, A.; Choi, K.Y.; Lee, K.Y.; Kwak, J.J. Intestinal tuberculosis: Clinicopathologic analysis and diagnosis by endoscopic biopsy. *Am. J. Gastroenterol.* **1998**, *93*, 606–609. [CrossRef] [PubMed]
114. Trinkle-Mulcahy, L.; Sleeman, J.E. The Cajal body and the nucleolus: "In a relationship" or "It's complicated"? *RNA Biol.* **2017**, *14*, 739–751. [CrossRef] [PubMed]
115. Berkovits, B.D.; Mayr, C. Alternative 3′ UTRs act as scaffolds to regulate membrane protein localization. *Nature* **2015**, *522*, 363–367. [CrossRef] [PubMed]
116. Lianoglou, S.; Garg, V.; Yang, J.L.; Leslie, C.S.; Mayr, C. Ubiquitously transcribed genes use alternative polyadenylation to achieve tissue-specific expression. *Genes Dev.* **2013**, *27*, 2380–2396. [CrossRef] [PubMed]
117. Lu, H.; Yu, D.; Hansen, A.S.; Ganguly, S.; Liu, R.; Heckert, A.; Darzacq, X.; Zhou, Q. Phase-separation mechanism for C-terminal hyperphosphorylation of RNA polymerase II. *Nature* **2018**, *558*, 318–323. [CrossRef] [PubMed]
118. Egloff, S. CDK9 keeps RNA polymerase II on track. *Cell Mol. Life Sci.* **2021**, *78*, 5543–5567. [CrossRef] [PubMed]
119. Cho, W.K.; Spille, J.H.; Hecht, M.; Lee, C.; Li, C.; Grube, V.; Cisse, I.I. Mediator and RNA polymerase II clusters associate in transcription-dependent condensates. *Science* **2018**, *361*, 412–415. [CrossRef]
120. Burke, K.A.; Janke, A.M.; Rhine, C.L.; Fawzi, N.L. Residue-by-Residue View of In Vitro FUS Granules that Bind the C-Terminal Domain of RNA Polymerase II. *Mol. Cell* **2015**, *60*, 231–241. [CrossRef]
121. Janke, A.M.; Seo, D.H.; Rahmanian, V.; Conicella, A.E.; Mathews, K.L.; Burke, K.A.; Mittal, J.; Fawzi, N.L. Lysines in the RNA Polymerase II C-Terminal Domain Contribute to TAF15 Fibril Recruitment. *Biochemistry* **2018**, *57*, 2549–2563. [CrossRef]
122. Boija, A.; Klein, I.A.; Sabari, B.R.; Dall'Agnese, A.; Coffey, E.L.; Zamudio, A.V.; Li, C.H.; Shrinivas, K.; Manteiga, J.C.; Hannett, N.M.; et al. Transcription Factors Activate Genes through the Phase-Separation Capacity of Their Activation Domains. *Cell* **2018**, *175*, 1842–1855.e16. [CrossRef]

123. Cai, D.; Feliciano, D.; Dong, P.; Flores, E.; Gruebele, M.; Porat-Shliom, N.; Sukenik, S.; Liu, Z.; Lippincott-Schwartz, J. Phase separation of YAP reorganizes genome topology for long-term YAP target gene expression. *Nat. Cell Biol.* **2019**, *21*, 1578–1589. [CrossRef] [PubMed]
124. Zou, S.; Tong, Q.; Liu, B.; Huang, W.; Tian, Y.; Fu, X. Targeting STAT3 in Cancer Immunotherapy. *Mol. Cancer* **2020**, *19*, 145. [CrossRef] [PubMed]
125. Herrmann, A.; Sommer, U.; Pranada, A.L.; Giese, B.; Kuster, A.; Haan, S.; Becker, W.; Heinrich, P.C.; Muller-Newen, G. STAT3 is enriched in nuclear bodies. *J. Cell Sci.* **2004**, *117*, 339–349. [CrossRef] [PubMed]
126. Sehgal, P.B. Interleukin-6 at the Host-Tumor Interface: STAT3 in Biomolecular Condensates in Cancer Cells. *Cells* **2022**, *11*, 1164. [CrossRef] [PubMed]
127. Sehgal, P.B.; Westley, J.; Lerea, K.M.; DiSenso-Browne, S.; Etlinger, J.D. Biomolecular condensates in cell biology and virology: Phase-separated membraneless organelles (MLOs). *Anal. Biochem.* **2020**, *597*, 113691. [CrossRef]
128. Nikolic, J.; Le Bars, R.; Lama, Z.; Scrima, N.; Lagaudriere-Gesbert, C.; Gaudin, Y.; Blondel, D. Negri bodies are viral factories with properties of liquid organelles. *Nat. Commun.* **2017**, *8*, 58. [CrossRef]
129. Geiger, F.; Acker, J.; Papa, G.; Wang, X.; Arter, W.E.; Saar, K.L.; Erkamp, N.A.; Qi, R.; Bravo, J.P.; Strauss, S.; et al. Liquid-liquid phase separation underpins the formation of replication factories in rotaviruses. *EMBO J.* **2021**, *40*, e107711. [CrossRef]
130. Lahaye, X.; Vidy, A.; Pomier, C.; Obiang, L.; Harper, F.; Gaudin, Y.; Blondel, D. Functional characterization of Negri bodies (NBs) in rabies virus-infected cells: Evidence that NBs are sites of viral transcription and replication. *J. Virol.* **2009**, *83*, 7948–7958. [CrossRef]
131. Hoenen, T.; Shabman, R.S.; Groseth, A.; Herwig, A.; Weber, M.; Schudt, G.; Dolnik, O.; Basler, C.F.; Becker, S.; Feldmann, H. Inclusion bodies are a site of ebolavirus replication. *J. Virol.* **2012**, *86*, 11779–11788. [CrossRef]
132. Zhou, Y.; Su, J.M.; Samuel, C.E.; Ma, D. Measles Virus Forms Inclusion Bodies with Properties of Liquid Organelles. *J. Virol.* **2019**, *93*, e00948-19. [CrossRef]
133. Alenquer, M.; Vale-Costa, S.; Etibor, T.A.; Ferreira, F.; Sousa, A.L.; Amorim, M.J. Influenza a virus ribonucleoproteins form liquid organelles at endoplasmic reticulum exit sites. *Nat. Commun.* **2019**, *10*, 1629. [CrossRef] [PubMed]
134. Rincheval, V.; Lelek, M.; Gault, E.; Bouillier, C.; Sitterlin, D.; Blouquit-Laye, S.; Galloux, M.; Zimmer, C.; Eleouet, J.F.; Rameix-Welti, M.A. Functional organization of cytoplasmic inclusion bodies in cells infected by respiratory syncytial virus. *Nat. Commun.* **2017**, *8*, 563. [CrossRef] [PubMed]
135. Fouquet, B.; Nikolic, J.; Larrous, F.; Bourhy, H.; Wirblich, C.; Lagaudriere-Gesbert, C.; Blondel, D. Focal adhesion kinase is involved in rabies virus infection through its interaction with viral phosphoprotein P. *J. Virol.* **2015**, *89*, 1640–1651. [CrossRef] [PubMed]
136. Raab-Traub, N. Novel mechanisms of EBV-induced oncogenesis. *Curr. Opin. Virol.* **2012**, *2*, 453–458. [CrossRef]
137. Kaiser, C.; Laux, G.; Eick, D.; Jochner, N.; Bornkamm, G.W.; Kempkes, B. The proto-oncogene c-myc is a direct target gene of Epstein-Barr virus nuclear antigen 2. *J. Virol.* **1999**, *73*, 4481–4484. [CrossRef]
138. Peng, Q.; Wang, L.; Qin, Z.; Wang, J.; Zheng, X.; Wei, L.; Zhang, X.; Zhang, X.; Liu, C.; Li, Z.; et al. Phase Separation of Epstein-Barr Virus EBNA2 and Its Coactivator EBNALP Controls Gene Expression. *J. Virol.* **2020**, *94*, e01771-19. [CrossRef]
139. Pham, C.L.; Shanmugam, N.; Strange, M.; O'Carroll, A.; Brown, J.W.; Sierecki, E.; Gambin, Y.; Steain, M.; Sunde, M. Viral M45 and necroptosis-associated proteins form heteromeric amyloid assemblies. *EMBO Rep.* **2019**, *20*, e46518. [CrossRef]
140. Sehgal, P.B. Metastable biomolecular condensates of interferon-inducible antiviral Mx-family GTPases: A paradigm shift in the last three years. *J. Biosci.* **2021**, *46*, 1–14. [CrossRef]
141. Davis, D.; Yuan, H.; Liang, F.X.; Yang, Y.M.; Westley, J.; Petzold, C.; Dancel-Manning, K.; Deng, Y.; Sall, J.; Sehgal, P.B. Human Antiviral Protein MxA Forms Novel Metastable Membraneless Cytoplasmic Condensates Exhibiting Rapid Reversible Tonicity-Driven Phase Transitions. *J. Virol.* **2019**, *93*, e01014-19. [CrossRef]
142. Sehgal, P.B.; Yuan, H.; Scott, M.F.; Deng, Y.; Liang, F.X.; Mackiewicz, A. Murine GFP-Mx1 forms nuclear condensates and associates with cytoplasmic intermediate filaments: Novel antiviral activity against VSV. *J. Biol. Chem.* **2020**, *295*, 18023–18035. [CrossRef]
143. Gu, Z.C.; Wu, E.; Sailer, C.; Jando, J.; Styles, E.; Eisenkolb, I.; Kuschel, M.; Bitschar, K.; Wang, X.; Huang, L.; et al. Ubiquitin orchestrates proteasome dynamics between proliferation and quiescence in yeast. *Mol. Biol. Cell* **2017**, *28*, 2479–2491. [CrossRef] [PubMed]
144. Enenkel, C. The paradox of proteasome granules. *Curr. Genet.* **2018**, *64*, 137–140. [CrossRef] [PubMed]
145. Yedidi, R.S.; Fatehi, A.K.; Enenkel, C. Proteasome dynamics between proliferation and quiescence stages of Saccharomyces cerevisiae. *Crit. Rev. Biochem. Mol. Biol.* **2016**, *51*, 497–512. [CrossRef] [PubMed]
146. Lavut, A.; Raveh, D. Sequestration of highly expressed mRNAs in cytoplasmic granules, P-bodies, and stress granules enhances cell viability. *PLoS Genet.* **2012**, *8*, e1002527. [CrossRef] [PubMed]
147. Nostramo, R.; Xing, S.; Zhang, B.; Herman, P.K. Insights into the Role of P-Bodies and Stress Granules in Protein Quality Control. *Genetics* **2019**, *213*, 251–265. [CrossRef] [PubMed]
148. Li, H.; Leo, C.; Zhu, J.; Wu, X.; O'Neil, J.; Park, E.J.; Chen, J.D. Sequestration and inhibition of Daxx-mediated transcriptional repression by PML. *Mol. Cell. Biol.* **2000**, *20*, 1784–1796. [CrossRef] [PubMed]
149. Shastrula, P.K.; Sierra, I.; Deng, Z.; Keeney, F.; Hayden, J.E.; Lieberman, P.M.; Janicki, S.M. PML is recruited to heterochromatin during S phase and represses DAXX-mediated histone H3.3 chromatin assembly. *J. Cell Sci.* **2019**, *132*, jcs220970. [CrossRef]

150. Wang, M.; Bokros, M.; Theodoridis, P.R.; Lee, S. Nucleolar Sequestration: Remodeling Nucleoli Into Amyloid Bodies. *Front. Genet.* **2019**, *10*, 1179. [CrossRef]
151. Weber, J.D.; Taylor, L.J.; Roussel, M.F.; Sherr, C.J.; Bar-Sagi, D. Nucleolar Arf sequesters Mdm2 and activates p53. *Nat. Cell Biol.* **1999**, *1*, 20–26. [CrossRef]
152. Poyurovsky, M.V.; Jacq, X.; Ma, C.; Karni-Schmidt, O.; Parker, P.J.; Chalfie, M.; Manley, J.L.; Prives, C. Nucleotide binding by the Mdm2 RING domain facilitates Arf-independent Mdm2 nucleolar localization. *Mol. Cell* **2003**, *12*, 875–887. [CrossRef]
153. Mekhail, K.; Gunaratnam, L.; Bonicalzi, M.E.; Lee, S. HIF activation by pH-dependent nucleolar sequestration of VHL. *Nat. Cell Biol.* **2004**, *6*, 642–647. [CrossRef] [PubMed]
154. Arabi, A.; Rustum, C.; Hallberg, E.; Wright, A.P. Accumulation of c-Myc and proteasomes at the nucleoli of cells containing elevated c-Myc protein levels. *J. Cell Sci.* **2003**, *116*, 1707–1717. [CrossRef] [PubMed]
155. Wong, J.M.; Kusdra, L.; Collins, K. Subnuclear shuttling of human telomerase induced by transformation and DNA damage. *Nat. Cell Biol.* **2002**, *4*, 731–736. [CrossRef] [PubMed]
156. Shou, W.; Seol, J.H.; Shevchenko, A.; Baskerville, C.; Moazed, D.; Chen, Z.W.; Jang, J.; Shevchenko, A.; Charbonneau, H.; Deshaies, R.J. Exit from mitosis is triggered by Tem1-dependent release of the protein phosphatase Cdc14 from nucleolar RENT complex. *Cell* **1999**, *97*, 233–244. [CrossRef]
157. Mehta, S.; Zhang, J. Liquid-liquid phase separation drives cellular function and dysfunction in cancer. *Nat. Rev. Cancer* **2022**, *22*, 239–252. [CrossRef]
158. Yamaguchi, A.; Takanashi, K. FUS interacts with nuclear matrix-associated protein SAFB1 as well as Matrin3 to regulate splicing and ligand-mediated transcription. *Sci. Rep.* **2016**, *6*, 35195. [CrossRef]
159. Murthy, A.C.; Dignon, G.L.; Kan, Y.; Zerze, G.H.; Parekh, S.H.; Mittal, J.; Fawzi, N.L. Molecular interactions underlying liquid-liquid phase separation of the FUS low-complexity domain. *Nat. Struct. Mol. Biol.* **2019**, *26*, 637–648. [CrossRef]
160. Rhoads, S.N.; Monahan, Z.T.; Yee, D.S.; Shewmaker, F.P. The Role of Post-Translational Modifications on Prion-like Aggregation and Liquid-Phase Separation of FUS. *Int. J. Mol. Sci.* **2018**, *19*, 886. [CrossRef]
161. Chook, Y.M.; Suel, K.E. Nuclear import by karyopherin-betas: Recognition and inhibition. *Biochim. Biophys. Acta* **2011**, *1813*, 1593–1606. [CrossRef]
162. Ishigaki, S.; Sobue, G. Importance of Functional Loss of FUS in FTLD/ALS. *Front. Mol. Biosci.* **2018**, *5*, 44. [CrossRef]
163. Nomura, T.; Watanabe, S.; Kaneko, K.; Yamanaka, K.; Nukina, N.; Furukawa, Y. Intranuclear aggregation of mutant FUS/TLS as a molecular pathomechanism of amyotrophic lateral sclerosis. *J. Biol. Chem.* **2014**, *289*, 1192–1202. [CrossRef] [PubMed]
164. Dormann, D.; Madl, T.; Valori, C.F.; Bentmann, E.; Tahirovic, S.; Abou-Ajram, C.; Kremmer, E.; Ansorge, O.; Mackenzie, I.R.; Neumann, M.; et al. Arginine methylation next to the PY-NLS modulates Transportin binding and nuclear import of FUS. *EMBO J.* **2012**, *31*, 4258–4275. [CrossRef] [PubMed]
165. Hofweber, M.; Hutten, S.; Bourgeois, B.; Spreitzer, E.; Niedner-Boblenz, A.; Schifferer, M.; Ruepp, M.D.; Simons, M.; Niessing, D.; Madl, T.; et al. Phase Separation of FUS Is Suppressed by Its Nuclear Import Receptor and Arginine Methylation. *Cell* **2018**, *173*, 706–719.e713. [CrossRef] [PubMed]
166. Weingarten, M.D.; Lockwood, A.H.; Hwo, S.Y.; Kirschner, M.W. A protein factor essential for microtubule assembly. *Proc. Natl. Acad. Sci. USA* **1975**, *72*, 1858–1862. [CrossRef] [PubMed]
167. Rai, S.K.; Savastano, A.; Singh, P.; Mukhopadhyay, S.; Zweckstetter, M. Liquid-liquid phase separation of tau: From molecular biophysics to physiology and disease. *Protein Sci.* **2021**, *30*, 1294–1314. [CrossRef] [PubMed]
168. Arendt, T.; Stieler, J.T.; Holzer, M. Tau and tauopathies. *Brain Res. Bull.* **2016**, *126*, 238–292. [CrossRef] [PubMed]
169. Martin, L.; Latypova, X.; Terro, F. Post-translational modifications of tau protein: Implications for Alzheimer's disease. *Neurochem. Int.* **2011**, *58*, 458–471. [CrossRef] [PubMed]
170. Savastano, A.; Flores, D.; Kadavath, H.; Biernat, J.; Mandelkow, E.; Zweckstetter, M. Disease-Associated Tau Phosphorylation Hinders Tubulin Assembly within Tau Condensates. *Angew. Chem. Int. Ed. Engl.* **2021**, *60*, 726–730. [CrossRef]
171. Ukmar-Godec, T.; Hutten, S.; Grieshop, M.P.; Rezaei-Ghaleh, N.; Cima-Omori, M.S.; Biernat, J.; Mandelkow, E.; Soding, J.; Dormann, D.; Zweckstetter, M. Lysine/RNA-interactions drive and regulate biomolecular condensation. *Nat. Commun.* **2019**, *10*, 2909. [CrossRef]
172. Ferreon, J.C.; Jain, A.; Choi, K.J.; Tsoi, P.S.; MacKenzie, K.R.; Jung, S.Y.; Ferreon, A.C. Acetylation Disfavors Tau Phase Separation. *Int. J. Mol. Sci.* **2018**, *19*, 1360. [CrossRef]
173. Ou, S.H.; Wu, F.; Harrich, D.; Garcia-Martinez, L.F.; Gaynor, R.B. Cloning and characterization of a novel cellular protein, TDP-43, that binds to human immunodeficiency virus type 1 TAR DNA sequence motifs. *J. Virol.* **1995**, *69*, 3584–3596. [CrossRef] [PubMed]
174. Buratti, E.; Dörk, T.; Zuccato, E.; Pagani, F.; Romano, M.; Baralle, F.E. Nuclear factor TDP-43 and SR proteins promote in vitro and in vivo CFTR exon 9 skipping. *EMBO J.* **2001**, *20*, 1774–1784. [CrossRef] [PubMed]
175. Prasad, A.; Bharathi, V.; Sivalingam, V.; Girdhar, A.; Patel, B.K. Molecular Mechanisms of TDP-43 Misfolding and Pathology in Amyotrophic Lateral Sclerosis. *Front. Mol. Neurosci.* **2019**, *12*, 25. [CrossRef]
176. Dhakal, S.; Wyant, C.E.; George, H.E.; Morgan, S.E.; Rangachari, V. Prion-like C-Terminal Domain of TDP-43 and alpha-Synuclein Interact Synergistically to Generate Neurotoxic Hybrid Fibrils. *J. Mol. Biol.* **2021**, *433*, 166953. [CrossRef] [PubMed]
177. Nonaka, T.; Hasegawa, M. Prion-like properties of assembled TDP-43. *Curr. Opin. Neurobiol.* **2020**, *61*, 23–28. [CrossRef] [PubMed]
178. Suk, T.R.; Rousseaux, M.W.C. The role of TDP-43 mislocalization in amyotrophic lateral sclerosis. *Mol. Neurodegener.* **2020**, *15*, 45. [CrossRef] [PubMed]

179. Neumann, M.; Kwong, L.K.; Lee, E.B.; Kremmer, E.; Flatley, A.; Xu, Y.; Forman, M.S.; Troost, D.; Kretzschmar, H.A.; Trojanowski, J.Q.; et al. Phosphorylation of S409/410 of TDP-43 is a consistent feature in all sporadic and familial forms of TDP-43 proteinopathies. *Acta Neuropathol.* **2009**, *117*, 137–149. [CrossRef]
180. Cohen, T.J.; Hwang, A.W.; Restrepo, C.R.; Yuan, C.X.; Trojanowski, J.Q.; Lee, V.M. An acetylation switch controls TDP-43 function and aggregation propensity. *Nat. Commun.* **2015**, *6*, 5845. [CrossRef]
181. Hebron, M.L.; Lonskaya, I.; Sharpe, K.; Weerasinghe, P.P.; Algarzae, N.K.; Shekoyan, A.R.; Moussa, C.E. Parkin ubiquitinates Tar-DNA binding protein-43 (TDP-43) and promotes its cytosolic accumulation via interaction with histone deacetylase 6 (HDAC6). *J. Biol. Chem.* **2013**, *288*, 4103–4115. [CrossRef]
182. Babinchak, W.M.; Haider, R.; Dumm, B.K.; Sarkar, P.; Surewicz, K.; Choi, J.K.; Surewicz, W.K. The role of liquid-liquid phase separation in aggregation of the TDP-43 low-complexity domain. *J. Biol. Chem.* **2019**, *294*, 6306–6317. [CrossRef]
183. Anselmi, M.; Calligari, P.; Hub, J.S.; Tartaglia, M.; Bocchinfuso, G.; Stella, L. Structural Determinants of Phosphopeptide Binding to the N-Terminal Src Homology 2 Domain of the SHP2 Phosphatase. *J. Chem. Inf. Model.* **2020**, *60*, 3157–3171. [CrossRef] [PubMed]
184. Tajan, M.; de Rocca Serra, A.; Valet, P.; Edouard, T.; Yart, A. SHP2 sails from physiology to pathology. *Eur. J. Med. Genet.* **2015**, *58*, 509–525. [CrossRef] [PubMed]
185. Tartaglia, M.; Gelb, B.D. Noonan syndrome and related disorders: Genetics and pathogenesis. *Annu. Rev. Genom. Hum. Genet.* **2005**, *6*, 45–68. [CrossRef] [PubMed]
186. Digilio, M.C.; Conti, E.; Sarkozy, A.; Mingarelli, R.; Dottorini, T.; Marino, B.; Pizzuti, A.; Dallapiccola, B. Grouping of multiple-lentigines/LEOPARD and Noonan syndromes on the *PTPN11* gene. *Am. J. Hum. Genet.* **2002**, *71*, 389–394. [CrossRef] [PubMed]
187. Miyamoto, D.; Miyamoto, M.; Takahashi, A.; Yomogita, Y.; Higashi, H.; Kondo, S.; Hatakeyama, M. Isolation of a distinct class of gain-of-function SHP-2 mutants with oncogenic RAS-like transforming activity from solid tumors. *Oncogene* **2008**, *27*, 3508–3515. [CrossRef] [PubMed]
188. Yu, Z.H.; Xu, J.; Walls, C.D.; Chen, L.; Zhang, S.; Zhang, R.; Wu, L.; Wang, L.; Liu, S.; Zhang, Z.Y. Structural and mechanistic insights into LEOPARD syndrome-associated SHP2 mutations. *J. Biol. Chem.* **2013**, *288*, 10472–10482. [CrossRef] [PubMed]
189. Yu, Z.H.; Zhang, R.Y.; Walls, C.D.; Chen, L.; Zhang, S.; Wu, L.; Liu, S.; Zhang, Z.Y. Molecular basis of gain-of-function LEOPARD syndrome-associated SHP2 mutations. *Biochemistry* **2014**, *53*, 4136–4151. [CrossRef]
190. Zhu, G.; Xie, J.; Kong, W.; Xie, J.; Li, Y.; Du, L.; Zheng, Q.; Sun, L.; Guan, M.; Li, H.; et al. Phase Separation of Disease-Associated SHP2 Mutants Underlies MAPK Hyperactivation. *Cell* **2020**, *183*, 490–502.e18. [CrossRef]
191. Boopathy, G.T.K.; Hong, W. Role of Hippo Pathway-YAP/TAZ Signaling in Angiogenesis. *Front. Cell Dev. Biol.* **2019**, *7*, 49. [CrossRef]
192. Masliantsev, K.; Karayan-Tapon, L.; Guichet, P.O. Hippo Signaling Pathway in Gliomas. *Cells* **2021**, *10*, 184. [CrossRef]
193. Franklin, J.M.; Guan, K.L. YAP/TAZ phase separation for transcription. *Nat. Cell Biol.* **2020**, *22*, 357–358. [CrossRef] [PubMed]
194. Kim, E.; Kang, J.G.; Kang, M.J.; Park, J.H.; Kim, Y.J.; Kweon, T.H.; Lee, H.W.; Jho, E.H.; Lee, Y.H.; Kim, S.I.; et al. O-GlcNAcylation on LATS2 disrupts the Hippo pathway by inhibiting its activity. *Proc. Natl. Acad. Sci. USA* **2020**, *117*, 14259–14269. [CrossRef] [PubMed]
195. Harvey, K.F.; Zhang, X.; Thomas, D.M. The Hippo pathway and human cancer. *Nat. Rev. Cancer* **2013**, *13*, 246–257. [CrossRef] [PubMed]

International Journal of *Molecular Sciences*

MDPI

Article

Free Cholesterol Accelerates Aβ Self-Assembly on Membranes at Physiological Concentration

Mohtadin Hashemi [1,†], Siddhartha Banerjee [1,2,†] and Yuri L. Lyubchenko [1,*]

1 Department of Pharmaceutical Sciences, University of Nebraska Medical Center, 986025 Nebraska Medical Center, Omaha, NE 68198-6025, USA; mohtadin.hashemi@unmc.edu (M.H.); sbanerjee6@ua.edu (S.B.)
2 Department of Chemistry and Biochemistry, The University of Alabama, Shelby Hall, Tuscaloosa, AL 35487, USA
* Correspondence: ylyubchenko@unmc.edu
† These authors contributed equally.

Abstract: The effects of membranes on the early-stage aggregation of amyloid β (Aβ) have come to light as potential mechanisms by which neurotoxic species are formed in Alzheimer's disease. We have shown that direct Aβ-membrane interactions dramatically enhance the Aβ aggregation, allowing for oligomer assembly at physiologically low concentrations of the monomer. Membrane composition is also a crucial factor in this process. Our results showed that apart from phospholipids composition, cholesterol in membranes significantly enhances the aggregation kinetics. It has been reported that free cholesterol is present in plaques. Here we report that free cholesterol, along with its presence inside the membrane, further accelerate the aggregation process by producing aggregates more rapidly and of significantly larger sizes. These aggregates, which are formed on the lipid bilayer, are able to dissociate from the surface and accumulate in the bulk solution; the presence of free cholesterol accelerates this dissociation as well. All-atom molecular dynamics simulations show that cholesterol binds Aβ monomers and significantly changes the conformational sampling of Aβ monomer; more than doubling the fraction of low-energy conformations compared to those in the absence of cholesterol, which can contribute to the aggregation process. The results indicate that Aβ-lipid interaction is an important factor in the disease prone amyloid assembly process.

Keywords: Alzheimer's disease; amyloid aggregation; lipid bilayer; cholesterol; time-lapse AFM imaging; molecular dynamics

Citation: Hashemi, M.; Banerjee, S.; Lyubchenko, Y.L. Free Cholesterol Accelerates Aβ Self-Assembly on Membranes at Physiological Concentration. *Int. J. Mol. Sci.* **2022**, *23*, 2803. https://doi.org/10.3390/ijms23052803

Academic Editor: Andrea Cavalli

Received: 11 February 2022
Accepted: 1 March 2022
Published: 3 March 2022

1. Introduction

The self-assembly of amyloid β (Aβ) is a process that results in the production of neurotoxic oligomer and fibrillar aggregates in Alzheimer's disease [1,2]. Understanding the mechanism by which these aggregates are formed has been the major focus of research in Alzheimer's disease and other fatal neurodegenerative diseases [3,4]. However, in the majority of in vitro studies, the Aβ concentrations used are several orders of magnitude higher than the physiologically relevant concentrations [5,6]; no aggregation is observed at the physiological low nanomolar concentration of Aβ. This suggests that the aggregation of Aβ in vivo utilizes pathways different from those probed by in vitro experiments.

Recently, an alternative aggregation mechanism has been discovered, allowing for the aggregation to occur at the physiologically relevant concentrations of Aβ [7,8]. This is the on-surface aggregation pathway, in which interactions with a surface act as a catalyst for the aggregation process. The model for the on-surface aggregation process suggests that the self-assembly of Aβ oligomers is initiated by the interaction of amyloid proteins with the cellular membrane. The membrane catalyzes amyloid aggregation by stabilizing an aggregation-prone conformation.

Cell membranes consist of a large variety of lipids, suggesting that numerous factors may contribute to the on-membrane aggregation of amyloids. Indeed, recent publications revealed the role of such lipids as cholesterol (Chol), sphingomyelins, and gangliosides on the formation of Aβ fibrils on membrane surfaces [9–11]. A very recent publication [12] demonstrated that Chol in the lipid bilayer significantly enhances the aggregation of Aβ(1-42) at nanomolar monomer concentration. Importantly, computer modeling showed that Aβ(1-42) has an elevated affinity to Chol-containing membranes, adopting a set of aggregation-prone conformations. These studies led to an aggregation model with membranes playing a critical role in triggering the aggregation process and hence, the disease state. Within this model, the membrane composition is a factor controlling the aggregation process, so a change in membrane composition can shift the ratio between monomeric and aggregated states of Aβ. This hypothesis is further strengthened by the data regarding the contribution of Chol, sphingomyelins, and gangliosides to the neurotoxicity of Aβ aggregates [13–15], which also highlights these lipids as prime candidates for possible disease defining parameters.

While phospholipids are the major constituent of the cellular lipid bilayer, Chol is the second most abundant lipid and provides stability to the cellular membrane. Importantly, recent findings show higher level of plasma Chol in Alzheimer's disease patients compared to healthy controls [16]. Furthermore, Chol has been identified to be present in plaques in a 1:1 ratio with Aβ [17,18]. Other studies revealed that feeding a Chol-enriched diet to rats resulted in the enhancement of APP, Aβ, and p-tau in the cortex region, which was associated with cognitive problems [19]. In a different study, it was observed that a Chol-rich diet increased the brain Chol level and resulted in motor function impairment [20]. Furthermore, neuronal Chol content has been linked with age, with higher Chol concentration being found in mature neurons compared to younger [21]. Together these results clearly connect Chol with disease development; however, the molecular mechanism of how Chol affects disease development remains unknown.

Aggregates extracted from patient brains have revealed the existence of oligomer-lipid ensembles, pointing to possible direct interaction of free lipids with Aβ [22,23]. Additionally, recent studies [24] have reported assemblies of Aβ(1-42) monomers with Chol. These reports lead us to posit that free lipids affect the aggregation of amyloid proteins. Here we tested the hypothesis on the role of free Chol in the aggregation of Aβ, at the physiologically relevant nanomolar concentration. Time-lapse Atomic Force Microscopy (AFM) was applied to monitor the *in-situ* formation of Aβ(1-42) aggregates on supported lipid bilayers in the presence of free Chol. These studies revealed that Aβ(1-42) aggregates are formed more rapidly on the lipid bilayer in presence of free Chol. Furthermore, the aggregation kinetics of Aβ in the presence of free Chol is greatest on bilayers containing Chol. Moreover, in the presence of free Chol, aggregates accumulate more rapidly in the bulk above the membrane bilayer. Altogether, these studies revealed a critical role of free Chol on the disease-prone aggregation of Aβ(1-42), suggesting that Chol can be a trigger of the aggregation process.

2. Results

2.1. Rapid Appearance of Aggregates in Presence of Free Cholesterol

The role of free Chol in the aggregation of Aβ(1-42) was investigated on a supported lipid bilayer surface. Briefly, a mixed lipid bilayer (PC-PS), containing 1-palmitoyl-2-oleoyl-glycero-3-phosphocholine (PC) and 1-palmitoyl-2-oleoyl-sn-glycero-3-phospho-L-serine (PS), was prepared as described earlier [12]. Then, 10 nM Aβ(1-42) monomer solution with and without 100 nM Chol was deposited on the bilayer and time-lapse AFM imaging was performed to visualize the on-surface aggregation process.

Figure 1a shows the lipid bilayer surface before the addition of Aβ solution. The surface is smooth and homogeneous, with no aggregate-like features or trapped vesicles, which is critical for monitoring the on-membrane aggregation events [25–27]. Aggregates were detected 1 h after the addition of the Aβ solution and continued growing in numbers in the subsequent time-points of 3 h and 5 h (Figure 1b,c). To quantify the aggregation process, the volume of the aggregates, at each timepoint, was measured (Figure 1d). The plot shows that the mean aggregate volume increases as a function of incubation time on the PC-PS lipid bilayer.

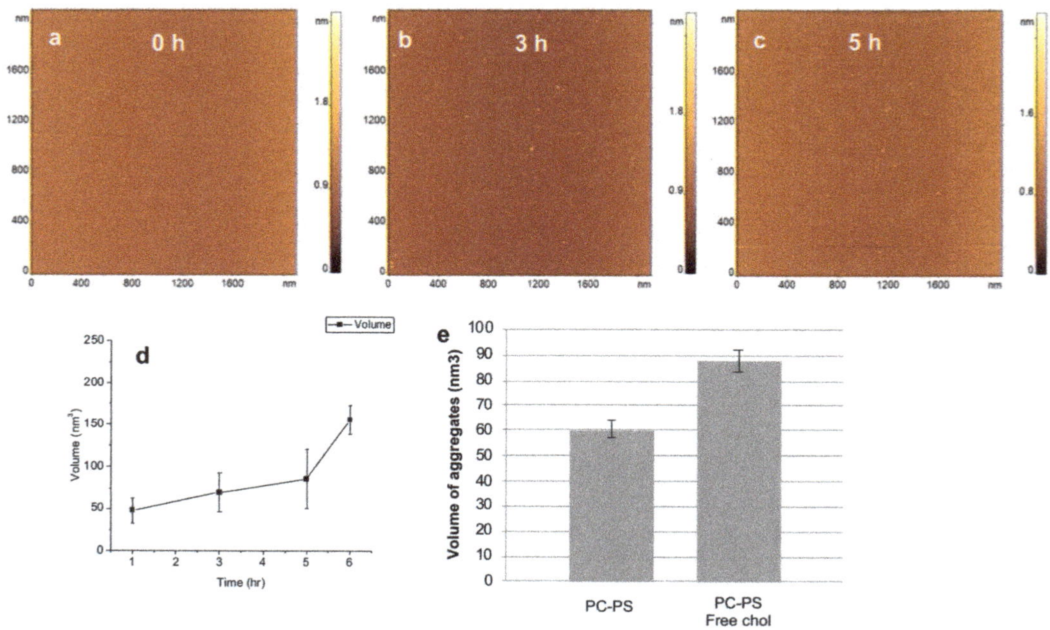

Figure 1. Aggregation of 10 nM Aβ(1-42), in the presence of 100 nM Chol, on PC-PS lipid bilayer. (**a**) AFM image of the bilayer surface before addition of Aβ(1-42)-Chol solution. (**b**,**c**) AFM images of the same area of the lipid bilayer 3 h and 5 h after addition of Aβ(1-42)-Chol solution. (**d**) Evolution of Aβ(1-42) aggregate volume with time. (**e**) Comparison of Aβ(1-42) aggregate volumes after 5 h incubation in the presence of PC-PS bilayer and PC-PS bilayer with Chol in solution. The volume of aggregates is significantly larger ($p < 0.0001$, *t*-test) in presence of free Chol.

As a control, we performed aggregation experiments by incubating 10 nM Aβ(1-42) on the PC-PS bilayer without Chol in solution. Comparison of the volume of aggregates formed after 5 h incubation, with and without Chol present in the solution, is shown in Figure 1e. It is evident that aggregates are significantly larger when free Chol is present in the solution during aggregation, compared to only the Aβ(1-42) in solution.

2.2. Acceleration of Aβ(1-42) Aggregation by Cholesterol inside Membrane

To understand if the bilayer composition is important during aggregation with free Chol in solution, we assembled a mixed bilayer with Chol, PC-PS-Chol bilayer, and followed the aggregation of Aβ in the presence of free Chol on this bilayer. Representative time-lapse AFM imaging data are shown in Figure 2 and Figure S1. Initially, the bilayer surface is smooth, Figure S1a. Aggregates appear within 30 min of Aβ-Chol solution addition; a few are highlighted with white arrows in Figure S1b. After 2 h of incubation, the lipid bilayer surface shows a significant number of large aggregates (Figure 2a). Quantitative volume measurements for the two time-points show the change in aggregate size (Figure S1c,d). The aggregate size increased approximately 4 times, from ~65 nm^3 to ~272 nm^3, between 30 min and 2 h.

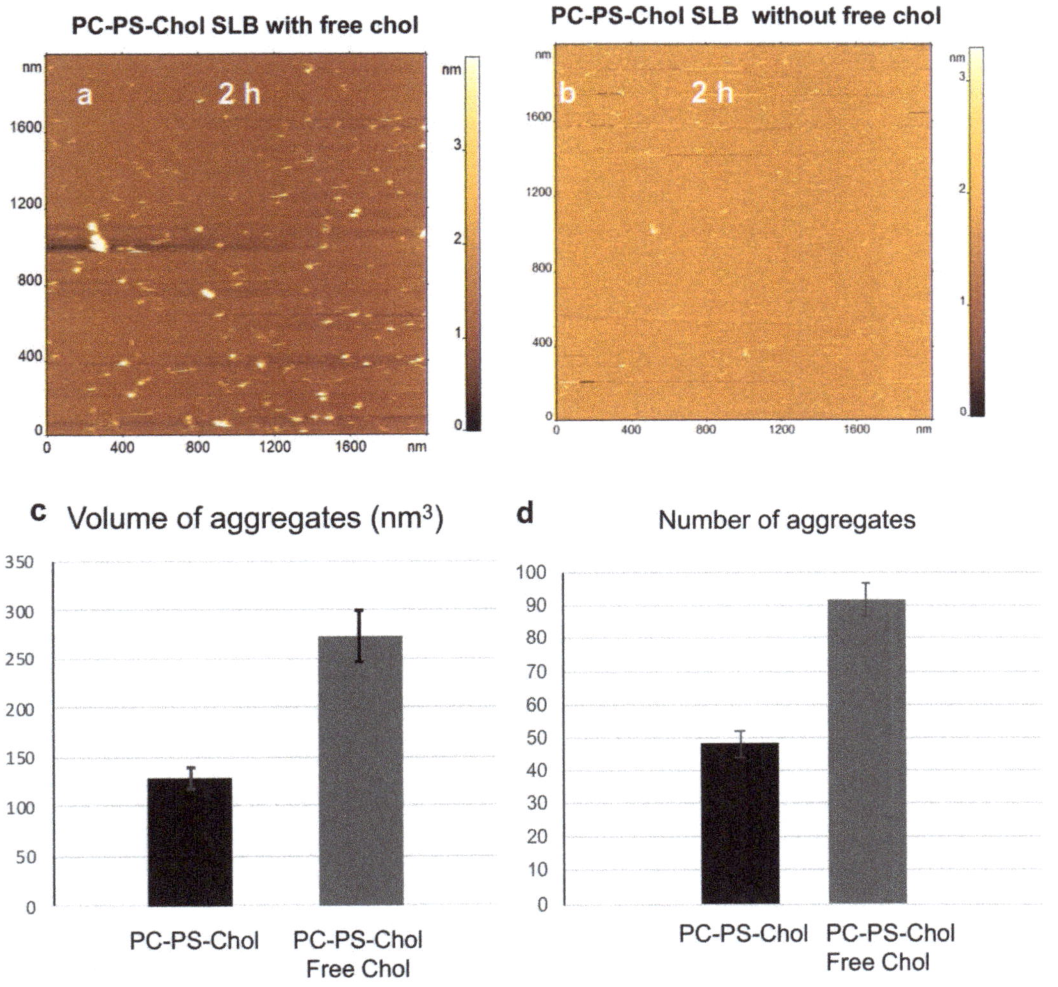

Figure 2. Aggregation of 10 nM Aβ(1-42) on PC-PS-Chol bilayer. (**a**) AFM image of the PC-PS-Chol lipid bilayer after 2 h incubation with 10 nM Aβ42 and 100 nM Chol in the solution. (**b**) AFM image of similar aggregation experiment as (**a**), except the absence of 100 nM Chol in the solution. (**c**) Comparison of the on-bilayer aggregate volumes in the two aggregation experiments. Data is the mean value of aggregate volumes, obtained through Gaussian fits. Presence of free Chol significantly increases ($p = 0.001$, t-test) oligomer volume. (**d**) Comparison of the number of aggregates formed on the lipid bilayers in the presence and absence of Chol in solution; presence of free Chol leads to significantly more oligomers ($p = 0.003$, t-test). For (**c**) and (**d**) the error bars represent the standard error of the mean.

We then performed aggregation experiments with only Aβ(1-42) in solution in the presence of a PC-PS-Chol bilayer, Figure 2b. Visually it is evident that greater number of aggregates are present when free Chol is in the solution. Quantitative analysis of the two experiments shows that the volume as well as the total number of aggregates are significantly greater when Aβ(1-42) aggregates in presence of free Chol in solution, Figure 2c,d.

To validate the observations and to test whether Chol itself can form aggregate-like features on the bilayer surface, we performed time-lapse experiments on the PC-PS-Chol bilayer in presence of Chol only. Figure S2a shows a large area of the bilayer surface prior to addition of Chol solution. Figure S2b, shows a zoom of the same area after 2 h incubation with Chol solution. Figure S2b,c shows another area on the bilayer surface after 2 h incubation with Chol solution; there are no aggregates or aggregate-like features on the surface of the bilayer. These observations clearly demonstrate that the aggregates, which were observed on the bilayer surface, were indeed self-assembled Aβ(1-42) oligomers and that Chol inside the membrane works in synergy with free Chol, catalyzing the self-assembly of amyloid oligomers.

2.3. Dynamics of Aβ(1-42) Aggregation in Presence of Free Cholesterol

After 2 h aggregation of Aβ-Chol solution on the PC-PS-Chol bilayer, the surface is practically covered with aggregates, Figure 2a. However, at 3 h significantly fewer aggregates are observed, Figure S3a. While the number of aggregates become fewer with increased aggregation time, their volumes increase, Figure S3b–d. Volume measurements of the aggregates after 1 h incubation show, Figure S3b, that the aggregate volumes are centered around 74 nm^3. As the aggregates become larger at 3 h, the distribution changes, and a peak around 293 nm^3 becomes prominent. Larger aggregates also appear, Figure S3c. At the 4 h incubation point the aggregates are significantly larger, with a peak around 397 nm^3, Figure S3d.

Previous studies [12] have shown that aggregates are capable of dissociating from the bilayer surface. Aggregates in the presence of free Chol show similar behavior, and the findings suggest that the presence of Chol in the solution accelerates the dissociation of aggregates. This phenomenon was tested by characterizing the accumulation of aggregates in the bulk solution above the bilayer using AFM. In these experiments, 10 nM Aβ(1-42) with 100 nM Chol solution was incubated on top of PC-PS-Chol bilayer surface. At certain time intervals an aliquot was taken from the bulk solution above the bilayer, deposited onto APS-functionalized mica, and characterized using AFM imaging. The data is assembled in Figure 3. Aggregates, accumulated in the bulk solution above the bilayer, were detected after 3 h, Figure 3a, and become more prominent after 6 h, Figure 3b. At the same time, control experiments conducted with Aβ(1-42) and Chol without the bilayer present show a negligible number of aggregates, Figure 3c. Volumes of the aggregates were also analyzed and show that the average size of the aggregates increases over time, Figure 3d. These results show that the aggregates, which dissociate from the surface, do accumulate in the bulk solution, increasing the level of soluble aggregates. The data also show that, compared with the control experiments, in which 10 nM Aβ(1-42) and 100 nM Chol were incubated without the bilayer, the presence of the bilayer leads to statistically significant more accumulation of aggregates in the bulk solution.

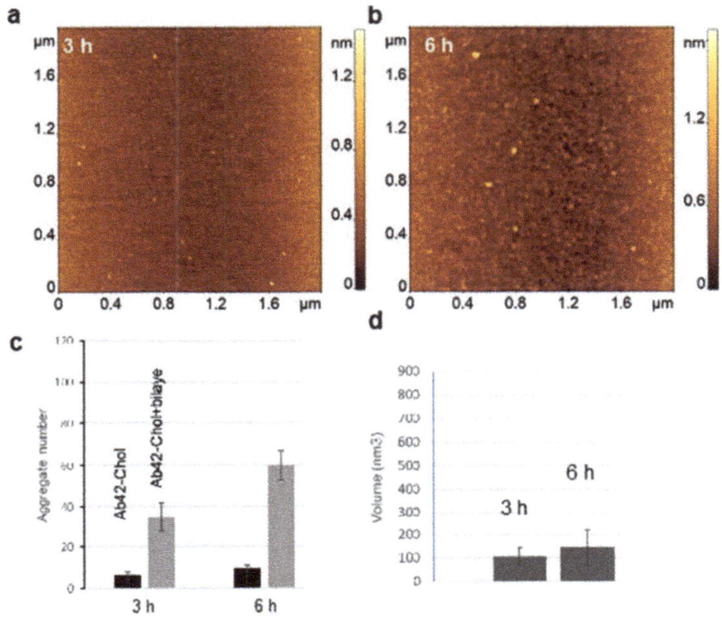

Figure 3. Aβ(1-42) aggregate desorption from PC-PS-Chol lipid bilayer in presence of free Chol. (**a**,**b**) AFM images of aggregates from aliquots taken from the solution above the PC-PS-Chol bilayer while 10 nM Aβ(1-42) and 100 nM Chol was incubating. Samples were taken 3 h and 6 h after addition of Aβ(1-42)-Chol solution. (**c**) Comparison of aggregates after 3 h and 6 h incubation of Aβ(1-42)-Chol in the absence and presence of PC-PS-Chol bilayer. Presence of free Chol significantly increases number of desorbed oligomers, furthermore the increase from 3 h to 6 h time point is also significant ($p = 0.009$, *t*-test). (**d**) Comparison of aggregate volumes formed in presence of free Chol, depicted in (**c**).

2.4. Computer Simulation of Interactions of Aβ(1-42) with Free Cholesterol

We used all-atom molecular dynamics simulations to elucidate the interaction of free Chol with Aβ(1-42) monomers. Briefly, monomeric Aβ(1-42) was placed in an explicit water box, and NaCl ions were used to neutralize the system charge and keep the ionic strength at a physiologically relevant concentration, 150 mM. Aβ(1-42) was placed at 4 nm from a single Chol molecule. Dynamics of Aβ(1-42) without Chol was simulated as a control. Five replicas of each simulation system were run for 10 μs, yielding a cumulative simulation time of 50 μs for each system.

The Aβ(1-42) monomer shows a rough free energy landscape (FEL), calculated using dihedral principle component analysis of the concatenated dataset, when in the presence of a single free Chol molecule, Figure 4a. The FEL contains well-separated energy minima in three distinct areas, two small areas to the upper and lower left, and a single, large, rough area to the right. The 10 lowest energy minima are highlighted in Figure 4a, and the representative structure for each cluster of said minima are also presented, showing the Chol molecule. These 10 clusters represent ~45.6% of the conformations sampled during the simulation. The number of protein residues in contact with Chol plotted versus the simulation time are given for each individual simulation run in Figure S4a–e. It is evident that the Chol molecule does not simultaneously interact with many residues of Aβ(1-42) at any given time. In fact, the majority of interactions occur through contacts with single residues. Quantitative analysis of these data show that specific regions of Aβ(1-42) are more likely to interact with the Chol molecule, Figure 4b. The contact probability for each residue, based on the combined 50 μs dataset, shows that residues 10 through 14 are most likely to interact with Chol, followed by residues 1–8 of the N-terminal region. Residues in the central hydrophobic region (CHC, residues 17–21) are also likely interaction partners, albeit with lower probability than the aforementioned regions.

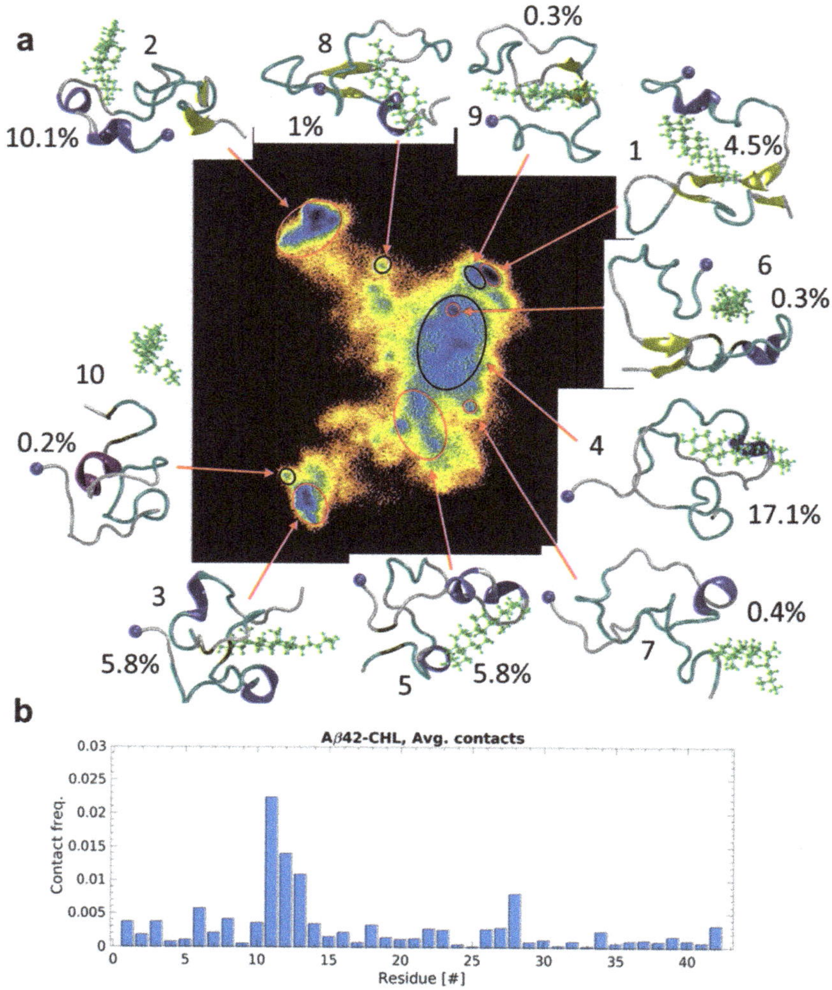

Figure 4. MD simulation of Aβ(1-42) interacting with Chol. (**a**) Free energy landscape based on dihedral principal component analysis of cumulative 50 μs simulation of Aβ(1-42) interacting with Chol. The 10 lowest energy minima are highlighted and the representative conformation of the Aβ(1-42) is shown. Percentages indicate the fraction of conformations relative to total number sampled during the simulations. Blue sphere denotes the N-terminal. (**b**) Average contact probability between residues of Aβ42 and the Chol molecule.

Aβ(1-42) monomer, in the absence of Chol, shows a dramatically different FEL, Figure 5, in which the deepest energy minimum is isolated and dominates by number of conformations (~11.7%) while the rest of the minima are scattered around a very rough area. Furthermore, the 10 lowest energy clusters only represent ~19.9% of the conformations sampled during the simulations. Comparing the evolution of secondary structure for the different simulations, Figures S5 and S6, shows that in both systems the Aβ(1-42) monomer is dominated by turn/bend conformations, with gradual increases in β-strand structure for each system. However, interactions with Chol seems to hinder the formation of long-lived β-strands, as in 3/5 of simulations β-strand appear and disappear more rapidly than in the control simulations without Chol, Figure S5 compared to Figure S6.

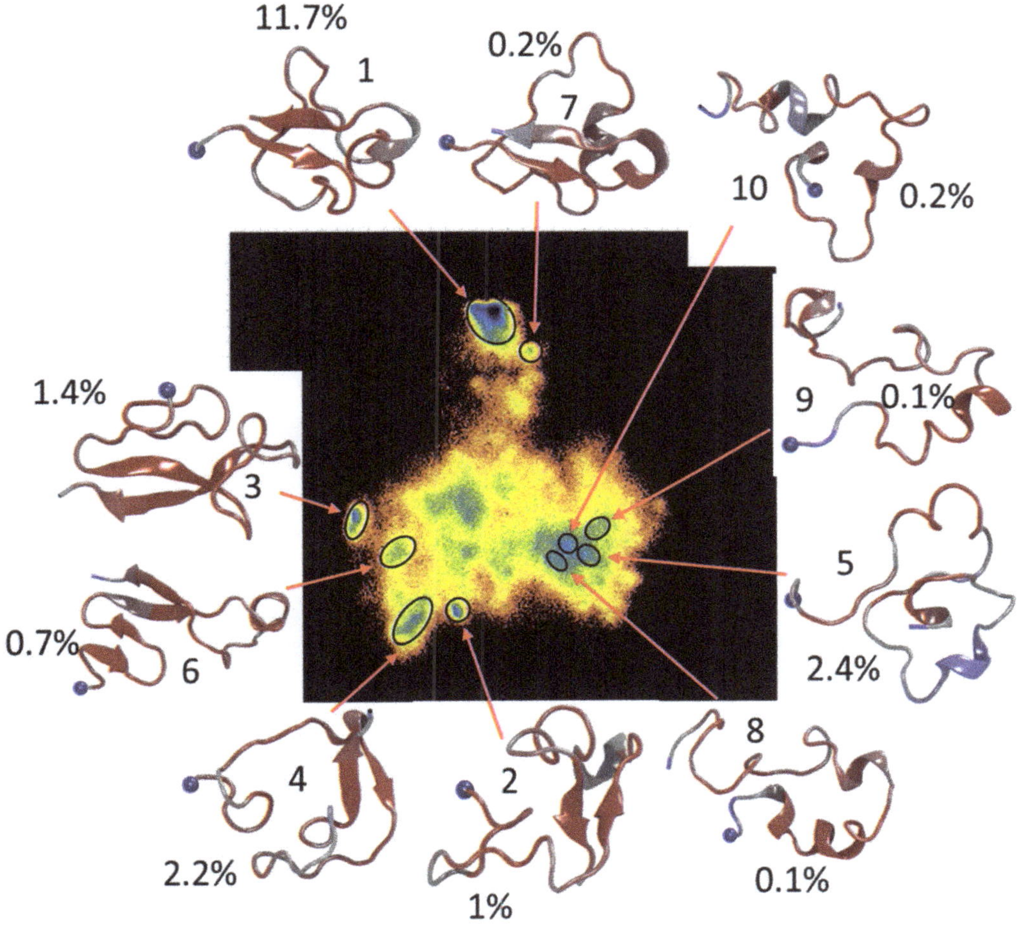

Figure 5. MD simulation of Aβ(1-42) monomer. Free energy landscape based on dihedral principal component analysis of cumulative 50 μs simulation of Aβ(1-42) monomer. The 10 lowest energy minima are highlighted and the representative conformation of the Aβ(1-42) is shown; colors indicate degree of fluctuation in structure, with red being highly conserved regions and blue being highly dynamic regions. Percentages indicate the fraction of conformations relative to total number sampled during the simulations. Blue sphere denotes the N-terminal.

3. Discussion

In our previous study, we have shown that the presence of Chol in the lipid bilayer facilitates aggregation of Aβ(1-42) leading to rapid formation of aggregates [12]. The number of aggregates formed in presence of Chol-containing bilayers was 6 times greater compared to the aggregates on bilayers devoid of Chol. These results revealed the critical role of Chol in the aggregation process. Here, we have shown that free Chol, in addition to Chol inside the lipid bilayer (PC-PS-Chol), has an accelerating effect on Aβ(1-42) aggregation.

Results unambiguously show that free Chol can further accelerate Aβ(1-42) aggregation, as the size and number of aggregates formed in presence of free Chol are greater compared to the experiments where it is absent (Figures 1 and 2). This enhanced effect of free Chol indicates the possibility of direct interaction between Chol and Aβ(1-42). Several studies have shown this type of direct binding, among them [28]. NMR studies have revealed Chol-binding regions of C99, which is the source of Aβ peptide generation due to the action of γ-secretase. The region encompassing residues 18–40 of Aβ(1-42) is observed to interact with Chol [29]. Furthermore, insertion studies of various length of peptide fragments such as Aβ(17-40), Aβ(22-35), Aβ(25-35) have shown that fragments containing residues 25–35 successfully penetrated the Chol containing monolayer [30].

The findings on direct binding of free Chol to Aβ monomers are in line with our all-atom simulations (Figures 4 and 5). Moreover, the energy landscapes qualitatively support the observation of increased dynamics in the Aβ molecule in the presence of Chol (Figure 4). The presence of Chol dramatically increases the sampling of the free energy landscape, but more importantly also increases the number of sampled low-energy conformations. The 10 lowest energy minima sampled by the Aβ(1-42) monomer, in presence of Chol, make up almost 46% of total conformations sampled during the 50 μs cumulative simulations. At the same time, in the absence of Chol, the 10 lowest minima make up almost 20% of the sampled conformations. This acceleration of conformational search may be the key for how Chol affects the aggregation. Indeed, comparing interactions with membranes with and without Chol showed that the Aβ(1-42) monomer experiences a similar increased sampling when Chol is present in the membrane [12]. Additionally, the affinity of the monomer to the membrane is also changed by Chol [12,31]. Furthermore, the simulations show that dimer formation on membranes with Chol inside occur almost 2X faster than on a similar membrane without Chol [12]. The effect of Chol on the free energy and conformational sampling has also been reported for Aβ dimers and trimers [32]. In addition to significant changes to the FEL, the authors also report that presence of Chol induces greater β-structure content in the dimers and trimers of the Aβ(1-42); they also report that dimer to trimer change in β-structure is also significant when Chol is present, going from 26% to 41% [32]. The discrepancy in fraction of β-structure secondary structure between monomer and oligomers can be explained by data obtained by Ono et al., in which different pure oligomers of defined sizes were compared [33]. They reported that oligomer size has a significant effect on the structure and that there is a significant alteration of the Aβ structure going from monomer to dimer.

Our results, demonstrating the accelerating effect of free Chol on Aβ(1-42) aggregation, directly suggest that interference or blocking of Chol-Aβ interaction may suppress spontaneous self-assembly of the protein and thereby reduce the early-stage toxic oligomers. Studies following this line of thought have shown promising results. Bexarotene, which binds to the Chol-binding domain of Aβ, poses a competition for Chol towards Aβ [34,35]. Treatment with nanomolar concentration of bexarotene prevented Aβ oligomer induced Ca^{2+} flux. These data indicate that the prevention of direct interaction of Chol with Aβ can significantly reduce the toxicity caused by the oligomers [34].

One of the important findings in the present study is the increased aggregate dynamics caused by the presence of free Chol (Figure 3 and Figure S3). The data shows that, although aggregates are rapidly formed on the surface, they are not firmly attached to the bilayer and can easily leave the surface spontaneously. This hypothesis is supported by a gradual accumulation of aggregates in the bulk solution above the membrane surface (Figure 3). These data clearly show that the bilayer surface, along with the presence of free Chol, can act as a highly efficient platform for producing oligomers, which then can either participate in further aggregation or act as toxic agents. Most notable, this efficient oligomer producing process occurs at physiologically low nanomolar concentrations of Aβ(1-42).

Another aspect of the oligomers formed in the presence of free Chol is their greater size compared to those formed in the absence of free Chol. Yasumoto et al. reported that low- (LMW) and high-molecular weight (HMW) oligomers use different pathways to damage neurons, with HMW being more neurotoxic and causing more direct damage to the membranes [36]. In particular, HMW oligomers caused significantly more membrane depolarization and impaired long-term potentiation. In the context of the current study, large oligomers, produced due to interactions with free Chol, that dissociated from the membrane surface may show similar mechanism of action as the HMW oligomers tested in the aforementioned study.

Overall, the present study shows that the presence of free Chol, along with in-membrane Chol, significantly accelerates the Aβ(1-42) aggregation. This process occurs at physiologically relevant conditions, including the low nanomolar protein concentration. These findings suggests that specific lipid-Aβ interactions are critical factors for the spontaneous formation of neurotoxic oligomers. These findings further extend our model on the critical role of membrane composition in the assembly of disease-prone amyloid aggregates [12]. Our new data suggest that free Chol facilitate the aggregation process of Aβ monomers. Importantly, there is a strong synergy between the in-membrane and free Chol in this membrane mediated catalysis of Aβ aggregation at physiologically relevant conditions. Note a recent publication [37], which found accumulation of free Chol in the brain for a neurovisceral Niemann-Pick type C (NPC) disease. These findings suggest that the effects of free Chol and other lipids may also be extended to other diseases. Further neurotoxic studies of nanoaggregates assembled on the membranes, in parallel with structural characterization of such aggregates, will pave the way for the development of novel diagnostic and therapeutic strategies for AD and can be extended to other neurodegenerative diseases associated with the formation of protein deposits.

4. Materials and Methods

4.1. Materials

Lipids were purchased from Avanti Polar Lipids, Inc. (Alabama, US). Aβ(1-42) was bought from AnaSpec (Fremont, CA, USA). Chloroform was procured from Sigma Aldrich Inc (St. Louis, MO, USA). The buffer solution that was used in this study is 20 mM HEPES, 150 mM NaCl, 10 mM $CaCl_2$, pH 7.4. All other chemicals, unless otherwise specified, were procured from Sigma at analytical chemistry grade or better.

4.2. Preparation of Supported Lipid Bilayer

PC-PS-Chol lipid bilayer was prepared on mica substrate as mentioned in the previous publication [12]. Briefly, POPC, POPS, and Chol vesicles were prepared by sonicating the mixture for 45 min until the mixture became clear and then deposited onto freshly cleaved mica surface attached to a glass slide. The slide was then incubated at 60 °C for 1 h. After the incubation, the sample was allowed to reach room temperature and then gently rinsed with a buffer containing 20 mM HEPES, 150 mM NaCl, pH 7.4. The bilayer was then imaged immediately by AFM in liquid.

4.3. Preparation of Aβ42 Protein Solution

The method for preparing the Aβ42 stock solution was kept similar to our previous publication [12]. Briefly, lyophilized Aβ(1-42) was dissolved in 100 μL of 1,1,1,3,3,3-hexafluoroisopropanol (HFIP) at room temperature with sonication. The HFIP was then evacuated completely in a vacufuge. Anhydrous DMSO was then added to prepare the stock solution, which was then kept at −20 °C. The stock solution was diluted in the buffer solution to prepare working solutions at the necessary concentrations. Working solutions were used immediately and leftover was discarded.

4.4. Time-Lapse AFM Imaging

Time-lapse data were obtained using an MFP-3D instrument (Asylum Research, Santa Barbara, CA, USA). AFM imaging, in buffer medium, was carried out in tapping mode using the cantilever "E" of MSNL probes (Bruker, Santa Barbara, CA, USA). The typical resonance frequency of the cantilever in buffer was 7–9 kHz with typical spring constants of ~0.1 N/m. Scan speed was typically between 1 to 2 Hz.

At the start of each time-lapse experiment the lipid bilayer was imaged to ensure a homogenous and smooth surface, devoid of any unruptured vesicles. Aβ solution was then added, and time-lapse imaging commenced in the same area of the bilayer. The cantilever was parked after recording each frame to ensure that no damage to the lipid bilayer surface occurred due to scanning.

4.5. AFM Data Analysis

The presented AFM images have undergone minimal processing. Flattening was applied to the images (fitted with 1st order polynomial) with FemtoScan software (Advanced Technologies Center, Moscow, Russia). Grain analysis tool in the software was applied to measure the volume of the oligomers. The volume data were plotted as histograms using Origin Pro software (OriginLab, Northampton, MA, USA) and fitted with Gaussian distribution. The mean value of the oligomer volume for each time point was determined using the peak value of the distribution and the error bars represent the standard deviation, unless otherwise mentioned.

4.6. Molecular Dynamics Simulations

To investigate the interaction of Aβ(1-42) monomer with Chol, we placed an Aβ(1-42) monomer (conformation taken from [38]) at 4 nm center-of-mass (CoM) from a single Chol molecule, solvated the system in TIP3P water, neutralized with NaCl counter ions, and maintained a final NaCl concentration of 150 mM. Protein was described using the Amber ff99SB-ILDN force field [39], while Chol was described using the lipid17 force field (an extension and refinement of lipid14 [40]). A control system with only Aβ(1-42) monomer was also created in a similar manner. The systems were then energy minimized, heated to 300 K, and run for 500 ps as NVT ensemble. Production simulations were run as an NPT ensemble for 10 μs; simulations for each system were repeated five times for a total of 50 μs for each system. Simulations were performed using a 2 fs integration time step. The simulations employed periodic boundary conditions with an isotropic pressure coupling at 1 bar, a constant temperature of 300 K, non-bonded interactions truncated at 10 Å, and electrostatic interactions treated using particle-mesh Ewald [41]. Simulations were performed using the Amber18 package [42].

4.7. Analysis of MD Trajectories

AmberTools20 suite of programs [43], Carma [44], and VMD [45] were used to analyze the obtained simulation trajectories. Graphs and mathematical analyses were obtained using MATLAB (MathWorks, Natick, MA, USA).

Supplementary Materials: The following supporting information can be downloaded at: https://www.mdpi.com/article/10.3390/ijms23052803/s1.

Author Contributions: Y.L.L., S.B. and M.H. designed the project. S.B. performed the AFM experiments. M.H. performed and analyzed the molecular dynamics simulations. All authors have read and agreed to the published version of the manuscript.

Funding: This research was funded by National Institutes of Health, grants GM096039 and GM118006 to Y.L.L.

Institutional Review Board Statement: Not applicable.

Informed Consent Statement: Not applicable.

Data Availability Statement: The data that support the findings of this study are available from the corresponding author upon reasonable request.

Acknowledgments: Anton 2 computer time was provided by the Pittsburgh Supercomputing Center (PSC) through Grant R01GM116961 from the National Institutes of Health. The Anton 2 machine at PSC was generously made available by D.E. Shaw Research. This work was completed utilizing the Holland Computing Center of the University of Nebraska, which receives support from the Nebraska Research Initiative. Authors thank Thomas D. Stormberg for proof reading.

Conflicts of Interest: The authors declare no conflict of interest.

References

1. Ross, C.A.; Poirier, M.A. Protein aggregation and neurodegenerative disease. *Nat. Med.* **2004**, *10*, S10–S17. [CrossRef] [PubMed]
2. Chiti, F.; Dobson, C.M. Protein Misfolding, Amyloid Formation, and Human Disease: A Summary of Progress Over the Last Decade. *Annu. Rev. Biochem.* **2017**, *86*, 27–68. [CrossRef] [PubMed]
3. Hardy, J.A.; Higgins, G.A. Alzheimer's disease: The amyloid cascade hypothesis. *Science* **1992**, *256*, 184. [CrossRef] [PubMed]
4. Hardy, J. Has the Amyloid Cascade Hypothesis for Alzheimers Disease been Proved? *Curr. Alzheimer Res.* **2006**, *3*, 71–73. [CrossRef] [PubMed]
5. Hu, X.; Crick, S.L.; Bu, G.; Frieden, C.; Pappu, R.V.; Lee, J.M. Amyloid seeds formed by cellular uptake, concentration, and aggregation of the amyloid-beta peptide. *Proc. Natl. Acad. Sci. USA* **2009**, *106*, 20324–20329. [CrossRef]
6. Grimmer, T.; Riemenschneider, M.; Förstl, H.; Henriksen, G.; Klunk, W.E.; Mathis, C.A.; Shiga, T.; Wester, H.-J.; Kurz, A.; Drzezga, A. Beta Amyloid in Alzheimer's Disease: Increased Deposition in Brain Is Reflected in Reduced Concentration in Cerebrospinal Fluid. *Biol. Psychiatry* **2009**, *65*, 927–934. [CrossRef]
7. Banerjee, S.; Hashemi, M.; Lv, Z.; Maity, S.; Rochet, J.C.; Lyubchenko, Y.L. A novel pathway for amyloids self-assembly in aggregates at nanomolar concentration mediated by the interaction with surfaces. *Sci. Rep.* **2017**, *7*, 45592. [CrossRef]
8. Pan, Y.; Banerjee, S.; Zagorski, K.; Shlyakhtenko, L.S.; Kolomeisky, A.B.; Lyubchenko, Y.L. Molecular Model for the Surface-Catalyzed Protein Self-Assembly. *J. Phys. Chem. B* **2019**, *124*, 366–372. [CrossRef]
9. Matsubara, T.; Nishihara, M.; Yasumori, H.; Nakai, M.; Yanagisawa, K.; Sato, T. Size and Shape of Amyloid Fibrils Induced by Ganglioside Nanoclusters: Role of Sialyl Oligosaccharide in Fibril Formation. *Langmuir* **2017**, *33*, 13874–13881. [CrossRef]
10. Chi, E.Y.; Frey, S.L.; Lee, K.Y.C. Ganglioside GM1-Mediated Amyloid-beta Fibrillogenesis and Membrane Disruption. *Biochemistry* **2007**, *46*, 1913–1924. [CrossRef]
11. Matsubara, T.; Yasumori, H.; Ito, K.; Shimoaka, T.; Hasegawa, T.; Sato, T. Amyloid-β fibrils assembled on ganglioside-enriched membranes contain both parallel β-sheets and turns. *J. Biol. Chem.* **2018**, *293*, 14146–14154. [CrossRef] [PubMed]
12. Banerjee, S.; Hashemi, M.; Zagorski, K.; Lyubchenko, Y.L. Cholesterol in Membranes Facilitates Aggregation of Amyloid β Protein at Physiologically Relevant Concentrations. *ACS Chem. Neurosci.* **2021**, *12*, 506–516. [CrossRef] [PubMed]
13. Hicks, D.A.; Nalivaeva, N.N.; Turner, A.J. Lipid Rafts and Alzheimer's Disease: Protein-Lipid Interactions and Perturbation of Signaling. *Front. Physiol.* **2012**, *3*, 189. [CrossRef] [PubMed]
14. Ahyayauch, H.; de la Arada, I.; Masserini, M.E.; Arrondo, J.L.R.; Goñi, F.M.; Alonso, A. The Binding of Aβ42 Peptide Monomers to Sphingomyelin/Cholesterol/Ganglioside Bilayers Assayed by Density Gradient Ultracentrifugation. *Int. J. Mol. Sci.* **2020**, *21*, 1674. [CrossRef]
15. Fantini, J.; Yahi, N. Molecular insights into amyloid regulation by membrane cholesterol and sphingolipids: Common mechanisms in neurodegenerative diseases. *Expert Rev. Mol. Med.* **2010**, *12*, e27. [CrossRef]
16. Popp, J.; Meichsner, S.; Kölsch, H.; Lewczuk, P.; Maier, W.; Kornhuber, J.; Jessen, F.; Lütjohann, D. Cerebral and extracerebral cholesterol metabolism and CSF markers of Alzheimer's disease. *Biochem. Pharmacol.* **2013**, *86*, 37–42. [CrossRef]
17. Panchal, M.; Loeper, J.; Cossec, J.-C.; Perruchini, C.; Lazar, A.; Pompon, D.; Duyckaerts, C. Enrichment of cholesterol in microdissected Alzheimer's disease senile plaques as assessed by mass spectrometry. *J. Lipid Res.* **2010**, *51*, 598–605. [CrossRef]
18. Gellermann, G.P.; Appel, T.R.; Davies, P.; Diekmann, S. Paired helical filaments contain small amounts of cholesterol, phosphatidylcholine and sphingolipids. *Biol. Chem.* **2006**, *387*, 1267–1274. [CrossRef]
19. Ullrich, C.; Pirchl, M.; Humpel, C. Hypercholesterolemia in rats impairs the cholinergic system and leads to memory deficits. *Mol. Cell. Neurosci.* **2010**, *45*, 408–417. [CrossRef]
20. Tansey, M.G.; Paul, R.; Choudhury, A.; Kumar, S.; Giri, A.; Sandhir, R.; Borah, A. Cholesterol contributes to dopamine-neuronal loss in MPTP mouse model of Parkinson's disease: Involvement of mitochondrial dysfunctions and oxidative stress. *PLoS ONE* **2017**, *12*, e0171285. [CrossRef]
21. Nicholson, A.M.; Ferreira, A. Increased Membrane Cholesterol Might Render Mature Hippocampal Neurons More Susceptible to β-Amyloid-Induced Calpain Activation and Tau Toxicity. *J. Neurosci.* **2009**, *29*, 4640–4651. [CrossRef] [PubMed]
22. Kiskis, J.; Fink, H.; Nyberg, L.; Thyr, J.; Li, J.-Y.; Enejder, A. Plaque-associated lipids in Alzheimer's diseased brain tissue visualized by nonlinear microscopy. *Sci. Rep.* **2015**, *5*, 13489. [CrossRef] [PubMed]
23. Martins, I.C.; Kuperstein, I.; Wilkinson, H.; Maes, E.; Vanbrabant, M.; Jonckheere, W.; Van Gelder, P.; Hartmann, D.; D'Hooge, R.; De Strooper, B.; et al. Lipids revert inert Aβ amyloid fibrils to neurotoxic protofibrils that affect learning in mice. *EMBO* **2007**, *27*, 224–233. [CrossRef] [PubMed]
24. Sparr, E.; Linse, S. Lipid-protein interactions in amyloid formation. *Biochim. Biophys. Acta* **2019**, *1867*, 455–457. [CrossRef] [PubMed]

25. Banerjee, S.; Hashemi, M.; Zagorski, K.; Lyubchenko, Y.L. Interaction of Aβ42 with Membranes Triggers the Self-Assembly into Oligomers. *Int. J. Mol. Sci.* **2020**, *21*, 1129. [CrossRef]
26. Lv, Z.; Banerjee, S.; Zagorski, K.; Lyubchenko, Y.L. Supported Lipid Bilayers for Atomic Force Microscopy Studies. *Methods Mol. Biol.* **2018**, *1814*, 129–143. [CrossRef]
27. Lv, Z.; Hashemi, M.; Banerjee, S.; Zagorski, K.; Rochet, J.-C.; Lyubchenko, Y.L. Assembly of α-synuclein aggregates on phospholipid bilayers. *Biochim. Biophys. Acta* **2019**, *1867*, 802–812. [CrossRef]
28. Di Scala, C.; Chahinian, H.; Yahi, N.; Garmy, N.; Fantini, J. Interaction of Alzheimer's β-amyloid peptides with cholesterol: Mechanistic insights into amyloid pore formation. *Biochemistry* **2014**, *53*, 4489–4502. [CrossRef]
29. Barrett Paul, J.; Song, Y.; Van Horn Wade, D.; Hustedt Eric, J.; Schafer Johanna, M.; Hadziselimovic, A.; Beel Andrew, J.; Sanders Charles, R. The Amyloid Precursor Protein Has a Flexible Transmembrane Domain and Binds Cholesterol. *Science* **2012**, *336*, 1168–1171. [CrossRef]
30. Di Scala, C.; Yahi, N.; Lelièvre, C.; Garmy, N.; Chahinian, H.; Fantini, J. Biochemical Identification of a Linear Cholesterol-Binding Domain within Alzheimer's β Amyloid Peptide. *ACS Chem. Neurosci.* **2013**, *4*, 509–517. [CrossRef]
31. Yu, X.; Zheng, J. Cholesterol Promotes the Interaction of Alzheimer β-Amyloid Monomer with Lipid Bilayer. *J. Mol. Biol.* **2012**, *421*, 561–571. [CrossRef] [PubMed]
32. Ngo, S.T.; Nguyen, P.H.; Derreumaux, P. Cholesterol Molecules Alter the Energy Landscape of Small Aβ1–42 Oligomers. *J. Phys. Chem. B* **2021**, *125*, 2299–2307. [CrossRef] [PubMed]
33. Ono, K.; Condron, M.M.; Teplow, D.B. Structure-neurotoxicity relationships of amyloid β-protein oligomers. *Proc. Natl. Acad. Sci. USA* **2009**, *106*, 14745–14750. [CrossRef] [PubMed]
34. Fantini, J.; Di Scala, C.; Yahi, N.; Troadec, J.-D.; Sadelli, K.; Chahinian, H.; Garmy, N. Bexarotene Blocks Calcium-Permeable Ion Channels Formed by Neurotoxic Alzheimer's β-Amyloid Peptides. *ACS Chem. Neurosci.* **2014**, *5*, 216–224. [CrossRef] [PubMed]
35. Cramer Paige, E.; Cirrito John, R.; Wesson Daniel, W.; Lee, C.Y.D.; Karlo, J.C.; Zinn Adriana, E.; Casali Brad, T.; Restivo Jessica, L.; Goebel Whitney, D.; James Michael, J.; et al. ApoE-Directed Therapeutics Rapidly Clear β-Amyloid and Reverse Deficits in AD Mouse Models. *Science* **2012**, *335*, 1503–1506. [CrossRef] [PubMed]
36. Yasumoto, T.; Takamura, Y.; Tsuji, M.; Watanabe-Nakayama, T.; Imamura, K.; Inoue, H.; Nakamura, S.; Inoue, T.; Kimura, A.; Yano, S.; et al. High molecular weight amyloid β 1-42 oligomers induce neurotoxicity via plasma membrane damage. *FASEB J.* **2019**, *33*, 9220–9234. [CrossRef] [PubMed]
37. Treiber-Held, S.; Distl, R.; Meske, V.; Albert, F.; Ohm, T.G. Spatial and temporal distribution of intracellular free cholesterol in brains of a Niemann–Pick type C mouse model showing hyperphosphorylated tau protein. Implications for Alzheimer's disease. *J. Pathol.* **2003**, *200*, 95–103. [CrossRef]
38. Zhang, Y.; Hashemi, M.; Lv, Z.; Lyubchenko, Y.L. Self-assembly of the full-length amyloid Abeta42 protein in dimers. *Nanoscale* **2016**, *8*, 18928–18937. [CrossRef]
39. Lindorff-Larsen, K.; Piana, S.; Palmo, K.; Maragakis, P.; Klepeis, J.L.; Dror, R.O.; Shaw, D.E. Improved side-chain torsion potentials for the Amber ff99SB protein force field. *Proteins* **2010**, *78*, 1950–1958. [CrossRef]
40. Dickson, C.J.; Madej, B.D.; Skjevik, Å.A.; Betz, R.M.; Teigen, K.; Gould, I.R.; Walker, R.C. Lipid14: The Amber Lipid Force Field. *J. Chem. Theory Comput.* **2014**, *10*, 865–879. [CrossRef]
41. Darden, T.; York, D.; Pedersen, L. Particle mesh Ewald: An N·log(N) method for Ewald sums in large systems. *J. Chem. Phys.* **1993**, *98*, 10089–10092. [CrossRef]
42. Case, D.A.; Aktulga, H.M.; Belfon, K.; Ben-Shalom, I.Y.; Brozell, S.R.; Cerutti, D.S.; Cheatham, T.E., III; Cisneros, G.A.; Cruzeiro, V.W.D.; Darden, T.A.; et al. *Amber 1*; University of California: San Francisco, CA, USA, 2018.
43. Case, D.A.; Aktulga, H.M.; Belfon, K.; Ben-Shalom, I.Y.; Brozell, S.R.; Cerutti, D.S.; Cheatham, T.E., III; Cisneros, G.A.; Cruzeiro, V.W.D.; Darden, T.A.; et al. *Amber 20*; University of California: San Francisco, CA, USA, 2020.
44. Glykos, N.M. Software news and updates carma: A molecular dynamics analysis program. *J. Comput. Chem.* **2006**, *27*, 1765–1768. [CrossRef] [PubMed]
45. Humphrey, W.; Dalke, A.; Schulten, K. VMD: Visual molecular dynamics. *J. Mol. Graph.* **1996**, *14*, 33–38. [CrossRef]

MDPI
St. Alban-Anlage 66
4052 Basel
Switzerland
www.mdpi.com

International Journal of Molecular Sciences Editorial Office
E-mail: ijms@mdpi.com
www.mdpi.com/journal/ijms

www.ingramcontent.com/pod-product-compliance
Lightning Source LLC
Chambersburg PA
CBHW061510210926
51850CB00008B/1876
9783036589312